T0241568

Lecture Notes in Computer Science 716

Andrew U. Frank   Irene Campari (Eds.)

Spatial Information Theory
A Theoretical Basis for GIS

European Conference, COSIT'93
Marciana Marina, Elba Island, Italy
September 19-22, 1993
Proceedings

Springer-Verlag

Berlin Heidelberg New York
London Paris Tokyo
Hong Kong Barcelona
Budapest

Series Editors

Gerhard Goos
Universität Karlsruhe
Postfach 69 80
Vincenz-Priessnitz-Straße 1
D-76131 Karlsruhe, Germany

Juris Hartmanis
Cornell University
Department of Computer Science
4130 Upson Hall
Ithaca, NY 14853, USA

Volume Editors

Andrew U. Frank
Irene Campari
Department for Geo-Information E127.1, Technical University of Vienna
Gusshausstraße 27-29, A-1040 Vienna, Austria

frank@geoinfo.tuwien.ac.at
campari@geoinfo.tuwien.ac.at

CR Subject Classification (1991): I.2-3, H.2, E.1-2, E.5, I.5-6, J.2

ISBN 3-540-57207-4 Springer-Verlag Berlin Heidelberg New York
ISBN 0-387-57207-4 Springer-Verlag New York Berlin Heidelberg

© Springer-Verlag Berlin Heidelberg 1993
Printed in Germany

Typesetting: Camera-ready by authors
Printing and binding: Druckhaus Beltz, Hemsbach/Bergstr.
45/3140-543210 - Printed on acid-free paper

Foreword

This volume collects the papers presented at the European Conference on Spatial Information Theory (COSIT'93), held in Marciana Marina on the island of Elba (Italy) in September 1993. Spatial Information Theory collects disciplinary topics and interdisciplinary issues that deal with the conceptualization and formalization of large-scale (geographic) space. It contributes towards a consistent theoretical basis for Geographic Information Systems (GIS).

Advances in computer technology and information science and also geography are applied to the practical problem of collecting, managing and presenting spatial data and have produced Geographic Information Systems. GISs are widely used in administration, planning, and science in many different countries, and for a wide variety of application areas. The unifying concept for the GIS is the relation of all information to space, which is realized differently in different applications and cultures. Spatial Information Theory attempts to discover the universally valid principles and to understand the differences of the particular solution. Research results are relevant for GIS, but are distributed in many disciplines and contacts between researchers are therefore hindered. At the same time, development of GIS is limited by the lack of a sound theoretical base.

COSIT'93 follows the international conference "GIS: From Space to Territory. Theories and Methods of Spatio-Temporal Reasoning" that took place in Pisa in September 1992*. That conference brought together experts from different disciplines, most notably computer science, geography, cognitive science and linguistics and was focused on spatial and temporal reasoning about geographic space. This event has established an interdisciplinary dialog within the international scientific community which has continued since and has led to the organization of COSIT'93.

The call for papers, mostly distributed by electronic mail, resulted in over 60 full papers submitted. They were of very high quality and covered a broad field of different disciplines. Each paper was distributed for review to four members of the program committee or other experts in the field. The program chairs then selected the 32 best papers based on the reviewers' assessment to be presented at COSIT'93 and to be included in the proceedings. Comments from the reviewers were sent back to the authors to help them in producing the final copy. We are grateful for the collaborative efforts of the authors and reviewers that allowed us to get this volume ready for the conference.

We thank all people who helped in organizing the conference. In particular the members of the program committee and the additional reviewers contributed generously. Sincere thanks also to Nahid Nayyeri from the ARA Congressi for the organizational and administrative support.

July 1993
Andrew U. Frank
Irene Campari

---

* The proceedings have been published by Springer-Verlag as Lecture Notes in Computer Science Volume 639.

Program Committee Chairs

Andrew U. Frank (Austria)
Irene Campari (Austria)

Program Committee Co-Chair

David Mark (USA)

Scientific Committee

Norbert Bartelme (Austria)
Benedetto Biagi (Italy)
Flavio Bonfatti (Italy)
Peter Burrough (The Netherlands)
Kai-Uwe Carstensen (Germany)
Nick Chrisman (USA)
Helen Couclelis (USA)
David Cowen (USA)
Leila De Floriani (Italy)
Costancio De Castro (Spain)
Pierre Dumolard (France)
Max Egenhofer (USA)
Giorgio Faconti (Italy)
Giacomo Ferrari (Italy)
Manfred Fischer (Austria)
Ubaldo Formentini (Italy)
Christian Freksa (Germany)
Dieter Fritsch (Germany)
Paolo Ghelardoni (Italy)
Chris Gold (Canada)
Reg Golledge (USA)
Mike Goodchild (USA)
Georg Gottlob (Austria)
Dietmar Grünreich (Germany)
Giuseppe Guarrasi (Italy)
Oliver Günter (Germany)
John Herring (USA)
Stephen Hirtle (USA)

Erland Jungert (Sweden)
Bob Jacobson (USA)
Fritz Kelnhofer (Austria)
Milan Konecny (Czechoslovakia)
Karl Kraus (Austria)
Werner Kuhn (Austria)
Ewald Lang (Germany)
Robert Laurini (France)
Duane Marble (USA)
Jan Masser (UK)
Mark Monmonnier (USA)
Jean-Claude Muller (The Netherlands)
John O'Callaghan (Australia)
Harlan Onsrud (USA)
Stan Openshaw (UK)
Giuseppe Papagno (Italy)
Edoardo Politano (Italy)
Giuseppe Pozzana (Italy)
Alina Potrikovska (Poland)
Francois Salgé (France)
Michel Scholl (France)
Hans-Jörg Schek (Switzerland)
Timos Sellis (Greece)
Joseph Strobl (Austria)
M. Tinacci Mossello (Italy)
A Min Tjoa (Austria)

Additional Referees

Kate Beard	(USA)
Mark Biekkens	(The Netherlands)
Kurt Brassel	(Switzerland)
Carola Eschenbach	(Germany)
Daniel Hernandez	(Germany)
Karen Kemp	(Austria)
Benjamin Kuipers	(USA)
Robert McMaster	(USA)
Georg Nagy	(USA)
Simone Pribbenow	(Germany)
Enrico Puppo	(Italy)
Ralf Röhrig	(Germany)
Terence Smith	(USA)
Heinz Stanek	(Austria)
Waldo Tobler	(USA)
Kai Zimmermann	(Germany)
Michael Worboys	(UK)

Sponsorship

Technische Universität Wien

Consiglio Nazionale delle Ricerche - Istituto CNUCE, Pisa

Amministrazione Provinciale di Livorno

IRPET-Istituto Regionale per la Programmazione Economica della Toscana

Comune di Marciana Marina

Associazione Industriali di Livorno

Cassa di Risparmio di Livorno

INTERGRAPH Corporation

Organizing Committee

Chair: Ubaldo Formentini (Università di Pisa, Italy)

Nahid Nayyeri
ARA Congressi (Livorno, Italy)

Table of Contents

V. Temporal Reasoning

VI. Data Models for Spatial and Temporal Data

VII. Cultural Differences in Spatial Cognition

VIII. Scales in Geographic Space

IX. User-Interface

X. Spatial Analysis

XI. Spatial Reasoning

XII. Posters

A Cognitive Model for the Process of Multimodal, Incremental Route Descriptions

Wolfgang Maaß

Graduiertenkolleg Kognitionswissenschaft
Universität des Saarlandes
Im Stadtwald 15
D-6600 Saarbrücken 11, Germany

E-Mail : maass@cs.uni-sb.de
Phone: (+49 681) 302-3496
Fax: (+49 681) 302-4421

Area of submission: Cognitive Science, Spatial and temporal reasoning in geographic space

Abstract. In normal life we often give *route descriptions* to inform someone about a specific route. They can be divided into the classes of *complete* and *incremental* route descriptions. We propose that the multimodal, incremental route description process consists of a wayfinding, a waypresentation and a control process and is based on a *segmentation hierarchy*. The presentation of a route description is characterized by the use of different presentation modi like speech, gesture and graphics. We propose a computational model for the multimodal, incremental route description process.

1 Introduction

Route Decriptions are common actions in normal life. They can be divided into two classes: *complete* and *incremental route descriptions (IRD)*. Normally, we use complete route descriptions to give the entire route all at once. The problem for the questioner is that he has to remember a lot of details all at the same time. On his way to his destination point he normally cannot ask the same person for more details. What we usual want are *incremental* route descriptions like those given by a codriver, who ideally gives timley route informations.

Psychology experiments have been conducted concerning the human ability to process and represent information about spatial environment ([Tolman 48; Piaget et al. 60; Siegel & White 75; Haury et al. 92]). The abstract representation format is called *cognitive map* ([Hartl 90]). In this work we are not concerned with the mental model problem. However we want to present a model of the *multimodal, incremental route description process*[1]. Furthermore we give some remarks about the possiblity to remember experiences made by route descriptions received in the past. We explore how to arrive from a spatial representation of a route [2] at a *multimodal incremental route description* (MIRD).

Route knowledge and its connections to other human abilities has been investigated in psychology by different views ([Blades 91a; Garling 89; Hayes-Roth & Hayes-Roth 79]). Piaget differentiated between landmark and route knowledge ([Piaget et al. 60]), whereas Siegel and White propose that landmarks are the preliminary stage of learning routes ([Siegel & White 75], for a good review see [Blades 91a]). Complete route descriptions are well examined for psychological ([Streeter et al. 85; Thorndyke & Goldin 83]) and linguistic reasons ([Klein 79; Wunderlich & Reinelt 82]), but there is little known about incremental route descriptions.

The main process is usually divided into two subprocesses. The first subprocess is the wayfinding process, that modells the human ability to learn and remember a route through the environment ([Blades 91b]). The second subprocess is called the waypresentation process. In general, both processes depend on the knowledge about the environment, on the questioner, and some external parameters like weather and time[3]. Approaches that integrate knowledge about the questioner and external influences to construct an adequate and adaptive model of the cognitive process of incremental route descriptions do not exist.

In general, route descriptions can be clearly defined to enable a questioner to find his path from a starting point to a destination point following some constraints. In this case many linguistic examinations about the structure of complete route descriptions are available ([Wunderlich & Reinelt 82; Klein 79; Meier et al. 88; Habel 87]). Wunderlich and Klein have divided the presentation into four phases: introdution, mainpart, verification and final phase. The sentence structures in complete route descriptions are very restricted and sometimes schematic. There are a number of computational models for the generation of textual route descriptions. Habel proposes a three-phase model that is a special case of the architecture presented in [Hoeppner et al. 90].

In contrast to complete route descriptions, temporal aspects are very central for MIRD. If one wants to present spatial informations one has to fix the temporal structure of the whole process. Therfore one has to decide on the start

---

[1] As the spatial basis for incremental route descriptions we use a 3-dimensional map representation.

[2] Multimodality means the presentation of information in different modes, like natural language, gestures, and graphics.

[3] From now on, these external parameters and the knowledge about the questioner will be called constraints.

of information retrieval and the beginning of the presentation of an information unit. A general constraint for a MIRD is that the information is presented timely.

A description like the following example combines various description levels:

> You must follow the Goethestraße until you reach a red house on the right side, where you turn right into the Schillerstraße.

In this description one level is used that is defined by paths (e.g. Goethe-straße, Schillerstraße) and one level that is defined by landmarks (e.g. red house). Usually, we change the level of description to give the questioner a unique orientation in his actual spatio-temporal environment. As the example shows, descriptions are not only at a very basic level (*... red house on the right side, where you turn right ...*), but sometimes we give some more abstract information (*... follow the Goethestraße*) to give the questioner a global orientation. Beside the topological and topographical information that is relevant for the route descriptions, we often use explanations *why* we have choosen a specific route segment, but not an alternative one (*Now, we drive down the Schillerstreet, because the Mozartstreet is a one-way street in the other direction*). For explanations we have to appeal to information that are relevant for the wayfinding process before.

2 A Cognitive Model for the Multimodal, Incremental Route Description Process

We use the term *cognitive model* in the sense of Pylyshyn:

> In order that a computer program be viewed as a literal model of co-gnition, the program must correspond to the process people actually perform at a sufficiently fine and theoretically motivated level of resolution. ([Pylyshyn 84] p. xv)

In this sense we want to present a model for the MIRD-process. First we give a general question of the whole process in order to delimit our model from others. Then the representation structure is introduced, followed by the subprocesses and their interactions.

Klein and v. Stutterheim propose that a text is used to answer an explicit or implicit question, which they call *quaestio* of a text ([Klein & v.Stutterheim 87]). The more precise the quaestio can be formulated the better is the structure of the text. So route descriptions are very interesting because of their precise quaestio:

> What is the way from X to Y?

One of the main criterions is the *adequacy* of route descriptions. Adequacy, in the sense that the description is similar to those given by humans ([Habel 87]) is intentionaly not very clearly defined. It reflects what Pylyshyn means with that *the eventual successes of cognitive science, if they come at all, will have to explain a varity of empirical phenomena* ([Pylyshyn 84] p.272). So the main task is to give a model that incorporates all known processes which seem to be relevant for the MIRD-process and associated representation structures. In general the route description process consists of the *wayfinding* and the *route presentation process*. What both processes have to achieve is relativly clear but the connection between them is almost undefined. For this reason we presume another process that is called the *control process*. The main task of this process is the coordination of process interactions. In *complete route description* there exist two competing strategies: planning in advance and stepwise planning ([Klein 79] p.7). In cause of the incrementality, the MIRD-process can be characterized as a stepwise planning process. At the startingpoint there is only a *rough path* to the destination point and only some segments of the whole route are consciously present. We assume that a path from the starting point to the destination point can be splitted into segments. Furthermore we presume that in MIRD in every moment of the description only the actual and the next segment are used ([Meier et al. 88] p.19).

Our general architecture of a route description process consists of a wayfinding and a presentation process with abilities to switch and interact between both processes, which is done by the control process. The central knowledge structure could be seen as a hierachical extension of the *primary plan*[4] introduced by Klein ([Klein 79]).

2.1 Segmentation Hierarchy

One of the first hierarchical models for a cognitive map is the approach suggested by Pailhous, ([Pailhous 70]), who organized the spatial knowledge into a *basic network* and a *secondary network*. It seems to be widely accepted that for different uses of cognitive maps we need different views or layers. One possible approach is the idea of overlaying many thematic maps of the same space ([Gluck 91] p.126).

In our model we propose a *segmentation hierarchy* that is similar to Kuipers differentiation into the hierarchy of regions. A *segment* is a vertical part of the hierarchy, that consists of two spatio-temporal entities, e.g landmarks, and their connection by a finite spatio-temporal unit (paths). The union of all segments of a specific route description is called the *segmentation hierarchy*. Segments are defined concerning a specific level that depends on various parameters[5].

A segment is the basis for all processes that are relevant for the MIRD-process. We will present an example to clearify what is meant by the segmentation hierarchy. The problem lies in presenting a route from the *starting point*

---

[4] The primary plan is part of a cognitive map that contains the starting and the destination point.

[5] If we go by feet important landmarks are the segment-constituing spatio-temporal entities which define the segementation level. If we go by car we use the level of turning point landmarks.

S to the destination point Z (see fig. 1). On the choosen route there are three turning point landmarks (A, B, C). Between S and A is one landmark, a gate on the right side, and between A and B are two landmarks, namely a house on the right side and a tower on the left side of the street. First we predict that the rough path[6] from S to Z is known by experience[7]. It is also known that the questioner is familiar with the segments SA, BC, and CZ but he does not know the segment AB.

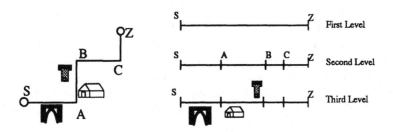

Fig. 1. Complete Segmentation Hierarchy

The second level of the hierarchy represents the differentiation between the starting point S, the landmarks A, B, C, and the destination Z. In this example, the landmarks A, B, C, are turning points where we have to turn right or left. It is widely accepted that turning-point landmarks are very important for temporal-spatial instructions, so they are used in nearly every model of route descriptions ([Klein 79; Wunderlich & Reinelt 82; Habel 87; Meier et al. 88; Hoeppner et al. 90]). It can be deduced that turning-point landmarks are very high up the segmentation hierarchy.

The next level is more detailed than the second level. In the third level, all objects that we know by normal perception, like houses, trees, human beings, animals etc., are integrated. The difference between the second and the third level is that we switch from *large scale space* to *small scale space*. In large scale space we look at a spatial environment where we cannot see the destination point from the starting point. In small scale space the destination point is visible from the starting point. It is not necessary to percept the landmark C from the starting point. But if we say that on the right side is a tree when we pass by it, it is necessary that we then see the tree.

On this level we want to show how the experience of the questioner can be integrated. As we assume for our example the questioner knows the segments SA, BC, and CZ. But he is not so familiar with the spatial environment that he

---

[6] It seems that the relative position of the starting to the destination point, and the distance between both are mainly responsible for the determination of the rough path.

[7] If the route is not known the waysearching process will be started.

knows the complete route, especially the connections between these segments. So information from the second level has to be used in describing the segments SA, BS, and CZ and from the third level for segment AB. For the description of segment AB, we integrate the house at the right side of the road and the tower at the left side, but the gate in segment SA is not mentioned.

Now we have the whole segmentation hierarchy for the complete route SZ. The hierarchy is incrementally generated (see example 1) and determined by two general phases. In the first phase a rough spatial path is examined. It reflects the common strategy that if we want to find a path on a map in an unkown area we will look at a general map where the starting point and the destination point are integrated. There we determine a rough path (usually using our finger) between both points. Generally it does not matter if the roads we determine in that way are not obviously connected. In this case we use the general assumption that in the western world all roads are somehow connected[8]. A model for this rough path is the first level of the segmentation hierarchy.

In the second phase, we refine the first level so that we find the segment SA and AB where we follow Klein ([Klein 79]) who has mentioned that in normal route descriptions we only use the actual and the next segment at one time[9].

Every time when we change the actual segment we have to determine the next. For example, if we turn left at A, we leave SA and reach AB that becomes the actual segment. Now, we have to determine the segment BC to be able to give the transition from segment AB to BC.

The whole segmentation hierarchy is incrementally generated on the way from the starting point to the destination point. This reflects the assumption that the whole description is in detail not known in the beginning. Description elements like a tree or a house are integrated as they are perceived. Another assumption is that humans can easily choose a new path if they have followed the wrong way. In the proposed model a new rough path, the new actual and the new next segment have to be determined to proceed with the description.

2.2 Wayfinding Process

Klein divides the planning of the primary plan for complete route descriptions into two techniques: advanced and stepwise planning. These techniques can be used in combination. Advanced planning, means that the route description process is mostly sequential. In contrast, stepwise route planning, implies that there is an interaction between the wayfinding and the route presentation process. It is therefore meaningful to assume a combination of both techniques ([Klein 79]).

Klein does not say a lot on how to find a way. He only reports of a person who shows a behaviour that can be outlined as a trial and error technique. Habel distinguishes between *route finding* and *route knowing* which we denote with *way search* and *experience-based wayselection* ([Habel 87]). Gluck remarks, that

---

[8] Here we do not take into account that mountains, vallies and seas are important restrictions (see the barrier effect ([Kosslyn et al. 74]).

[9] At that time segments BC, and CZ are still not computed.

the potential optimizing functions that are relevant for the wayfinding process, are not restricted to minimal distance or even minimal effort. He also mentions, however, that there is no model that accounts for them ([Gluck 91]). The wayfinding process modelled in the TOUR model is a simple and slow approach to find a way ([Kuipers 77]). In the TRAVELLER model an incomplete wayfinding process is used, that can not find partial ways in the cognitive map ([Leiser & Zilbershatz 89]).

To also give adequate descriptions, the wayfinding process has to integrate the spatial representation, information about the questioner and external constraints. The spatial representation, is in the map-based approach a topographical representation of the spatial environment. The adequacy of a description, is mainly based on the information about the questioner and some external parameters. There is a wide difference between an 8 and an 80 years old person, as much as between a tourist and a stockbroker. Also the weather (e.g. rainy, dry) and the time (e.g. rush hour, night) are important.

We have to distinguish between the selection of a path that is known by experience (experience-based wayselection) and the search for a path by using a map ([Elliott & Lesk 82]). Kuipers ([Kuipers 77; Kuipers 78]) in his TOUR model and Leiser/Zilbershatz ([Leiser & Zilbershatz 89]) in their TRAVELLER model mainly use an experience-based approach. Both models are based on a graph network which represents a mental model of the spatial environment and concentrate on how to integrate new knowledge about the *external* world into it. Leiser and Zilbershatz divide route knowledge into a basic network for the main routes and a secondary network for the other routes. This division is based on experiments done by Pailhous ([Pailhous 69]) and Chase ([Chase 82]).

The waysearching process does not seem to be as well understood as the experience-based wayselection process[10]. In most of the relevant works, the route that has to be presented is almost known or generated with some straightforward algorithm[11]. But the waysearch process has also to observe various kinds of constraints. If the person goes by car, the search process has to consider roads for cars. It is also relevant to consider the weather, e.g. if it rains you will prefer to show someone a covered path. There is much evidence that the cognitive ability of wayfinding combines the experience-based wayselection process and the waysearch processes. This is meaningful because it is not usual that we know a large scale space route without searching for some path segments. It seems that a combination of both processes is what we do if we are looking for a route.

If there is no experience that could be used, the path for this segment has to be searched. In figure 1 the segments SA, BC, and CZ are known by experience on the second level. A familiar segment is relevant for the wayfinding process if it connects two spatio-temporal entities on the prospected path under

---

[10] It seems to be very unlikely that heuristic search algorithm, like A*, are similiar to human search ability in maps.

[11] The algorithm are mainly based on recursive procedures or some kind of standard search algorithm like A* [Habel 87; Carstensen 91].

consideration of external constraints.

The output of the wayfinding process is a *segmentation hierarchy fragment* (SHF). The rough path is in some kind completly determined because you always have an unspecific idea of the way[12]. At any position the actual segment and all following familiar segments are considered until there is a segment in which the path has to be searched for. The complete hierarchy instantiates the *segmentation hierarchy*.

2.3 Control Process

The SHF is an abstract semantic representation of the objects for the path from one spatio-temporal entity to the next. There is no information attached that assigns items of the SHF to a presentation mode, because we suppose that the wayfinding and the presentation processes are independent[13] from each other. This assumption is based on neurological results that show that presentation processes, for example the speaking process, are principally independent ([Springer & Deutsch 88] p.3) from each other. The coordination of the presentation processes is done by a central process which we call the *control process*[14].

The assignment of items of the SHF to a presentation mode can be divided into two main techniques. First, there are schematic descriptions of route segments. For example, if you want to tell someone that he has to go up a street until he reaches a crossing, you use some kind of presentation like:

You have to go up the street [point with a gesture along the street][15] until you reach a crossing.

On the other side, we have to consider unusual descriptions, that are mostly unusual in cause of the spatio-temporal environment. For example, if you want to describe someone how to hillclimb to the next position.

You have to put your right food on that ledge [point to the ledge] and than grep with your right hand the crack one meter right above the ledge [point to the crack] .

It seems to be clear that the usage of schemas depends on how common we are with the spatio-temporal environment. If you are a good hillclimber you will

---

[12] This depends on the distance and the way you move. E.g. when you go by car and you want to go from Berlin to Saarbrücken, then you first look for the longest Autobahn segments between both cities. An important assumption is that there always exist an acceptable path from every point in Berlin to that Autobahn segment and from that to Saarbrücken.

[13] Presenation modes are any independent information mode, like *natural language, gestures, graphics etc.*

[14] Such a central control process is also supported by neurological results. E.g. the Thalamus seems to coordinate some aspects of speech and motorical skills ([Springer & Deutsch 88] p.109).

[15] The brackets denote gesture actions.

not think about the hillclimbing description. On the other side, if you live in the forest you may have some problems to give the first description. Beside the user and environment dependencies, external parameters, like the actual time, are important for MIRD. For example a landmark may be relevant in the night because it is illuminated, but not during the day.

It seems that we have no problems to switch between the presentation modes if we want to inform the questioner about, for example, a red house on the right side of a street. We can do this with a verbal expression, with a pointing gesture or a line drawing. So it seems to be evident that there is a central representation structure. When the control process assigns items of the SHF to presentation modes, it has to consider some problems with the classification of the items, like the *cross-referentiality problem*. This means that an item can be parallely presented in different modes like the crack in the second example. To solve this problem the control process assigns the crack item to the verbal and the gesture mode and sends this item to both presentation processes.

2.4 Presentation Processes

The control process distributes the route information to be presented according to the individual strength of each presentation mode. For presentation processes we use approaches presented in ([Herzog et al. 93; Maaß et al. 93]). The presentation processes are mainly the verbal, the gesture and little more less the graphical process. The control process passes a semantic representation structures of the intended information to the processes. For explanation of the main points, we use the following simple example:

```
MOVE( MODE    (verbal)
      TYPE    (by-feed)
      ACTUAL (type(street),   time(t1),   name(Goethestraße))
      RIGHT   (type(house),   time(t2),   attr(red))
      LEFT    (type(tower),   time(t3),   attr(big))
      UNTIL   (type(crossing), time(t1, t4))
      NEXT    (type(street),  time(t4),   name(Schillerstrasse)))
```

The semantic structure provides global information, like the presentation mode and the way the questioner moves. It also contains information about the actual spatial segment, ACTUAL(type(street), time(t1), name(Goethestraße)), and the following segment, NEXT(type(street), time(t1), name(Schillerstrasse)). If it is necessary to give more detailed descriptions, the details are integrated in this structure, like RIGHT(type(house), time(t2), attr(red)) and LEFT(type(tower), time(t3), attr(big)). Another important item of the structure is the next spatio-temporal entity, UNTIL(type(crossing), time(t1, t4)), that defines the connection to the next segment. The timemarker t_i fixes the moment when the items are presented. Items with the same timemarker are presented at the same time in one coherent expression.

With this semantic structure the text process generates a verbal expression like:

t1: Go down the Goethestraße, until you reach a crossing.
t2: Pass the red house on the right side
t3: Pass the big tower on the left side
t4: At the crossing, turn right into the Schillerstrasse.

3 Conclusion

The research in route descriptions leads to the conclusion that we have to assume three subprocesses: a wayfinding, a presentation process, and a control process. Central for our approach is the segmentation hierachy, that is incrementally generated, and the integration of various external constraints, which influence all subprocesses.

We presented a global framework, that has to be refined, perhaps modified, and compared with results of psychological examinations in the future. Especially the interaction between the subprocesses and the representation structure has to be examined in more detail. We have finished a 3-dimensional domain model and the realization of new display techniques (2-dimensional projections, perspective views, animation). Our current work is concerned with the seamless coordination of the presentation modes.

4 Example

Here is the complete example used in the paper.

1. t1: Go along the street until you reach landmark A.
   [point with a gesture along the street]
   t2: Turn left into the Goethestraße. [Point left]
   (First SHF)
2. t3: Go down the Goethestraße, until you reach a crossing.
   [point straight ahead]
   t4: Pass the red house on the right side.
   t5: Pass the big tower on the left side.
   t6: At the crossing, turn right into the Schillerstraße. [point right]
   (Second SHF)
3. t7: Follow the Schillerstrasse. [point straight ahead]
   t8: Turn left behind landmark C. [point right]
   (Third SHF)
4. t9: Go down this street until you reach your destination point.
   [point straight ahead]
   t10: Here you see your destination point.
   (Final SHF)

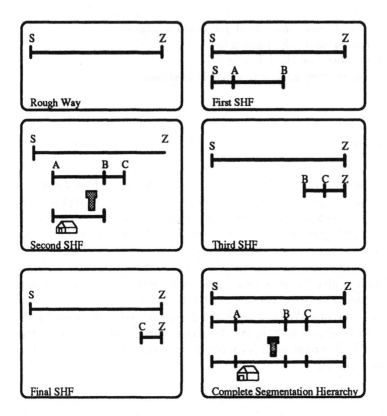

Fig. 2. Incremental construction of the *segment hierarchy*

References

[Blades 91a] M. **Blades**. *The Development of the Abilities Required to Understand Spatial Representations*. In: D. M. Mark and A. U. Frank (eds.), Cognitive and Linguistic Aspects of Geographic Space, pp. 81–115. Dordrecht: Kluwer, 1991.

[Blades 91b] M. **Blades**. *Wayfinding Theory and Research: The Need for a New Approach*. In: D. M. Mark and A. U. Frank (eds.), Cognitive and Linguistic Aspects of Geographic Space, pp. 137–165. Dordrecht: Kluwer, 1991.

[Carstensen 91] K.-U. **Carstensen**. *Aspekte der Generierung von Wegbeschreibungen*. Master's thesis, Universität Osnabrück, 1991.

[Chase 82] W.G. **Chase**. *Spatial Representations of Taxi Drivers*. In: D.R. Rogers and J.A. Sloboda (eds.), Acquisition of Symbolic Skills. New York: Plenum, 1982.

[Elliott & Lesk 82] R. J. **Elliott** and M. E. **Lesk**. *Route Finding in Street Maps by Computers and People*. In: Proc. of AAAI-82, pp. 258–261, Pittsburgh, PA, 1982.

[Garling 89] T. **Garling**. *The role of cognitive maps in spatial decisions*. Journal of Environmental Psychology, 9:269–278, 1989.

12

[Gluck 91] M. **Gluck**. *Making sense of Human Wayfinding: Review of Cognitive and Linguistic Knowledge for Personal Navigation with a new Research Direction.* In: D.M. Mark and A.U. Frank (eds.), Cognitive and Linguistic Aspects of Geographic Space, pp. 117–135. Netherlands: Kluwer Academic Publishers, 1991.

[Habel 87] Ch. **Habel**. *Prozedurale Aspekte der Wegplanung und Wegbeschreibung.* LILOG-Report 17, IBM, Stuttgart, 1987.

[Hartl 90] A. **Hartl**. *Kognitive Karten und kognitives Kartieren.* In: C. Freksa and C. Habel (eds.), Repräsentation und Verarbeitung räumlichen Wissens, pp. 34–46. Springer, 1990.

[Haury et al. 92] Ch. **Haury**, H.M. **Engelbert**, R.**Graf**, and Th. **Herrmann**. *Lokalisationssequenzen auf der Basis von Karten- und Straßenwissen: Erste Erprobung einer Experimentalanordnung.* Bericht 47, SFB 245, Heidelberg/Mannheim, August 1992.

[Hayes-Roth & Hayes-Roth 79] B. **Hayes-Roth** and F. **Hayes-Roth**. *A cognitive model of planning.* Cognitive Science, 3:275–310, 1979.

[Herzog et al. 93] G. **Herzog**, W. **Maaß**, and P. **Wazinski**. *VITRA GUIDE: Utilisation du Langage Naturel et de Représentation Graphiques pour la Description d'Iténitaires.* In: Colloque Interdisciplinaire du Commitée National Images et Langages: Multimodalitée et Modélisation Cognitive", Paris, 1993.

[Hoeppner et al. 90] W. **Hoeppner**, M. **Carstensen**, and U. **Rhein**. *Wegauskünfte: Die Interdependenz von Such- und Beschreiungsprozessen.* In: C. Freksa and C. Habel (eds.), Informatik Fachberichte 245, pp. 221–234. Springer, 1990.

[Klein & v.Stutterheim 87] W. **Klein** and C. **v.Stutterheim**. *Quaestion und referentielle Bewegung in Erzählungen.* Linguistische Berichte, 109:163–183, 1987.

[Klein 79] W. **Klein**. *Wegauskünfte.* LiLi, 9(33):9–57, 1979.

[Kosslyn et al. 74] S. **Kosslyn**, H.L. **Pick jr.**, and G.R. **Fariello**. *Cognitive Maps in Children and Men.* Child Development, 45:707–716, 1974.

[Kuipers 77] B. **Kuipers**. *Representing Knowledge of Large-Scale Space.* PhD thesis, MIT AI Lab, Cambridge, MA, 1977. TR-418.

[Kuipers 78] B. **Kuipers**. *Modelling Spatial Knowledge.* Cognitive Science, 2:129–153, 1978.

[Leiser & Zilbershatz 89] D. **Leiser** and A. **Zilbershatz**. *THE TRAVELLER: A Computational Model of Spatial Network Learning.* Environmental and Behaviour, 21(4):435–463, 1989.

[Maaß et al. 93] W. **Maaß**, P. **Wazinski**, and G. **Herzog**. *VITRA GUIDE: Multimodal Route Descriptions for Computer Assisted Vehicle Navigation.* In: Proc. of the Sixth International Conference on Industrial and Engineering Applications on Artificial Intelligence and Expert Systems, Edinburgh, U.K., June 1-4,1993.

[Meier et al. 88] J. **Meier**, D. **Metzing**, T. **Polzin**, P. **Ruhrberg**, H. **Rutz**, und M. **Vollmer**. *Generierung von Wegbeschreibungen.* KoLiBri Arbeitsbericht 9, Fakultät für Linguistik und Literaturwissenschaft, Universität Bielefeld, 1988.

[Pailhous 69] J. **Pailhous**. *Representation de l'espace urbain et cheminements.* Le Travail Humain, pp. 32–87, 1969.

[Pailhous 70] J. **Pailhous**. *La Representation de l'Espace Urbain - L'example du Chauffeur de Taxi.* Presses Universitaires de France, 1970.

[Piaget et al. 60] J. **Piaget**, B. **Inhelder**, and A. **Szemenska**. *The Child's Conception of Gometry.* New York: Basic Books, 1960.

[Pylyshyn 84] Z. W. **Pylyshyn**. *Computation and Cognition.* Cambridge, MA: MIT Press, 1984.

[Siegel & White 75] A. W. **Siegel** and S. H. **White**. *The Development of Spatial Representation of Large-Scale Environments*. In: W. Reese (ed.), Advances in Child Developement and Behaviour. New York: Academic Press, 1975.

[Springer & Deutsch 88] Sally P. **Springer** and G. **Deutsch**. *Linkes - rechtes Gehirn: funktionelle Asymmetrien*. Heidelberg: Spektrum der Wissenschaft, 2 edition, 1988.

[Streeter et al. 85] L.A. **Streeter**, D. **Vitello**, and S.A. **Wonsiewicz**. *How to Tell People Where to Go: Comparing Navigational Aids*. International Journal of Man-Machine Studies, 22:549–562, 1985.

[Thorndyke & Goldin 83] P. W. **Thorndyke** and S. E. **Goldin**. *Spatial Learning and Reasoning Skill*. In: H. L. Pick and L. P. Acredolo (eds.), Spatial Orientation: Theory, Research, and Application, pp. 195–217. New York, London: Plenum, 1983.

[Tolman 48] E.C. **Tolman**. *Cognitive Maps in Rats ans Men*. Psychological Review, 55:189–208, 1948.

[Wunderlich & Reinelt 82] D. **Wunderlich** and R. **Reinelt**. *How to Get There From Here*. In: R. J. Jarvella and W. Klein (eds.), Speech, Place, and Action, pp. 183–201. Chichester: Wiley, 1982.

COGNITIVE MAPS, COGNITIVE COLLAGES, AND SPATIAL MENTAL MODELS

BARBARA TVERSKY

DEPARTMENT OF PSYCHOLOGY, STANFORD UNIVERSITY
STANFORD, CA 94305-2130 USA

Abstract. Although *cognitive map* is a popular metaphor for people's mental representations of environments, as it is typically conceived, it is often too restrictive. Two other metaphors for mental representations are proposed and supported. *Cognitive collages* are consistent with research demonstrating systematic errors in memory and judgment of environmental knowledge. Yet, for some simple or well-known environments, people seem to have coherent representations of the coarse spatial relations among elements. These *spatial mental models* allow inference and perspective-taking but may not allow accurate metric judgments.

1 Introduction

1.1 Cognitive Maps

There is a popular view that people's mental representations of environments are embodied in "cognitive maps." Like many useful concepts, the term *cognitive map* has many senses, leading to inevitable misunderstandings. One prevalent sense is that cognitive maps are maplike mental constructs that can be mentally inspected. They are presumed to be learned by gradually acquiring elements of the world, first *landmarks*, pointlike elements, then *routes*, linelike elements, and finally unifying the landmarks and routes with metric *survey* information. The appeal of this view is manifold. As *cognitive,* they are presumed to differ from "true" maps of the environment. Social scientists from many disciplines would be quick to bring forth evidence for that. As *maps,* they are presumed to be coherent wholes that reflect spatial relations among elements. As mental constructs available to mental inspection, cognitive maps are presumed to be like real maps available to real inspection, as well as like images, which, according to the classical view of mental imagery, are like internalized perceptions.

1.2 Constructionist View

In this paper, I will present evidence not compatible with the view of mental representations of environments as cognitive maps. I will discuss two alternative *constructionist* views of the mental representations underlying people's knowledge of environments. According to the constructionist view, people acquire disparate pieces of knowledge about environments, knowledge that they use when asked to remember an environment, describe a route, sketch a map, or make a judgment about location, direction, or distance. The separate pieces include recollections of journeys, memories of maps, recall of verbal (aural or written) directions and facts, and more. As for any human memory task, it is possible that not all t he relevant stored information will be retrieved when needed.

1.3 Cognitive Collages

In many instances, especially for environments not known in detail, the information relevant to memory or judgment may be in different forms, some of them not maplike at all. Some of the information may be systematically distorted as well. It is unlikely that the pieces of information can or will be organized into a single, coherent maplike cognitive structure. In these cases, rather than resembling maps, people's internal representations seem to be more like *collages*. Collages are thematic overlays of multimedia from different points of view. They lack the coherence of maps, but do contain figures, partial information, and differing perspectives. In the second section, I will review some of the evidence for the notion that cognitive collage is often a more appropriate metaphor for environmental knowledge than cognitive map. That evidence shows that memory and judgment are systematically distorted and potentially contradictory, thus not easily reconcilable in a maplike structure.

1.4 Spatial Mental Models

In other situations, especially where environments are simple or well-learned, people seem to have quite accurate mental representations of spatial layouts. On close examination, these representations capture the categorical spatial relations among elements coherently, allowing perspective-taking, reorientation, and spatial inferences. In contrast to cognitive maps and cognitive collages, these have been termed *spatial mental models*. Unlike cognitive maps, they may not preserve metric information. Unlike cognitive collages, they do preserve coarse spatial relations coherently. These are relations that are easily comprehended from language as well as from direct experience. In the third section I will review some evidence for the success of language in inducing coherent mental representations of the categorical spatial relations in environments.

2 Systematic Errors in Memory for Environments

2.1 Hierarchical Representations of Space

When students at U.C. San Diego were asked to draw the direction between San Diego and Reno, they incorrectly indicated that San Diego was west of Reno [36]. Indeed, it is surprising to learn that Reno is in fact west of San Diego. After all, California is on the western coast of the United States, and Reno is far inland, in Nevada. A glance at a map reveals that the coast of California, far from running north-south, in fact cuts eastward as it cuts southward. Stevens and Coupe attributed their findings to hierarchical representations of space. People do not remember the absolute locations of cities. Instead, they remember the states cities are part of, and the relative locations of the states. Then they infer the relative locations of cities from the locations of their superset states.

   Since this (shall I call it "landmark?") study, other evidence for hierarchical representations of geographic knowledge has accumulated. Hierarchical organization has been found to distort distance judgments as well as direction judgments [11].

Hirtle and Jonides asked one group of students at the University of Michigan in Ann Arbor to form subjective groups of buildings in town. They grouped the buildings according to function, commercial or educational. Another group of students was asked to judge distances between pairs of buildings. Distances between functional groupings were overestimated relative to distances within functional groupings. Chase found that a detailed hierarchical organization distinguished experienced taxi drivers from novices [4]. Other studies have demonstrated that people impose a hierarchy on what is in reality a flat two-dimensional display, and that that affects judgment and memory for environments [for example, 12, 23, 24, 25, 44; for a brief review, see 42 and 43]. Of course there are no hierarchies in maps, so this widespread cognitive phenomenon already introduces a distorting factor difficult to reconcile with maps.

2.2 Cognitive Perspective

Experienced hikers know that distances between nearby landmarks appear relatively larger than distances between faraway landmarks, though it is difficult to make adequate compensation for that. A similar phenomenon occurs in making distance judgments from memory. Holyoak and Mah [14] asked one group of students to imagine themselves on the East Coast of the United States, and another group to imagine themselves on the West Coast of the United States. Both groups were then asked to estimate the distances between pairs of U. S. cities along an east-west axis, for example, San Francisco and Salt Lake City, New York City and Pittsburgh. The students given a West Coast perspective overestimated the distances between the westerly pairs relative to the easterly pairs, and the students given an East Coast perspective did the opposite. Thus, the vantage point assigned for making the judgments systematically distorted the judgments.

2.3 Cognitive Reference Points

When I am out of state and asked where I live, I usually answer, "Near San Francisco." If I am closer to home, I will answer, "On Stanford campus," or "Off Stanford Avenue," or "Next door to the ----s." In other words, rather than giving an exact location, I convey where I live relative to a reference point [see 8] that I believe my questioner will know. Not only do we describe less-known locations relative to better-known landmarks, we also seem to remember them that way. As is often the case in memory, we describe situations to ourselves just as we would describe them to others.

Remembering less prominent locations relative to landmarks induces a distortion that is particularly intractable for metric maps, namely asymmetric distance. Sadalla, Burroughs, and Staplin [32] have found that people judge the distance from an ordinary building to a landmark to be *smaller* than the distance from a landmark to an ordinary building.

2.4 Alignment

Remembering one spatial location with respect to another leads to direction distortions as well. Two nearly-aligned locations tend to be grouped, in a Gestalt sense, in memory, and then remembered as more closely aligned than they actually were [41]. Students were given two maps of the Americas, one a correct map, and the other, a map in which South America was moved westward with respect to North America, so that the two Americas were more closely aligned. A significant majority of the students thought the altered map was the correct one. Another group of students selected a world map in which the Americas were moved northward relative to Europe and Africa in preference to a correct map of the world. In the preferred incorrect map, the United States was more closely aligned with Europe and South America with Africa.

Alignment errors in memory were also obtained for judgments of directions between cities, for example, students incorrectly thought that Boston was west of Rio de Janeiro and that Rome was south of Philadelphia. Alignment was also observed in memory for local environments most likely learned from navigation rather than maps, in memory for artificial countries and cities, and in memory for blobs not interpreted as maps.

2.5 Rotation

Remembering a spatial location relative to a frame of reference can also lead to direction distortions [41]. Think of a situation where the orientation of a land mass is not quite the same as the orientation of its frame of reference. A good example is South America, which appears to be tilted in a north-south east-west frame. In fact, when students were given cutouts of South America and asked to place them correctly with respect to the canonical directions, most students uprighted South America. Similar errors appeared for the San Francisco Bay Area, the environment immediately surrounding the students, and for artificial maps and blobs as well, in our work as well as that of others [4,21, 22].

2.6 Other Systematic Errors

This is by no means a complete catalog of systematic errors in memory and judgment of environments. Irregular geographic features may be regularized. For example, Parisians straighten out the Seine [27], and Americans seem to straighten out the Canadian border [36 as interpreted by 43]. Turns and angles are regularized to right angles [3, 13, 29, 34]. Distance judgments are arguably more complex than direction judgments. They are rarely known directly, so they seem to entail use of a number of surrogates that may yield distortions. Distances have been judged longer when a route has barriers or detours [7, 18, 30], when a route has more turns or nodes [33, 35], and when a route has more clutter [40].

2.7 Cognitive Collages

Thus, a number of different factors, hierarchical representations, cognitive perspectives, cognitive reference points, alignment to other locations, rotation to a frame of reference, regularization of geographic features, and more, can systematically distort memory and judgment of environments. On the whole, each empirical study has isolated the effects of a single factor, but in real cases, many factors may be operative. There is no guarantee that the distorting factors are consistent; in fact, it seems easy to construct cases where one factor would distort in one direction and another factor in another direction. The distortions alone are incompatible with a metric mental map, and inconsistent distortions make mental maps an even less satisfactory explanation. Of course, not all our spatial knowledge is distorted. Some of it may be quite accurate. But even so, it is unlikely to be complete, so that problems arise when trying to put it all together, especially if some of the information is erroneous and if the information from different sources is not compatible.

The inconsistencies, however, seem to provide a mechanism to reduce error. When subjects are asked for more information from an environment, it turns out that their judgments become more accurate [1, 2, 26]. This could happen because when confronted with their own inconsistencies, people retrieve additional information that allows them to reconcile the inconsistencies in the correct direction. It could also happen if there were a large number of unreliable judgments, with the majority going toward the correct. Figures can emerge from collages. In many real world situations, however, people are asked for partial information and may not use other information as a corrective.

2.8 Two Basic Relations

The situation is not always as chaotic as I've implied. Most of us manage to find our ways most of the time, either because an environment is familiar, or because we use maps or instructions or environmental cues or all of the above or more. For some well-learned environments, large-scale or small, people's knowledge can be well-organized and systematic. In those cases, the knowledge often has the form of locating elements relative to one another from a point of view or of locating an element relative to a higher order environmental feature or reference frame. Interestingly, the systematic errors described depend on these basic relations. The errors attributable to cognitive perspective, cognitive reference points, and alignment rest on representing landmarks relative to one another from a vantage point. The errors attributable to hierarchical organization and to rotation are based in representing a landmark relative to a higher order feature, a region or a frame of reference. Although much of human knowledge about space, including systematic errors, can be reduced to these two relations, some cannot.

These two simple relations can form a foundation for spatial knowledge from which memory and judgment are constructed. Although the relations can be quite coarse, they can also be refined by adding constraints imposed by other spatial relations. Significantly, these relations also form the basis for spatial language used in descriptions of environments. Because the spatial relations between elements or between an element and a reference frame can be expressed by the many disparate formats that convey environmental knowledge, the relations provide a means to integrate spatial information from different formats.

3 Spatial Descriptions

One of the major functions of language is to convey experience vicariously. Anyone who has laughed out loud reading a novel or felt their heart beat rapidly reading a mystery knows that. Describing space effectively must have been an early use of language, in order to tell others where to find food and where to avoid danger. Although modern-day spatial language can convey locations of landmarks with great accuracy using formal systems designed for that purpose, everyday spatial language is not very precise. Typical spatial expressions like "next to," "between," "to the left of," "in front of," "east of," and "on top of" describe spatial relations at coarse levels of precision, but their frequency in the language suggests that they are easily produced and readily understood. These expressions convey the relations between elements. Expressions like "within," "contains," "divides," "borders," and "curves" convey the relations between elements and reference frames. [For more discussion of spatial language, see for example, 6, 9, 15, 19, 28, 31.]

3.1 Comprehending Route and Survey Descriptions

Taylor and I have been interested in the nature of the spatial information that language alone can impart [37, 38]. Thus far, we have only investigated those spatial expressions that seem to be readily produced and understood. Spatial descriptions normally assume a perspective, explicit or implicit. An informal survey of guidebooks indicated that descriptions of environments take one of two perspectives. A *route* perspective takes readers on a mental tour of the environment, describing landmarks with respect to the (mentally) changing position of the reader in terms of the reader's front, back, left, and right. A *survey* perspective gives readers a bird's eye view, and describes landmarks relative to one another in terms of north, south, east, and west. These two perspectives have parallels with two major means of learning about environments, the first through exploration, and the second through maps. They also have parallels to a distinction made in knowledge representation that is both popular and controversial, namely, procedural and declarative.

Design. In our first set of experiments [38], students studied either a route or a survey description of each of four environments. Two of the environments were large-scale, one county-sized and the other a small town, and two were smaller, a zoo and a convention center. The environments contained about a dozen landmarks. After studying the descriptions, students responded true or false to a series of statements: verbatim statements taken from both the perspective read and the other perspective and inference statements from both perspectives. The inference statements contained information that was not explicitly stated in either text, but could be inferred from information in either text. If perspective was encoded in the mental representations, then inference statements from the read perspective should be verified more quickly than inference statements from the other perspective. After responding to the statements, students drew maps of the environments.

Results. From only studying the descriptions, students were able to produce maps that were nearly error-free, indicating that language alone was sufficient to accurately convey coarse spatial relations. The speed and accuracy to answer the true/false questions suggested that readers formed at least two mental representations of the text, one of the language of the text, and another of the situation described by the text, that is the spatial relations among the landmarks. We termed the latter a spatial mental model [cf. 16] to distinguish it from an image. Responses to verbatim statements were faster and more accurate than responses to inference statements. Presumably, verbatim statements were verified against a representation of the language of the descriptions, but inference statements had to be verified against a representation of the situation, a spatial mental model. Even though responses we refaster and more accurate to verbatim statements, the overall level of responding to inference statements was high. Subjects were able to verify spatial relations not specifically stated in the text, further support for the creation of spatial mental models. Responses to inference statements from the read perspective were neither faster nor more accurate than responses to inference statements from the other perspective, for both perspectives. This result was obtained in four separate experiments, including one where the students read only a single description and did not know they would be asked to draw maps.

Spatial Mental Models vs. Images. In this situation, where subjects studied coherent spatial descriptions of relatively simple environments, perspective did not seem to be encoded in the spatial mental models. Rather, the spatial mental models constructed seemed to be more abstract than either perspective. These spatial mental models appeared to capture the spatial relations among landmarks in a perspective-free manner, allowing the taking of either perspective with equal ease. As such, these spatial mental models are akin to an architect's model or a structural description of an object. They have no prescribed perspective, but permit many perspectives to be taken on them. Thus, spatial mental models are more abstract than images, which are restricted to a specific point of view [see 10, 17].

3.2 Producing Spatial Descriptions

Descriptions composed of the simple spatial relations between landmarks and between landmarks and reference frames were successful in inducing coherent spatial mental representations. It is then natural to ask what is the nature of the spatial descriptions that ordinary people spontaneously produce. In two experiments, Taylor and I [39] gave students maps to study, and asked them to write descriptions of the environments from memory. A compass rose appeared in each of the maps, allowing orientation with respect to the canonical axes, north-south and east-west. In a third stud y, we asked subjects to write descriptions of familiar environments they had learned from experience.

Survey, Route, and Mixed Descriptions. The descriptions subjects produced indicated that subjects regarded the maps as environments, and not as marks on pieces of paper. Perspective was scored using the definitions of route and survey described previously. As our intuitions suggested, descriptions used either route or survey perspectives, or a combination of both. No other style of description emerged. In the mixed perspective descriptions. either one perspective was used for parts of an environment and the other perspective for other parts, or both perspectives were used simultaneously for at least part of the descriptions. Across a wide variety of environments, survey, route, and mixed descriptions were obtained, their relative frequency depending in part on features of the environments. This was despite widespread claims that most spatial descriptions take a consistent perspective, specifically, a route perspective [for review, see 20]. The descriptions that subjects wrote from memory were quite accurate. They allowed a naive group of subjects to place nearly all the landmarks correctly [37].

Basic Relations and Coherence. A detailed analysis of the words, phrases, and clauses used in the descriptions revealed that the essence of a route description was describing the locations of landmarks relative to a single referent with a known perspective, in this case, the moving position of the reader. The essence of a survey description was describing the location of a landmark relative to the location of another landmark from a fixed perspective. These parallel the two basic relations described earlier. The situation is slightly more complex, however. In route descriptions, although the referent was constant, the orientation and location of the referent kept changing. Readers had to keep track of that orientation and location relative to the canonical frame of reference. Both our own and our subjects' route descriptions oriented readers with respect to north-south east-west. In survey descriptions, the referent kept changing, but the orientation was constant. Route descriptions, then, establish coherence by relating all landmarks to a single referent. They are complicated by the task of keeping track of the orientation of the referent. Survey descriptions establish coherence by using a single orientation, but they are complicated by changing the referent element. When either type of information is consistent and complete, as it was in the descriptions we wrote and in many of the descriptions subjects wrote, the individual pieces of information can be integrated into a coherent representation of the spatial relations among the landmarks independent of any specific perspective.

3.3 Spatial Mental Models

The integration of the relative locations of landmarks independent of perspective or orientation that occurs when people read spatial descriptions also seems to occur as people navigate the world. It would be inefficient to remember successive snapshots of the world because they would not allow recognition or navigation from other points of view. It makes more sense to isolate landmarks, and to remember their locations relative to one another and relative to a frame of reference so that recognition and way-finding are successful from different starting points. For simple, familiar environments, whether learned from direct experience, or learned vicariously through language, people can form coherent mental representations of the spatial relations among landmarks.

4 Conclusions

Despite its considerable appeal, as traditionally used, the "cognitive map" metaphor does not reflect the complexity and richness of environmental knowledge. That knowledge comes in a variety of forms, memory snippets of maps we've seen, routes we've taken, areas we've heard or read about, facts about distances or directions. It can also include knowledge of time zones and flying or driving times and climate. Even knowledge of historical conquests and linguistic families can be used to make inferences about spatial proximity. Some of that information may contain errors, systematic or random. When we need to remember or to make a judgment, we call on whatever information seems relevant. Because the snippets of information may be incomparable, we may have no way of integrating them. For those situations, *cognitive collage* is a more fitting metaphor for environmental knowledge.

Yet, there are areas that we seem to know quite well, either because they are familiar or simple or both. Even in those cases, metric knowledge can be schematic or distorted. What our knowledge seems to consist of in those cases is the coarse spatial relations among landmarks, what we have termed *spatial mental models*. Although spatial mental models may not allow accurate metric judgments, they do allow spatial perspective-taking and inferences about spatial locations. They are constructed from basic spatial relations, relations between elements with respect to a perspective or between an element and a frame of reference.

Those situations that are simple and that we know well are also easy to describe. Languages abound in expressions for categorical spatial relations. These expressions are readily produced and easily understood. Although many languages have adopted technical systems to convey metric information about location, orientation, and distance, this terminology is not widely used in everyday speech. When it is used in everyday situations, it is often used schematically. Apparently, descriptions using categorical spatial relations are sufficient for everyday uses.

Viewed as mental models or cognitive collages, environmental knowledge is not very different from other forms of knowledge. Just as for environments, there are areas of other knowledge where our information is consistent and integrated, but there are also areas where, because of incompleteness or incomparability or error, information cannot be consistent and integrated.

References

1. J. Baird: Studies of the cognitive representation of spatial relations: I. Overview. Journal of Experimental Psychology: General 108, 90-91 (1979)

2. J. Baird, A. Merril, J. Tannenbaum: Studies of the cognitive representations of spatial relations: II. A familiar environment. Journal of Experimental Psychology: General 108, 92-98 (1979)

3. R. W. Byrne: Memory for urban geography. Quarterly Journal of Experimental Psychology 31, 147-154 (1979)

4. W. G. Chase: Spatial representations of taxi drivers. In: Dr. R. Rogers, J. A. Sloboda (eds.): Acquisition of symbolic skills. New York: Plenum 1983

5. W. G. Chase, M. T. H. Chi: Cognitive skill: Implications for spatial skill in large-scale environments. In: J. H. Harvey (ed.): Cognition, social behavior, and the environment. Hillsdale, NJ: Erlbaum 1981

6. H. H. Clark: Space, time, semantics, and the child. In: T. E. Moore (ed.): Cognitive development and the acquisition of language. New York: Academic Press 1973, pp. 27-63

7. R. Cohen, L. M. Baldwin, R. C. Sherman: Cognitive maps of a naturalistic setting. Child Development 49, 1216-1218 (1978)

8. H. Couclelis, R. G. Golledge, N. Gale, W. Tobler: Exploring the anchor-point hypothesis of spatial cognition. Journal of Environmental Psychology 7, 99-122 (1987)

9. C. J. Fillmore: Santa Cruz lectures on deixis. Bloomington: Indiana University Linguistics Club 1975

10. R. A. Finke, R. N. Shepard: Visual functions of mental imagery. In: K. R. Boff, L. Kaufman, J. P. Thomas (eds.): Handbook of perception and human performance. New York: Wiley 1986

11. S. C. Hirtle, J. Jonides: Evidence of hierarchies in cognitive maps. Memory and Cognition 13, 208-217 (1985)

12. S. C. Hirtle, M. F. Mascolo: The effect of semantic clustering on the memory of spatial locations. Journal of Experimental Psychology: Learning, Memory and Cognition 12, 181-189 (1986)

13. S. C. Hirtle, M. F. Mascolo: The heuristics of spatial cognition. Proceedings of the 13th annual conference of the cognitive science society. Hillsdale, N: Erlbaum 1992, pp. 629-634

14. K. J. Holyoak, W. A. Mah: Cognitive reference points in judgments of symbolic magnitude. Cognitive Psychology 14, 328-352 (1982)

15. R. Jackendoff, B. Landau: Spatial language and spatial cognition. In: D. J. Napoli, J. A. Kegl (eds.), Bridges between psychology and linguistics: A Swarthmore Festschrift for Lila Gleitman. Hillsdale, NJ: Erlbaum 1991), pp. 145-169

16. P. N. Johnson-Laird: Mental models. Cambridge, MA: Harvard University Press 1983

17. S. M. Kosslyn: Image and mind. Cambridge, MA: Harvard University Press 1980

18. S. M. Kosslyn, H. L. Pick, G. R. Fariello: Cognitive maps in children and men. Child Development 45, 707-716 (1974)

19. G. Lakoff: Women, fire, and dangerous things: What categories reveal about the mind. Chicago: University of Chicago Press 1987

20. W. J. M. Levelt: Speaking. Cambridge, MA: MIT Press 1989

21. R. Lloyd: Cognitive maps: Encoding and decoding information. Annals of the Association of American Geographers 79, 101-124 (1989)

22. R. Lloyd, C. Heivly: Systematic distortions in urban cognitive maps. Annals of the Association of American Geographers 77, 191-207 (1987)

23. R. H. Maki: Categorization and distance effects with spatial linear orders. Journal of Experimental Psychology: Human Learning and Memory 7, 15-32 (1981)

24. T. P. McNamara: Mental representations of spatial relations. Cognitive Psychology 18, 87-121 (1986).

25. T. P. McNamara, J. K. Hardy, S. C. Hirtle: Subjective hierarchies in spatial memory. Journal of Experimental Psychology: Learning, Memory, and Cognition 15, 211-227 (1989)

26. A. Merril, J. Baird: Studies of the cognitive representations of spatial relations: III. A hypothetical environment. Journal of Experimental Psychology: General 108, 99-106 (1979)

27. S. Milgram, D. Jodelet: Psychological maps of Paris. In: H. M. Proshansky, W. H. Ittelson, L. G. Rivlin (eds.): Environmental psychology. Second Edition. New York: Holt, Rinehart and Winston 1976, pp. 104-124

28. G. A. Miller, P. N. Johnson-Laird: Language and perception. Cambridge, MA: Harvard University Press 1976

29. I. Moar, G. H. Bower: Inconsistency in spatial knowledge. Memory and Cognition 11, 107-113 (1983)

30. N. Newcombe, L. S. Liben: Barrier effects in the cognitive maps of children and adults. Journal of Experimental Child Psychology 34, 46-58 (1982)

31. G. Retz-Schmidt: Various view on spatial prepositions. AI Magazine 9, 95-105 (Summer 1988)

32. E. K. Sadalla, W. J. Burroughs, L. J. Staplin: Reference points in spatial cognition. Journal of Experimental Psychology: Human Learning and Memory 5, 516-528 (1980)

33. E. K. Sadalla, S. G. Magel: The perception of traversed distance. Environment and Behavior 12, 65-79 (1980)

34. E. K. Sadalla, D. R. Montello: Remembering changes in direction. Environment and Behavior 21, 346-363 (1989)

35. E. K. Sadalla, L. J. Staplin: The perception of traversed distance: Intersections. Environment and Behavior 12, 167-182 (1980)

36. A. Stevens, P. Coupe: Distortions in judged spatial relations. Cognitive Psychology 13, 422-437 (1978)

37. H. A. Taylor, B. Tversky, Descriptions and depictions of environments. Memory and Cognition 20, 483-496 (1992)

38. H. A. Taylor, B. Tversky: Spatial mental models derived from survey and route descriptions. Journal of Memory and Language 31, 261-282 (1992)

39. H. A. Taylor, B. Tversky: Perspective in spatial descriptions (in preparation)

40. P. W. Thorndyke: Distance estimation from cognitive maps. Cognitive Psychology 13, 526-550 (1981)

41. B. Tversky, Distortions in memory for maps. Cognitive Psychology 13, 407-433 (1981)

42. B. Tversky: Distortions in memory for visual displays. In: S. R. Ellis (ed.), M. K. Kaiser, A. Grunwald (assoc. eds.): Pictorial communication in virtual and real environments. London: Taylor and Francis 1991, pp. 61-75

43. B. Tversky: Distortions in cognitive maps. Geoforum 23, 131-138 (1992)

44. R. N. Wilton: Knowledge of spatial relations: The specification of information used in making inferences. Quarterly Journal of Experimental Psychology, 31, 133-146 (1979)

Preparation of this paper and the research described were supported by the Air Force Office of Scientific Research, Air Force Systems Command, USAF, under grant or cooperative agreement number, AFOSR 89-0076.

A Logical Framework for Reasoning about Space

Laure VIEU

Laboratoire d'Intelligence Artificielle,
INRA, BP 27, 31326 Castanet-Tolosan cedex, FRANCE
email: vieu@toulouse.inra.fr

Abstract. In this paper, we present a theory of space as a framework for spatial reasoning. We believe this formalism is useful for representing geographic space, at least when two constraints are present: a necessity to reason qualitatively over spatial information, and a lack of precise, homogeneous spatial data. This theory is based on mereology, an axiomatic theory of part-whole relation. It includes a formalization of topological concepts as well as some geometric notions, namely distance and orientation. It can be extended to a theory of space-time.

1 Introduction

In this paper, we present a theory of space as a framework for spatial reasoning. This logical formalism was first developed as a part of a system for understanding natural language expressions referring to space [2]; thus, it suits the cognitive structure of space particularly well. Since it represents space symbolically, this formalism can also be used as a framework for qualitative spatial reasoning. Actually, a system based on a theory of the same kind has been developed in qualitative physics for modeling physical systems such as a force pump [19]. Our belief is that this formalism is also useful for representing geographic space, at least when two constraints are present: a necessity to reason over spatial information, and a lack of precise, homogeneous spatial data.

We are now developing our theory within a project aiming at setting up artificial intelligence (AI henceforth) methods for modeling and simulating natural phenomena evolving in space, with the perspective of environmental protection [4]. To ground our research on real problems, we studied three concrete applications: forest fires, avalanches and concentrated flow erosion. The latter will be taken as an illustration in this paper. This special kind of erosion occurs on cultivated lands characterized by moderate topographic and climatic conditions. Simulating this phenomena implies predicting the location of the so-called ephemeral hydrographic network generated in a given watershed by a given rainfall, and evaluating the water and soil quantities circulating in each channel of this network [17]. As a consequence, the project requires developing spatial and qualitative reasoning methods [23].

After justifying its relevance for this project in section 2, section 3 describes the theory of space on which our system, used for both representing spatial information and reasoning over it, is based.

2 A case for a symbolic representational formalism

Reasoning over spatial information first requires choosing a formalism to represent this information. This formalism should be appropriate in two respects: it has to faithfully render the available data and make the kind of reasoning needed possible.

Spatial reasoning is an AI discipline which has been mainly developed in robotics and vision. In both cases, data are measures given by sensors; they may be incomplete but generally, their granularity is homogeneous. In addition, in robotics and especially in route planning, reasoning results must be numerical inputs to effectors.

For these reasons, many works in spatial reasoning are based on space representations of "absolute" type (i.e. reference frames where each object is given a position by means of coordinates) and computational geometry algorithms. Spatial reasoning proper occurs at a high level, for instance when heuristics are applied in path finding.

However, it is sometimes needed to describe space "symbolically", that is, by spatial relations between objects. This is the case in natural language processing and other fields linked to cognitive science, because mental space is, at least partly, symbolic [8]. It is the case in man-machine interfaces, for instance when the output of calculations has to be readily intelligible. This requirement appears also more generally, when precise geometric calculations are unworkable owing to lack of data or knowledge, and when the reasoning process is essentially based on the discrimination between qualitatively different spatial situations. These two cases occur in qualitative physics [7, 9, 19] as well as in our problem of physical process modeling. For instance, since the regions concerned with concentrated flow erosion present rather low slope angles, no numerical data are precise enough to calculate the actual runoff route.

Often though, representational formalisms used for describing space relationally ("languages" of spatial relations) are conceived as a layer on top of absolute-type space representations, i.e. spatial relations are extracted from a numerical space. In general, these formalisms cannot be used independently because some properties of space are only assumed (i.e., not explicit), being embedded in the numerical representation on which they are grounded. Reasoning over spatial relations is then difficult in these formalisms; this is the case, for instance, with geographical information systems (GIS).

A real symbolic approach [10, 11, 18, 19] uses from the start a relational space, integrating all the needed spatial properties as possible inferences from a combination of relations. In such an approach, which can be compared with interval-based representations of time [1], spatial reasoning is involved in the most basic tasks (e.g. transitivity of inclusion). It should be noted that even though some complex reasoning tasks may then be easier than in a numerical approach (e.g. predicting that a die thrown into a funnel will come out [7]), others are intractable: the relational approach cannot replace the absolute one when, for instance, a precise value of an angle or a distance has to be calculated from numerical data such as digitized images. On the other hand, a symbolic approach is more appropriate for dealing with imprecise data (e.g. the information "a is between b and c" can be exploited even when the distances are unknown) and with non-homogeneous data (e.g. distinct parts of space are described at different scales, that is, some parts are more detailed than others). This does not prevent a symbolic representation from being integrated with a numerical one. Indeed, we think that it is one of the most promising ways leading to spatial reasoning in geographic space.

In the remainder of this paper, we present a relational formalism for representing space and reasoning about it. To be as explicit and rigorous as possible, we give its theoretical framework, a theory of space expressed in higher-order logic with equality.

3 A Theory of Space

Unlike theories of time [3], relatively few theories have been proposed for modeling space. Ours is an extension of the theory proposed by B. L. Clarke [5, 6], on which [19] is also based. This theory comes in two main parts. The first describes mereological and topological relations between spatial entities. This part is adapted from [5], eliminating the fusion operator that made it second-order. The second, which is based on the first, builds the "points" to be considered in a given situation and then describes orientation and distance relations between these points. This part is an extended version of the theory described in [22].

3.1 Mereology and Topology

Mereology is a theory first introduced by Lesniewski early in this century as a component of a system proposed as an alternative to set theory [15]. Since then, mereology has repeatedly been taken up as a theory of part-whole relation in formal ontology [12, 14, 20]. In [5, 6, 21] it constitutes the basis of axiomatic theories for topology and geometry taking "individuals" or "bodies" as primitive entities instead of points. Individuals can there be seen as representing portions of space (or even space-time) determined by concrete objects, and need not be considered as sets of spatial points.

We present in this paper a theory of geometry based on the system proposed by Clarke in [5, 6] for modeling topology. This mereological / topological system is built on the sole primitive relation of "connection". This choice makes it quite distinct from "classical mereology" [14, 15] which is usually based on one of the following relations: "proper part", "overlap" or "discrete from".

In this theory, two individuals are "connected" when they share some part or when they are in contact. Connection is axiomatized as follows in [5]:
Connection is reflexive and symmetric:
A1 $\forall x\ C(x,x)$
A2 $\forall x\ \forall y\ (C(x,y) \Rightarrow C(y,x))$
Two individuals are (spatially) equal when they are connected with the same individuals:
A3 $\forall x\ \forall y\ (\forall z\ (C(z,x) \Leftrightarrow C(z,y)) \Rightarrow x=y)$
Several mereological relations are then defined:
D1 $P(x,y) \equiv_{def} \forall z\ (C(z,x) \Rightarrow C(z,y))$ "x is part of y"
D2 $PP(x,y) \equiv_{def} P(x,y) \wedge \neg P(y,x)$ "x is a proper part of y"
D3 $O(x,y) \equiv_{def} \exists z\ (P(z,x) \wedge P(z,y))$ "x overlaps y"

Contact, also called "external connection", tangential part and non tangential part are not relations belonging to classical mereology. Their definition was made possible by Clarke's choice of the connection relation as the primitive:
D4 $EC(x,y) \equiv_{def} C(x,y) \wedge \neg O(x,y)$ "x is externally connected to y"

D5 TP(x,y) $\equiv_{def}$ P(x,y) $\wedge$ $\exists$z (EC(z,x) $\wedge$ EC(z,y))   "x is a tangential part of y"
D6 NTP(x,y) $\equiv_{def}$ P(x,y) $\wedge$ $\neg\exists$z (EC(z,x) $\wedge$ EC(z,y)) "x is a non tangential part of y"

These relations can be represented pictorially as in Figure 1:

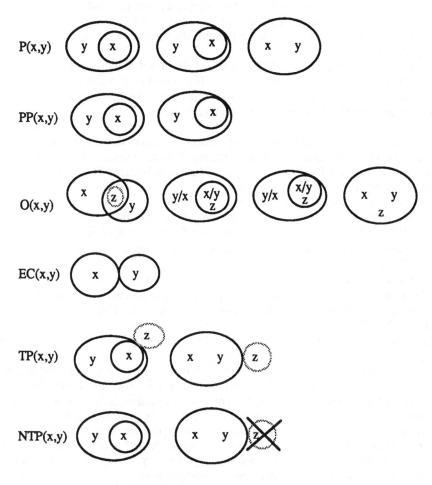

Fig. 1. Mereological relations.

Those definitions together with the three axioms confer the desired inferential properties to the relations: part, proper part and non tangential part are transitive, overlap and external connection are symmetric... Many properties are also obtained combining different relations; for instance we have:

$$\forall x \; \forall y \; \forall z \; ((P(x,y) \wedge O(x,z)) \Rightarrow O(y,z))$$

which states that an individual overlapping a part of another individual also overlaps the latter.

Classical mereology as well as Clarke's contain a fusion operator for summing up any collection of individuals into a new individual. It is principally used for defining Boolean operators such as intersection or complement. We felt that this general operator was here unnecessary. Indeed, it is possible to introduce the sum and the intersection of two individuals, the complement of an individual, and the universal individual without going second-order in this first part of the theory. We achieved this with the following axioms:

A4 $\forall x\, \forall y\, \exists z\, \forall u\, (C(u,z) \Leftrightarrow \exists v\, ((P(v,x) \vee P(v,y)) \wedge C(v,u)))$

this[1] individual z is noted x+y and represents the sum of x and y.

A5 $\forall x\, (\exists y\, \neg C(y,x) \Rightarrow \exists z\, \forall u\, (C(u,z) \Leftrightarrow \exists v\, (\neg C(v,x) \wedge C(v,u))))$

this individual y is noted -x and represents the complement of x.

A6 $\exists x\, \forall u\, C(u,x)$

this individual x is noted a* and represents the universal individual.

A7 $\forall x\, \forall y\, (O(x,y) \Rightarrow \exists z\, \forall u\, (C(u,z) \Leftrightarrow \exists v\, (P(v,x) \wedge P(v,y) \wedge C(v,u))))$

this individual z is noted x∩y and represents the intersection of x and y.

We get then a pseudo-Boolean algebra over individuals.[2]

The notion of non tangential part, stemming from the notion of external connection, enables Clarke to introduce the topological notions of interior, closure, and open or closed individual. Here again, Clarke's definitions based on the fusion operator are replaced by first-order axioms:

A8 $\forall x\, \exists y\, \forall u\, (C(u,y) \Leftrightarrow \exists v\, (NTP(v,x) \wedge C(v,u)))$

this individual y is noted ix and represents the interior of x.

A9 $\forall x\, (\exists y\, \neg C(y,x) \Rightarrow \exists z\, \forall u\, (C(u,z) \Leftrightarrow \exists v\, (\neg C(v,i(-x)) \wedge C(v,u))))$

this individual z is noted cx and represents the closure of x.[3]

D7 $OP(x) \equiv_{def} x=ix$ "x is open".

D8 $CL(x) \equiv_{def} x=cx$ "x is closed".

Two additional axioms are necessary to ensure that any individual possesses an interior and that the intersection of two open individuals is itself open:

A10 $\forall x\, \exists y\, NTP(y,x)$

A11 $\forall x\, \forall y\, (\forall z\, [(C(z,x) \Rightarrow O(z,x)) \wedge (C(z,y) \Rightarrow O(z,y))] \Rightarrow \forall z\, (C(z,x\cap y) \Rightarrow O(z,x\cap y)))$

These topological notions may not be directly useful by themselves; indeed, the difference between a closed individual and its interior (the former includes its "skin" or boundaries[4] while the latter doesn't) is not always intuitive. Still, it is important to prove that they can be introduced in this theory, for they guarantee that a number of important spatial concepts (e.g. connectivity, continuity) can in turn be defined. For example, they enable us to define a relation of "weak contact" to describe the situation in which two objects touch without sharing any boundary, as opposed to the external connection that can be seen as a relation of "strong contact":

D9 $WCONT(x,y) \equiv_{def} (\neg C(x,y) \wedge \forall z\, ((\neg z=a^* \wedge P(x,z) \wedge OP(z)) \Rightarrow C(cz,y)))$: "x is touching y"

An individual and its complement are not connected but do touch. Two closed individuals may touch, each being part of the other's complement. Weak contact has

---

[1]Note that axiom A3 guarantees the unicity of the individuals whose existence is asserted, and thus justifies the notations we introduce.

[2]It isn't a Boolean algebra because there is no null individual, otherwise any two individuals would overlap. Note that -x exists iff x≠a* and that x∩y exists iff O(x,y).

[3]Note that cx exists iff x≠a*.

[4]Note, though, that these "boundaries" cannot be handled directly as individuals in this theory, otherwise no difference between external connection and overlap could be made. After the introduction of points in the following section, boundaries may be defined as sets of boundary points, see [2, 22].

proved to be useful in analyzing the semantics of natural language's spatial relations [2] and should be more appropriate than external connection in modeling worlds evolving through time since the occurrence or disappearance of EC(x,y) alters x and y identity whereas WCONT doesn't.

3.2 Points and Geometry

Up to now, we have set up a theory of space without talking at all about spatial points. Set-theoretical and topological concepts such as inclusion and closure have been introduced in a non standard way (mereology aside). This presents significant advantages concerning representational issues for we need not having a function assigning particular sets of points to every spatial object.

Indeed, such an assignment would either be arbitrary or stay at a theoretical level (i.e. the points would be abstract entities not linked to any reality): In the first case, arbitrariness would occur when the exact positions of objects are not known, and moreover simply because a specific scale, or granularity, would have to be chosen. As explained in section 2 this can lead both to erroneous assumptions and to representational limitations. In the second case, one could think that doing so, we could simply use set theory and all of classical geometry. However, our position is that describing spatial relations directly between objects might be preferred for the sake of ontological simplicity. Actually, if the existence of points is just assumed, they cannot be manipulated as referring to anything real. For instance, it would be impossible to know whether a region is included in another by checking the belonging of all the points of the first to the second. So, in order to reason over relations such as inclusion, one would need to axiomatize their behavior, and in fact build a theory very similar to mereology.

Yet, some spatial notions are better grasped when referring to spatial points. This is the case for alignment and distance and, more generally, for all geometric notions resting on the concepts of straight line, circle, etc. Introducing points in the theory while avoiding the arbitrariness described above can be done by *defining* each point as a set of individuals, the exact opposite to the classical point of view. In this approach, the only points present in the representation are those we can "talk about" given the specific situation described; they recover an ontological and cognitive justification. In addition, the exact granularity needed is obtained automatically: a point may be subdivided into several points when further information is added at the mereological / topological level. There are two ways of conceiving point construction in this approach. One is to build maximal sets of nested individuals (for any two individuals of this set, one is part of the other) as put to use in [21]; the other is to apply the ultra-filter technique (for any two individuals, their intersection belongs to the set, any individual having as part a member of the set belongs to the set, and the set is maximal). In [6], Clarke gives a definition of points that matches more or less the ultra-filter construction. However, Clarke's definition is flawed: only very strange structures[5] are models for the resulting theory. We then give the following definitions for interior points (IPT) and boundary points (BPT):[6]

---

[5] As soon as there are two externally connected individuals, the closure of any other individual (universe excepted) must connect those two individuals [22].

[6] The Greek alphabet is used to quantify over sets of individuals; Λ is the null set.

31

D10 IPT(α) $\equiv_{\text{def}} \neg\ \alpha=\Lambda\ \wedge$ (a)
  $\forall x\ \forall y\ ((x\in\alpha \wedge y\in\alpha) \Rightarrow (O(x,y) \wedge x\cap y\in\alpha)) \wedge$ (b)
  $\forall x\ \forall y\ ((x\in\alpha \wedge P(x,y)) \Rightarrow y\in\alpha) \wedge$ (c)
  α maximal (i.e. $\forall\beta\ ((\beta$ verifies (a), (b) and (c)) $\wedge \alpha\subseteq\beta) \Rightarrow \alpha=\beta))$

D11 BPT(α) $\equiv_{\text{def}} \exists x\ \exists y\ (x\in\alpha \wedge y\in\alpha \wedge EC(x,y)) \wedge$ (a)
  $\forall x\ \forall y\ ((x\in\alpha \wedge y\in\alpha) \Rightarrow ((O(x,y) \wedge x\cap y\in\alpha) \vee$
  $\exists z\ \exists t\ (z\in\alpha \wedge t\in\alpha \wedge P(z,x) \wedge P(t,y) \wedge EC(z,t)))) \wedge$ (b)
  $\forall x\ \forall y\ ((x\in\alpha \wedge P(x,y)) \Rightarrow y\in\alpha) \wedge$ (c)
  α maximal

D12 PT(α) $\equiv_{\text{def}}$ IPT(α) $\vee$ BPT(α)   "α is a point"

Figure 2 illustrates these two kinds of points:

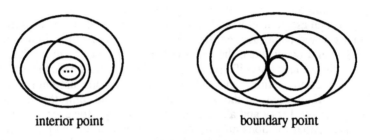

interior point   boundary point

Fig. 2. The two kinds of points.

For the sake of generality, we add an axiom stating the existence of points although it is needed only when there is an infinite number of individuals:
A12 $\forall x\ \forall y\ (C(x,y) \Rightarrow \exists\alpha\ (PT(\alpha) \wedge x\in\alpha \wedge y\in\alpha))$

Relations between points can now be introduced, so we can deal with relative distances and orientation. Two primitives are needed: "α is closer to β than to γ" ($K(\alpha,\beta,\gamma)$), adapted from [3], and "α precedes β" ($\alpha<\beta$). K is axiomatized as follows:
A13 $\forall\alpha\ \forall\beta\ \forall\gamma\ (K(\alpha,\beta,\gamma) \Rightarrow (PT(\alpha) \wedge PT(\beta) \wedge PT(\gamma)))$
K is transitive and irreflexive:
A14 $\forall\alpha\ \forall\beta\ \forall\gamma\ \forall\delta\ ((K(\alpha,\beta,\gamma) \wedge K(\alpha,\gamma,\delta)) \Rightarrow K(\alpha,\beta,\delta))$
A15 $\forall\alpha\ \forall\beta\ \forall\gamma\ \forall\delta\ ((K(\alpha,\beta,\gamma) \wedge K(\gamma,\alpha,\beta)) \Rightarrow K(\beta,\alpha,\gamma))$
A16 $\forall\alpha\ \forall\beta\ \neg K(\alpha,\beta,\beta)$
A17 $\forall\alpha\ \forall\beta\ (\alpha\neq\beta \Rightarrow K(\alpha,\alpha,\beta))$
Space is "connected" for distance:
A18 $\forall\alpha\ \forall\beta\ \forall\gamma\ \forall\delta\ (K(\alpha,\beta,\gamma) \Rightarrow (K(\alpha,\beta,\delta) \vee K(\alpha,\delta,\gamma)))$

From the relation K, it is possible to define a 3-place relation of equidistance (E):
D13 $E(\alpha,\beta,\gamma) \equiv_{\text{def}} \neg K(\alpha,\beta,\gamma) \wedge \neg K(\alpha,\gamma,\beta)$ "α is at an equal distance from β and γ", adding an axiom for triangular equality:
A22 $\forall\alpha\ \forall\beta\ \forall\gamma\ \forall\delta\ ((E(\alpha,\beta,\gamma) \wedge E(\gamma,\alpha,\beta)) \Rightarrow E(\beta,\alpha,\gamma))$.

Other distance-related definitions such as equidistance and order between pairs of points can also be introduced [22]. We then obtain a strict total order between equivalence classes of pairs of points.

Since points are defined as sets of individuals and conversely, the set of points "incident" in an individual can be reconstructed, distance notions between individuals are induced by the axioms on K. For instance, IK(x,y,z), "x is closer to y than to z", is defined as follows:

D14 $IK(x,y,z) \equiv_{def} \exists\alpha \exists\beta (PT(\alpha) \wedge PT(\beta) \wedge x\in\alpha \wedge y\in\beta \wedge \forall\gamma \forall\delta ((PT(\gamma) \wedge PT(\delta)$
$\wedge x\in\gamma \wedge z\in\delta) \Rightarrow INF(\alpha,\beta,\gamma,\delta)))$ — where $INF(\alpha,\beta,\gamma,\delta)$ stands for "the distance between α and β is lesser than the distance between γ and δ", i.e. it's the order relation between pairs of points mentioned above.

It is then easy to show that two connected individuals are at a "null" distance, that is, pairs of connected individuals form an equivalence class which is the smallest with respect to the order defined by IK. It can also be proved that pairs of individuals in weak contact form an equivalence class immediately larger than the null distance one.

$<$ is a strict partial order axiomatized as follows:

A19 $\forall\alpha \forall\beta \ \alpha<\beta \Rightarrow (PT(\alpha) \wedge PT(\beta))$

$<$ is irreflexive and transitive:

A20 $\forall\alpha \ \neg\alpha<\alpha$

A21 $\forall\alpha \forall\beta \forall\gamma ((\alpha<\beta \wedge \beta<\gamma) \Rightarrow \alpha<\gamma)$

$<$ determines a circuit-free oriented graph which can represent orientation from a particular point of view, for example, the upstream/downstream order. For instance, in our erosion application, the graph would represent the potential hydrographic network generated by surface runoff. It may be useful to deal simultaneously with several kind of orientations (e.g. north/south, east/west...); in that case, we introduce as many $<_i$ as needed. Alternatively, direction primitive elements could be used as a third argument of $<$; an algebra for combining such directions could then be axiomatized as done in [10].

Note that we are not restricted dimensionally in any way; this theory can be used for 2D space as well as 3D, as well as nD... As a matter of fact, this theory can be interpreted as being a theory of space-time, that is to say, individuals can be conceived as spatio-temporal (4D) entities. This is particularly useful for the erosion problem, where the watershed configuration evolves in time along with rainfalls and farming operations. All of the relations between individuals introduced above can be seen as spatio-temporal, but in addition, specific temporal relations are needed. For example, complete temporal precedence and temporal overlap between individuals can be added along the lines of [13]; see [22]. In that case, the spatial relations between (spatio-temporal) points have to be restricted to temporally concomitant (simultaneous) points.

From both relations K and $<$, we introduce alignment between three points and straight lines as maximal sets of points that are three by three aligned:

"α is on a line between β and γ":[7]

D14 $B(\alpha,\beta,\gamma) \equiv_{def} ((\beta<\alpha \wedge \alpha<\gamma) \vee (\gamma<\alpha \wedge \alpha<\beta)) \wedge \forall\delta (\alpha=\delta \vee K(\beta,\alpha,\delta) \vee K(\gamma,\alpha,\delta))$

"α, β and γ are on a line":

D15 $AL(\alpha,\beta,\gamma) \equiv_{def} B(\alpha,\beta,\gamma) \vee B(\beta,\alpha,\gamma) \vee B(\gamma,\alpha,\beta)$

"Γ[8] is a straight line":

D16 $SL(\Gamma) \equiv_{def} \Gamma\neq\Lambda \wedge \forall\alpha \forall\beta \forall\gamma ((\alpha\in\Gamma \wedge \beta\in\Gamma \wedge \gamma\in\Gamma \wedge \alpha\neq\beta \wedge \alpha\neq\gamma \wedge \beta\neq\gamma)$
$\Rightarrow (AL(\alpha,\beta,\gamma) \wedge \forall\delta (AL(\alpha,\beta,\delta) \Rightarrow \delta\in\Gamma)))$

---

[7]When there are several orientation relations $<_i$, we add as many clauses as needed so as to take into account each $<_i$: $((\beta<_i\alpha \wedge \alpha<_i\gamma) \vee (\gamma<_i\alpha \wedge \alpha<_i\beta) \vee (\beta<_{i+1}\alpha \wedge \alpha<_{i+1}\gamma) \vee ...)$.

[8]Upcase greek letters take for values sets of sets of individuals.

A last but very important axiom links alignment and equidistance:

A23 $\forall \alpha \; \forall \beta \; \forall \gamma \; \forall \delta \; \forall \epsilon \; ((AL(\gamma,\delta,\epsilon) \wedge E(\gamma,\alpha,\beta) \wedge E(\delta,\alpha,\beta)) \Rightarrow E(\epsilon,\alpha,\beta))$

4- Perspectives

This theory can be developed further; for instance, we can define perpendicularity between two straight lines [22], as well as introduce squares, circles, etc. and thus start a description of simple shapes. The previous section shows that many (naive) geometric concepts can be dealt with in a relational theory of space. However, relatively few works have been dedicated to this topic and some important issues still need to be addressed.

For one thing, points are here built on the basis of mereological and topological data only. This means we assume that the structure of space is completely determined by this kind of data, and that other data (such as distance and orientation) only describe this structure further, without altering it. This assumption has not been secured, though. We are now defining another version of this theory in which distance and orientation concepts are introduced by relations over individuals. This new version should make clear whether our assumption was true or not. Another advantage of this new version will be that the construction of points will then be postponed until all the data has been represented, and merely play the part of an interface between the theory and its models —as done in [13] for time. The points themselves will no longer be dealt with within the theory, so the theory will be first-order throughout and its computational tractability will benefit of the completeness of predicate logic. The difficulty of clearly expressing such concepts as straight lines or circles in terms of relations between individuals could be seen as a drawback, though. If this theory is to match the cognitive structure of space, it may be not a disadvantage, since these concepts are abstract geometric constructs.

Another thing is that distance here is relational, and we may have some numerical values we want to take advantage of. Being able to deal with hybrid reasoning is important in itself; what is more, it is decisive if we want to contemplate the perspective of an integration with GIS. Combining symbolic and numerical distances (as well as surface areas...) is a next step we are studying at the moment. It is not an easy task because it implies dealing with the relative imprecision numerical distances carry. More generally, if we were to use the theory as a language of spatial relations for a GIS,[9] we would have to take into account the inherent imprecision of the maps and the relative quality of heterogeneous data.

In parallel with the exploration of the two research directions mentioned above, we are on the way to implementing the representation and reasoning techniques that come with this theory. Although the erosion simulation is the motivating application that has a guiding role in the project, we try to build a system as independent as possible of the specificities of the erosion problem.

---

[9]For this kind of use, the definitions for points would have to be dropped, obviously: individuals would refer to sets of points in the maps. The axiomatization of the geometric relations between points could remain in order to take advantage of imprecise data in addition to the maps.

References

1. J. Allen: Maintaining knowledge about temporal intervals. Communications of the ACM 26(11), 832-843 (1983)
2. M. Aurnague, L. Vieu: A three-level approach to the semantics of space. In: C. Zelinsky-Wibbelt (ed.): The Semantics of Prepositions - From Mental Processing to Natural Language Processing. Berlin: Mouton de Gruyter, to appear in 1993
3. J. van Benthem: The logic of time. Dordrecht: Reidel 1983
4. L. Buisson, R. Martin-Clouaire, L. Vieu, J.-L. Wybo: Représentation de l'espace, dynamique qualitative et raisonnement spatial. Project of the "Programme Interdisciplinaire de Recherche - Environnement", CNRS 1992
5. B. Clarke: A calculus of individuals based on "connection". Notre Dame Journal of Formal Logic 22(3), 204-218 (1981)
6. B. Clarke: Individuals and points. Notre Dame Journal of Formal Logic 26(1), 61-75 (1985)
7. E. Davis: A logical framework for commonsense predictions of solid object behavior. Artificial Intelligence in Engineering 3(3), 125-140 (1988)
8. M. Denis: Image et cognition. Paris: Presses Universitaires de France 1989
9. K. Forbus: Qualitative reasoning about space and motion. In: D. Gentner, A. L. Stevens (eds.): Mental models. Hillsdale (NJ): Erlbaum 1983, pp. 53-73
10. A. Frank: Qualitative Spatial Reasoning about Distances and Directions in Geographic Space. Journal of Visual Languages and Computing 3, 343-371 (1992)
11. C. Freksa: Using Orientation Information for Qualitative Spatial Reasoning. In: A. Frank, I. Campari, U. Formentini (eds.): Theories and Methods of Spatio-Temporal Reasoning in Geographic Space, Proceedings of the International Conference GIS - From Space to Territory. Berlin: Springer 1992, pp. 162-178
12. N. Guarino, R. Poli (eds.): Proceedings of the International Workshop on Formal Ontology. Padova: LADSEB-CNR (Internal Report 01/93) 1993
13. H. Kamp: Events, instants and temporal reference. In : R. Bäuerle, U. Egli, A. von Stechow (eds.): Semantics from different points of view. Berlin: Springer 1979, pp. 376-417
14. H. Leonard, N. Goodman: The calculus of individuals and its uses. The Journal of Symbolic Logic 5, 45-55 (1940)
15. S. Lesniewski: O podstawach matematyki (On the foundations of mathematics). Przeglad filozoficzny (Philosophical Review) vols. 30-34, (1927-1931).
16. D. Lewis: Parts of Classes. Oxford (MA): Blackwell 1991
17. B. Ludwig: L'érosion par ruissellement concentré des terres cultivées du Nord du Bassin Parisien, Ph.D. dissertation. Strasbourg: Université Louis Pasteur 1992
18. A. Mukerjee, G. Joe: A Qualitative Model for Space. In: Proceedings of the Eighth National Conference on Artificial Intelligence, AAAI 90. Cambridge (MA): MIT Press 1990, pp. 721-727
19. D. Randell, A. Cohn: Modeling topological and metrical properties in physical processes. In: R. Brachman, H. Levesque, R. Reiter (eds.): Proceedings of the First International Conference on Principles of Knowledge Representation and Reasoning. San Mateo (CA): Morgan Kaufmann 1989, pp. 357-368
20. P. Simons: Parts - A study in ontology. Oxford: Clarendon Press 1987

21. A. Tarski: Les fondements de la géométrie des corps. In: A. Tarski: Logique, Sémantique, Métamathématique. Paris: Armand Colin 1972

22. L. Vieu: Sémantique des relations spatiales et inférences spatio-temporelles, Ph.D. dissertation. Toulouse: Université Paul Sabatier 1991

23. L. Vieu, R. Martin-Clouaire: Spatial and Qualitative Reasoning for Modelling Physical Processes. Submitted (20p)

Maintaining Qualitative Spatial Knowledge*

Daniel Hernández

Fakultät für Informatik
Technische Universität München
80290 Munich, Germany
e-mail: danher@informatik.tu-muenchen.de

Abstract. We present mechanisms used to maintain the consistency of a knowledge base of spatial information based on a qualitative representation of 2-D positions. These include the propagation heuristics used when inserting new relations as well as the reason maintenance mechanisms necessary to undo the effects of propagation when deleting a relation. Both take advantage of the rich structure of the spatial domain.

1   Introduction

For a representation to be of any use, we have to consider not only its constituents and how they correspond to what is being represented, but also the mechanisms operating on them. In this paper we look into the mechanisms that allow us to reason with qualitative representations of 2-D positions. These mechanisms are determined in part by the tasks for which qualitative reasoning is used, such as: Inferring knowledge implicit in the knowledge base; answering queries given partial knowledge and a specific context; maintaining various types of consistency; acquiring new knowledge; and, particularly in the case of spatial knowledge, building cognitive maps and visualizing qualitatively represented spatial situations.

Even though the qualitative approach has been extensively used for modeling physical phenomena (Bobrow 1984; Weld and de Kleer 1990), it is only recently that research on qualitative models of space has been undertaken. Allen (1983) introduced an interval-based temporal logic, in which knowledge about time is maintained qualitatively by storing comparative relations between intervals. Freksa (1992a) presents a generalization of Allen's temporal reasoning approach based on semi-intervals and introduces the notion of "conceptual neighborhood" of qualitative relations. There have been some prior efforts to extend Allen's temporal approach to spatial dimensions (Guesgen 1989; Mukerjee and Joe 1990). However, these extensions just use Cartesian tuples of the one-dimensional relations, loosing the "cognitive plausibility" that Allen's approach has in the temporal domain. Our representation of positions in 2-D space establishes different

---

* The project on which the work reported here is based has been funded by the German Ministry for Research and Technology (BMFT) under FKZ ITN9102B. The author is solely responsible for the contents of this publication.

qualitative relations for the two relevant dimensions topology and orientation. Related research on topological relations in the context of Geographic Information Systems has been done by Egenhofer (1989, 1991), Egenhofer and Al-Taha (1992), Egenhofer and Sharma (1993), and Smith and Park (1992). Kuipers and Levitt (1988) describe a series of influential systems for navigation and mapping in large-scale space. Of particular interest here is the QUALNAV model (see also Levitt and Lawton 1990), which includes a coordinate-free, topological representation of relative spatial location, and integrates metric knowledge of relative or absolute angles and distances. The symbolic projection schema introduced by Chang, Shi, and Yan (1987) in the context of pictorial databases represents two dimensional spatial arrangements by projecting them into two "2D-strings" along the vertical and horizontal axes. Various extensions of this model have been proposed including further operators and a combination with quad-trees (Chang and Li 1988), local operators to handle overlapping objects (Jungert 1988). Frank (1991) presents a qualitative algebra for reasoning about cardinal directions, which are easier to analyze then relative orientations because the frame of reference is fixed in space. Cohn, Cui, and Randell (1992) summarize a theory of space and time based on a calculus of individuals founded on "connection" and expressed in the many sorted logic LLAMA. A basic set of dyadic topological relations is defined using the primitive $C(x,y) =$ 'x connects with y', x and y being regions. Freksa (1992b) and Freksa and Zimmermann (1992) present an approach to qualitative spatial reasoning based on directional orientation information. They distinguish 15 possible positions and orientations of a point based on the left/straight/right distinction w.r.t. a vector ab as well as the front/neutral/back distinction w.r.t. the lines orthogonal to ab on the end points of a and b. For a general overview of recent literature in the area of spatial reasoning see, for example, McDermott (1992) and Topaloglou (1991).

In previous work we have explored various aspects of the qualitative representation of space (Hernández 1991) including mechanisms used to transform between different frames of reference (these transformations are necessary to obtain canonical reference frames, which are a pre-requisite for qualitative inference); methods for the efficient computation of composition tables for positional relations based on the structure of the relational domains, and "abstract maps", which allow the solution of some tasks by diagrammatical means. In (Hernández and Zimmermann 1992) we discuss a method for constraint relaxation that uses the structure of the relational domain to weaken constraints by including other neighboring relations in their disjunctive definitions, instead of retracting them as a whole. This approach leads faster to solutions of meaningfully modified sets of otherwise unsatisfiable constraints.

In the following section we first briefly introduce the representation model and concentrate in the later sections on the algorithms required to maintain the consistency of a qualitative knowledge base of spatial information. These include the propagation heuristics used when inserting new relations as well as the reason maintenance mechanisms required to undo the effects of propagation when deleting a relation.

2 Qualitative Representation of Positions in 2-D

We focus on 2-D projections of 3-D scenes. Two factors determine the qualitative position of objects in 2-D space: the relative orientation of objects to each other and the extension of the involved objects. Considering these factors independently from each other results in two classes of spatial relations:

- topological relations (ignore orientation)
- orientation relations (ignore extension, i.e., objects = points)

Our goal is to combine these two classes of relations to provide a model of orientation that accounts for extended objects. For this purpose we define a small set of spatial relations from the two relevant dimensions topology and orientation.

Topological[2] relations describe how the boundaries of the two objects relate. A complete set of topological relations can be derived from the combinatorial variations of the point set intersection of boundaries and interiors of the involved objects by imposing the constraints of physical space on them (Egenhofer and Franzosa 1991). The resulting set of eight mutually exclusive relations is: disjoint (d), tangent (t), overlaps (o), contains-at-border (c@b), included-at-border (i@b), contains (c), included (i), equal (=).

Orientation relations describe where the objects are placed relative to one another. The orientation dimension results from the transfer of distinguished reference axes from an observer to the reference object. There are various levels of hierarchically organized orientation relations of different granularities. The level with the eight distinctions most commonly used contains the following relations (abbreviations in parentheses) $front_3(f_3)$, $back_3(b_3)$, $left_3(l_3)$, $right_3(r_3)$, $left-back_3(lb_3)$, $right-back_3(rb_3)$, $left-front_3(lf_3)$, and $right-front_3(rf_3)$.

Relative orientations must be given w.r.t. a *reference frame*, which can be *intrinsic* (orientation given by some inherent property of the reference object), *extrinsic* (orientation imposed by external factors), or *deictic* (orientation imposed by point of view). When reasoning about orientations, the reference frame is *implicitly* assumed to be the intrinsic orientation of the parent object (i.e., the one containing the objects involved), unless *explicitly* stated otherwise. The relative position is given by a topological/orientation relation pair:

<primary_object, [topological,orientation], ref_object, ref_frame>

2.1 The Structure of the Topological and Orientation Domains

Topological relations have a fork-like neighboring structure, whereas orientations form a uniform circular neighborhood on each level. As we will show below taking advantage of this structure leads to more efficient propagation algorithms. Since we use pairs of topological and orientation relations to represent relative

---

[2] Previous work used the term "projection" instead of topological. This has been changed because of the possible confusion with the use of the word in "2-D projections of 3-D scenes" and in "projective spatial prepositions" (which are related to what we call orientation relations).

positions, it is interesting to look at their combined structure. Figure 1 gives a simplified overview of level 3 orientations and linearly adjacent topological relations. Figure 2 shows a partial detailed view of two neighboring level 2 orientations and all 8 topological relations.[3] Arcs between nodes denote neighboring topological/orientation pairs. In the simplified visualization the containment

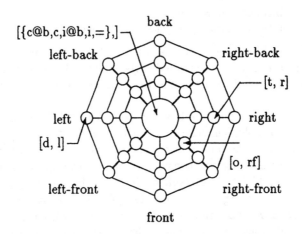

Fig. 1. Combined structure of topological and orientation relations (overview)

and equality topological relations have been merged in a single node. Figure 2 is intended to be seen as a perspective view of a side cut of the 3-D visualization of the combined structure. In order not to clutter the figure unnecessarily, only the two neighboring orientations **back**$_2$ and **right**$_2$ are shown, together with all the links connecting neighboring relation pairs. The c, =, and i nodes are in the middle of the structure, because, according to our conventions, they are not oriented. The oriented nodes (e.g., [i❶b,b$_2$]) are linked to neighboring topological nodes of the same orientation (e.g., [o,b$_2$], [i,b$_2$]), and to equivalent nodes of same topology in neighboring orientations (e.g., [i❶b,r$_2$], [i❶b,l$_2$]).

3   Reasoning with Qualitative Representations

While all spatial reasoning tasks in a qualitative representation can be formulated as constraint satisfaction problems, the general techniques known in the literature incur an unnecessary computational overhead.[4] One reason for this

---

[3] The figures actually omit the links for neighboring relation pairs that result from the simultaneous change of topological and orientation relations as in [d,f] → [t,rf].

[4] Comprehensive reviews of the constraint satisfaction literature can be found in, e.g., Mackworth (1987), Meseguer (1989), Kumar (1992).

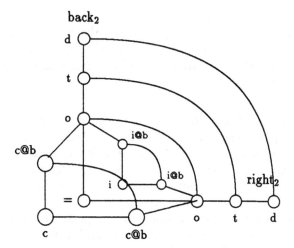

Fig. 2. Combined structure of topological and orientation relations (detail)

overhead is, that they try to achieve global consistency among all constraints, whereas many spatial reasoning tasks require only local consistency. Another reason is, that they ignore the rich structure of space, which further constraints the set of possible solutions.

A qualitative description of a spatial configuration in form of a set of positional relations can be represented as a constraint network, where nodes corresponding to objects are linked by arcs corresponding to the relative positional relations between two objects. Sets of relations (relsets, for short), corresponding conceptually to disjunctions of possible relations between two objects, are ubiquitous in the algorithms to be described. Disjunctions are a way of expressing uncertainty about the "real" relation between the objects. Whenever the number of different relations is relatively small, as is always the case with qualitative representations, a bit-string representation of sets is the usual choice.

The propagation of constraints in the network is necessary to check the consistency of the relations among adjacent objects, for example, after insertion of a new relation or deletion of a previously assumed relation. In the following subsections, we will describe algorithms for these two tasks that take advantage of the structure of space described in the previous section.

3.1 Inserting New Relations

Inserting a new relation between two objects[5] affects not only those two objects but might yield additional constraints on the relations between other objects in the scene through constraint propagation. Allen (1983) introduced an algorithm

---

[5] We assume the relation has been transformed to the canonical implicit frame of reference.

for updating an interval-based temporal network that is based on constraint propagation. This algorithm effectively computes the closure of the set of temporal assertions after each new insertion. In what follows, we first describe Allen's original algorithm and introduce then several important enhancements based on structural properties of space.

Allen's Propagation Algorithm. The first step in the process of inserting a new relation R_{ij} between i, j is intersecting the new relset with whatever relset was known before (this is here the universal relset, in case no relation had been previously inserted, which behaves as "identity" for the set intersection operation). In case this intersection results in a more constrained relset, the nodes i, j are placed in the queue for propagation.[6]

The computation of the closure is achieved by repeatedly calling a procedure PROPAGATE as long as there are entries in the queue. PROPAGATE does the main work by propagating the effects of the new constraint to "comparable" nodes (for now, assume all nodes in the network to be comparable). This is done by determining if the new relation between i and j can be used to constrain the relation between i and other nodes, or between those other nodes and j (Fig. 3). If one of these relations can indeed be constrained, then it is placed in the queue for further propagation. Furthermore, contradictions, characterized by an empty resulting relset, are signaled if found in this process. Contradictions will normally trigger a constraint relaxation process.

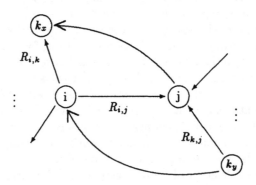

Fig. 3. Propagation algorithm (visualization)

Example. To illustrate the propagation algorithm, we use the following set of initial relations, which have already been transformed to a canonical frame of

---

[6] Note that, because the new link is obtained by intersection with the old one, it suffices to test if the new one is different from the old one, which might be cheaper.

reference (with possible corresponding verbal descriptions to the right):

1. $<T, [t,f], F>$ *The table (T) is at the window (F).*
2. $<S, \{[d,f],[t,f]\}, T>$ *In front of it there is a chair (S).*
3. $<W, \{[d,1],[d,r]\}, T>$ *A bookcase (W) is next to it.*

Adding R_{TF} creates the first link in the network (Fig. 4a). Adding R_{ST} triggers the computation of R_{SF} through composition of R_{ST}/R_{TF} (Fig. 4b). Adding R_{WT} leads to the computation of R_{WF}, and, if we allow inverting links, to $R_{WS} = R_{WT}/R_{TS}$ (Fig. 4c). Note, that the resulting relations in this last case are rather unspecific, because we do not know if the bookcase (W) is left or right of the table (T). Now suppose such information becomes indirectly available through a statement such as *"The bookcase (W) is to the right of the chair (S).",* i.e., $R_{WS} = [d,r]$. The intersection with the previously computed relset $[\{d,t\},\{1,1b,b,rb,r\}]$ results again in $[d,r]$. The propagation algorithm computes new values for $R_{WT} = R_{WS}/R_{ST}$ and $R_{WF} = R_{WS}/R_{SF}$ both equal to $[\{d,t\},\{r,rf,f\}]$. The intersection with previously computed relsets leads finally to $R_{WT} = [d,r]$ and $R_{WF} = [\{d,t\},\{r,rf,f\}]$ (Fig. 4d).

Exploiting the Structure of Space

Reference Objects. The predicate "comparable" is one place in Allen's algorithm that allows us to introduce modifications. Allen himself uses it to control propagation by introducing "reference intervals" to cluster sets of fully connected intervals, and defining comparable to be true only if the intervals share a reference interval, or one is the reference interval of the other. Similarly, we can limit the propagation by assigning parent objects (i.e., those resulting from hierarchical decomposition through containment) or functional clusters (for example, "dining-table-group" consisting of table, chairs, etc.) as "reference objects". However, we do not require full connectivity of objects with a common reference object. In most cases, it suffices if all objects are related to at least the central object in the group (e.g., the table in the dining-table-group).

Degrees of Coarseness. The next modification of the algorithm is not so straightforward. We want to take into account the fact that the information content of positional relations is not homogeneous. By information content we mean how much a relation constrains the relative position of objects. For example, a level 3 orientation is more constraining than a level 1 orientation, and t is more constraining than d. The specificity of a positional relation is, strictly speaking, a function of the relative size of the corresponding acceptance area.[7] However, establishing those areas is more involved than what is actually needed to differentiate among the relevant specificity classes. Thus, we use the "degree of coarseness" (doc), a number derived from the number of options left open by a relation (and confirmed by the "coarsening" factor of the resulting compositions), as a converse approximate measure. Small doc-values correspond to more

---

[7] The area in which a particular orientation is accepted as a valid description of the relative position of two objects is called "acceptance area".

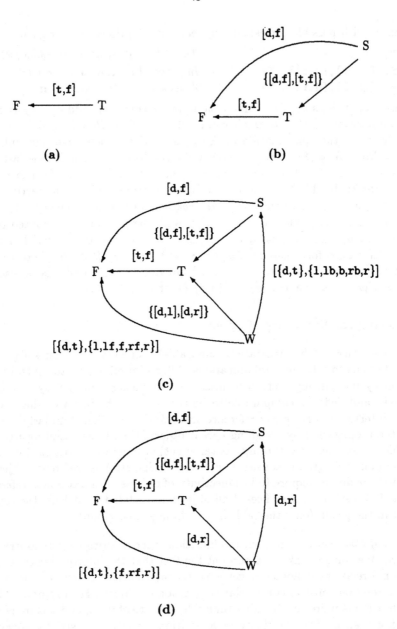

Fig. 4. Propagation algorithm (example)

constraining relations, while large doc-values correspond to less constraining relations.

characteristics			rel	doc
topological			=	1
	size&shape restriction		c, i	4
			i⊕b, c⊕b	2
oriented	boundary contact		t	3
			o	4
			d	8

Table 1. Degree of coarseness of topological relations (in the context of positional information)

Table 1 shows the degree of coarseness of topological relations in the context of positional information together with the factors contributing to the corresponding docs. = has the lowest doc, because it constrains the relative position of two objects to an unique value; c and i are not oriented but constrain the position of one of the objects to be within the boundaries of the other; t and o demand boundary contact between the objects, thus restricting their positions, whereas d has the highest doc and is useful only when used together with an orientation;[8] finally i⊕b and c⊕b are quite specific because they are oriented, demand boundary contact and have the size and shape restriction of the containment relations.

Table 2 lists the degree of coarseness of the most common orientation relations. It corresponds roughly to the number of basic sectors covered by each of the relations. For the case of extended objects, a further distinction between more specific corner orientations and less specific side orientations can be made, as shown here for the third level. The doc of a topological/orientation pair is the sum of the docs of its components. The doc of a relset is the sum of the docs of the set members.

The doc-values are used to control propagation in the following way: While adding a new relation (Fig. 5), its doc is compared against a pre-defined constant **maxdoc**, above which a relation is considered too unspecific to be worth adding. The value of **maxdoc** is application dependent. Setting it to half the sum of the docs of all possible relations at the granularity levels allowed has proven to be a useful heuristic. If the doc of the relset is below or equal to **maxdoc**, then it is inserted by combining it with the previously known relset. COMBINE can be

---

[8] The unspecificity of d should not be overrated: In most applications, the distance between objects is bounded by the extension of the parent object.

characteristics			rel	doc
orientation	level 3	corners	lf_3, rf_3, lb_3, rb_3	1
		sides	l_3, r_3, f_3, b_3	3
	level 2		l_2, r_2, f_2, b_2	4
	level 1		l_1, r_1	8
			f_1, b_1	8

Table 2. Degree of coarseness of the most common orientation relations (in the context of positional information)

```
To  ADD Rij
  begin
    if doc(Rij) > maxdoc
        then exit;
    Old ← N(i,j);
    N(i,j) ← Combine(N(i,j),Rij);
    If N(i,j) = ∅
        then Signal contradiction;
    if N(i,j) ≠ Old
        then add <i,j,doc(N(i,j))> to Agenda;
    Nodes ← Nodes ∪ {i,j};
  end;
```

Fig. 5. Adding a new relation

assumed for now to be equivalent to set intersection (it will be modified below). If the new combined relation is different from the previously known, it is placed on the agenda for further propagation. An agenda is a data structure to keep track of what to do next based on some sort order. It is usually implemented as an ordered list of queues. In this case, we use the doc of the relation to be propagated as entry key, allowing us to propagate first more specific relations, i.e., those with lower docs. Relations with equal docs are processed in a first-in-first-out manner.

The effects of new relations added to the network can be computed by calling COMPUTEEFFECTS (Fig. 6), which fetches the first entry from the agenda, double checks its doc to be below the limit (this is necessary because PROPAGATE also adds to the agenda, and could also be used for dynamic control through changing limits), and calls PROPAGATE.

PROPAGATE (Fig. 7) is essentially the same as in the original algorithm, except for COMBINE and the use of an agenda instead of a queue. Note, that the propagation algorithm assumes all relations, including semantically similar rela-

```
To  COMPUTEEFFECTS
   While Agenda is not empty do
      begin
         Get next <i,j,d> from Agenda;
         If d ≤ maxdoc
            then Propagate(i,j);
      end;
```

Fig. 6. Computing the effects of a new added relation

```
To  PROPAGATE i,j
   begin
      For each node k such that Comparable(i,k) do
         begin
            New ← Combine(N(i,k),Constraints(N(i,j),N(j,k)));
            If New = ∅
               then Signal contradiction;
            If New ≠ N(i,k)
               then add <i,k,doc(New)> to Agenda;
            N(i,k) ← New;
         end;
      For each node k such that Comparable(k,j) do
         begin
            New ← Combine(N(k,j),Constraints(N(k,i),N(i,j)));
            If New = ∅
               then Signal contradiction;
            If New ≠ N(k,j)
               then add <k,j,doc(New)> to Agenda;
            N(k,j) ← New;
         end;
   end;
```

Fig. 7. Weighted Propagation

tions at different levels of granularity (e.g., b_2 and b_3), to be mutually exclusive. Knowing about the hierarchical structure of the relational domains used, allows us to extend the algorithm to succeed in cases where the original algorithm would fail, and signal a contradiction. For example, the intersection of a set containing only $[t,l_2]$ and a set containing only $[t,l_3]$ shouldn't be "empty", signaling a contradiction, but rather lead to preferring $[t,l_3]$. This can be done by looking at the range representation of l_2 and l_3, instead of viewing them as unrelated relations. Furthermore, the intersection of sets containing neighboring relations, such as for example $\{[t,l_3]\}$ and $\{[t,lb_3]\}$, should not be "empty", but rather lead to a coarser relation such as $[t,l_2]$. These extensions are implemented by the modified COMBINE in Fig. 8, which checks to see if subsumed or neighboring relations are available, if the regular intersection of two relsets is empty. In the first case, the subsumed (more specific) relation is placed in the new combined set. In the second case, the "next common coarse relation" (nccr) is added to the combined set. This is a form of constraint relaxation embedded in the propagation process (Hernández and Zimmermann 1992).

```
To  COMBINE R1, R2
  begin
      temp ← R1 ∩ R2;
      if temp ≠ ∅
          then Return temp;
      C ← ∅;
      For each r1 ∈ R1
        For each r2 ∈ R2
          begin
              if subsumes(r1,r2) then C ← C ∪ r2;
              if neighbor(r1,r2) then C ← C ∪ nccr(r1,r2);
          end;
      Return C;
  end;
```

Fig. 8. Combining two relsets

Continuing our example, assume the initial relation between W and T to be $\{[d,l_2],[d,r_2]\}$, and suppose that we learn later that $R_{WT} = [d,rf_3]$. Instead of returning an empty set as an intersection of the two sets would, COMBINE recognizes that r_2 subsumes rf_3 and returns $[d,rf_3]$ as result. After propagation, the relations between W and F, and between W and S are constrained to be $\{[d,f],[d,rf]\}$, and $[\{d,t\}, \{b,rb,r\}]$, respectively.

Complexity. Allen (1983) and later Vilain, Kautz, and van Beek (1990) showed the algorithm to run in polynomial time w.r.t. the number of intervals in the database. $O(n^3)$ set operations are required in the worst case for the algorithm to run to completion for n intervals, because there are at most n^2 relations between n intervals, each of which can only be non-trivially updated (and correspondingly entered on the queue for propagation) a constant number of times (each update removes at least one of 13 possible relations). In turn each of the $O(n^2)$ propagations requires $O(n)$ set operations resulting in the said $O(n^3)$ set operations (which in a bit-string implementation of sets can be assumed to take constant time each).

As a consequence of the modified COMBINE, New ≠ N(i,k) in PROPAGATE being true does not imply **New** $\subset$ **N(i,k)**. However, the worst case complexity analysis of the original algorithm is not affected by this change, because every non-trivial update of a relset either removes at least one relation (intersection, original algorithm), or replaces one relation by a subsumed one, or one or more fine relations by a coarser one. The average case performance is greatly improved by the modifications described, particularly by the hierarchical decomposition, and by the preferential propagation of specific relations.

3.2 Deleting Relations

Deleting relations between two nodes is not just a matter of removing a link from the data structure representing the network. The consequences of the propagation of the constraint now being deleted must be taken back as well.

This requires a further modification of the insertion algorithm described in the previous section to maintain justifications for derived constraints. Instead of just modifying a link to contain the new constrained relset, we also record the link whose propagation led to the new constraint. In general, a justification is a list of the links that were used to derive the relation of the new link. To allow for multiple derivation paths, usually a list of justifications is maintained. At the same time a link has pointers to those links that it in turn served in deriving in a so called justificands list. Relations originally entered by the user (considered as "premises") have an empty justification list. This information is usually maintained in a separate "dependency network", where the links of the constraint network are the "nodes",[9] and the arcs connecting them represent the dependency structure. Fig. 9 shows a typical graphical representation of dependency networks, and illustrates the terminology introduced above (empty justifications marking premises are shown as solid rectangles). The dependency network can also serve as direct indexing and retrieval mechanism for sets of consistent relsets. The process operating on a dependency network is called "reason maintenance".

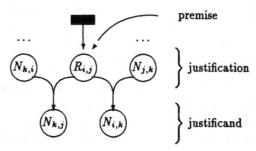

Fig. 9. Graphical representation of dependency networks

---

[9] Note, however, that each particular relset R_{ij} between objects i and j is recorded as a separate node of the dependency network.

Because they can operate independently from the problem solver (the constraint propagation algorithm, in our case), reason maintenance algorithms and the corresponding dependency networks have been developed as separate systems called "Reason Maintenance Systems" (RMS). RMSs were introduced in the late 70s in the context of computer-aided circuit analysis by Stallman and Sussman (1977), and first studied as independent systems by Doyle (1979).[10] Further variants were later developed by McAllester (1980), de Kleer (1986), and others.

Based on the dependency information recorded during the propagation of constraints, the RMS establishes which nodes of the dependency network are affected by the deletion of a link in the constraint network. Figure 10a shows the dependency network for the example from the previous section, after the first set of relations (the "premises" R_{FT}, R_{ST}, R_{WT}) and its consequences (R_{SF}, R_{WF}) have been established.[11] The one as superscript indicates that this particular relset is the first one assigned to that link. Figure 10b shows the next step in which R_{WS} is added as a premise, resulting in the derivation of new values for R_{WT}^2 and R_{WF}^2 (indicated by the two as superscript). Note that the old values R_{WT}^1 and R_{WF}^1 are cancelled out. Now suppose we originally got the wrong description and are forced to delete the relation between S and T. In that case, all the consequences derived by using that relation must be removed as well, as shown in Fig. 10c. Unfortunately, we are left only with the relations R_{TF}^1 and R_{WS}^1, even though we had previously the relations R_{WT}^1 and R_{WF}^1 that did not depend on the erased relation. The RMS mechanism, however, does usually reconsider the evidence in support of all nodes based ultimately on premises, when other values are erased, and would restore in this case the old values for R_{WT}^1 and R_{WF}^1 (Fig. 10d). Thus, maintaining dependencies not only allows us to retract the consequences of deleted relations, but also helps us to avoid repeated computations. Of course, there is always a tradeoff between the overhead of caching inferences and the costs of recomputing them, particularly when restrictive propagation policies apply.

Upon deletion of a link in the constraint network, reason maintenance is triggered, leading to the corresponding erasure of a node in the dependency network. The basic idea of reason maintenance is similar to the "garbage collection" mechanism in LISP systems: The status and support of all elements in the network are reset first, and then recomputed by a recursive mark procedure that—starting from the premises—finds all nodes with "well-founded" support. Nodes without such support are physically removed from the network. While the procedure illustrates the basic idea, it is usually not very efficient, because all nodes in the network are visited (only once, but anyway). Thus, other variants start from the erased node, and recursively mark all its justificands for erasure. However, because of the potential of circular dependencies, a somehow

---

[10] They were originally called "Truth Maintenance Systems", a somewhat misleading term, still often found in the literature. We prefer the name "Reason Maintenance Systems" following McDermott (1983).

[11] For better readability, we are omitting here the propagation through inverted links (e.g., $R_{WS} = R_{WT}/R_{TS}$).

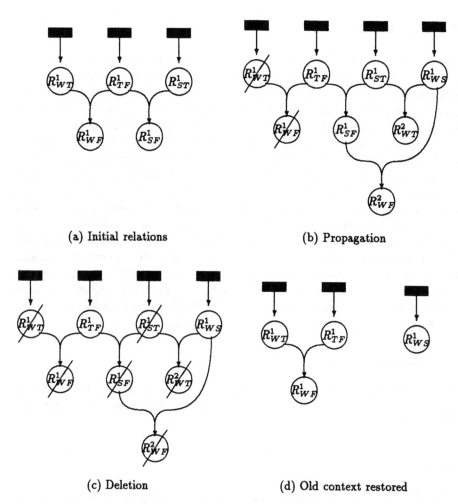

(a) Initial relations

(b) Propagation

(c) Deletion

(d) Old context restored

Fig. 10. Example dependency network

more involved procedure is needed (Hernández 1992). It resets the non-circular supporter/supportee links and assigns them a tentative status, assuming that all nodes marked inactive ($\neq$ {IN, OUT}) will ultimately be erased (OUT). If that happens not to be the case, i.e., if one such node becomes active (IN), because of a valid justification, all its supportees must be reconsidered, because they might have been assigned a new status based on the assumption that the supporter be OUT. Details of this algorithm can be found in (Hernández 1984), where several variants were implemented based on ideas in (Charniak et al. 1980) and (Doyle 1979).

4 Conclusion

Even though the solution techniques for the general constraint satisfaction problem available in the literature represent large efficiency improvements over the obvious backtracking algorithm, they are "limited by their generality". That is, being general, domain-independent techniques, they ignore the structure of the relational domain. Thus, in this paper we show that taking the structure of the richly constrained spatial domain into consideration leads to more efficient algorithms. One way of exploiting this structure is introducing heuristics to control the propagation of constraints by using the hierarchical and functional decomposition of space to limit constraints to physically adjacent objects. Spatial reasoning can be done at coarser or finer levels of that structure, depending on the kind of information available. In particular, if only coarse information is available, the reasoning process is less involved than if more details are known. Also, a weighting of positional relations according to their information content is used to avoid "information decay" in the network due to the propagation of weak relations. Retracting the consequences of previous propagations in order to maintain the consistency of the knowledge base requires keeping track of dependencies and a reason maintenance mechanism.

Acknowledgements

The research reported here has benefited from numerous discussions with W. Brauer, C. Freksa, C. Habel, S. Högg, D. Kobler, I. Schwarzer, K. Zimmermann and others.

References

Allen, J. F. (1983). Maintaining knowledge about temporal intervals. *Communications of the ACM*, *26*(11), 832–843.

American Association for Artificial Intelligence, 8th-AAAI (1990). *Proc. of the Eighth National Conference on Artificial Intelligence*, Menlo Park/Cambridge. AAAI Press/The MIT Press.

Bobrow, D. (Ed.). (1984). *Qualitative Reasoning about Physical Systems*. Elsevier Science Publishers, Amsterdam.

Chang, S.-K. and Li, Y. (1988). Representation of multi-resolution symbolic and binary pictures using 2DH-strings. In *Proceedings of the IEEE Workshop on Language for Automation*, pp. 190–195.

Chang, S.-K., Shi, Q.-Y., and Yan, C.-W. (1987). Iconic indexing by 2D-strings. *IEEE Transactions on Pattern Analysis and Machine Intelligence*, *9*(3), 413–427.

Charniak, E., Riesbeck, C. K., and McDermott, D. V. (1980). *Artificial Intelligence Programming*. Erlbaum, Hillsdale, N.J.

Cohn, A. G., Cui, Z., and Randell, D. A. (1992). Logical and computational aspects of spatial reasoning. In Pribbenow, S. and Schlieder, C. (Eds.),

Spatial Concepts: Connecting Cognitive Theories with Formal Representations. ECAI-92 Workshop.

Doyle, J. (1979). A truthmaintenance system. *Artificial Intelligence, 12,* 231–272.

Egenhofer, M. J. (1989). A formal definition of binary topological relationships. In Litwin, W. and Schek, H.-J. (Eds.), *Third International Conference on Foundations of Data Organization and Algorithms,* Vol. 367 of *Lecture Notes in Computer Science,* pp. 457–472. Springer-Verlag, Berlin.

Egenhofer, M. J. (1991). Reasoning about binary topological relations. In *Second Symposium on Large Spatial Databases,* Vol. 525 of *Lecture Notes in Computer Science.* Springer-Verlag.

Egenhofer, M. J. and Al-Taha, K. K. (1992). Reasoning about gradual changes of topological relationships. In Frank et al. (1992), pp. 196–219.

Egenhofer, M. J. and Franzosa, R. (1991). Point-set topological spatial relations. *International Journal of Geographical Information Systems, 5*(2), 161–174.

Egenhofer, M. J. and Sharma, J. (1993). Assessing the consistency of complete and incomplete topological information. *Geographical Systems, 1*(1). To appear.

Frank, A. U., Campari, I., and Formentini, U. (Eds.). (1992). *Theories and Methods of Spatio-Temporal Reasoning in Geographic Space. Intl. Conf. GIS—From Space to Territory. Pisa, Sept. 1992,* Vol. 639 of *Lecture Notes in Computer Science,* Berlin. Springer-Verlag.

Frank, A. U. (1991). Qualitative spatial reasoning with cardinal directions. In Kaindl, H. (Ed.), *7. Österreichische Artificial Intelligence Tagung.* Springer-Verlag.

Freksa, C. (1992a). Temporal reasoning based on semi-intervals. *Artificial Intelligence, 54,* 199–227.

Freksa, C. (1992b). Using orientation information for qualitative spatial reasoning. In Frank et al. (1992), pp. 162–178.

Freksa, C. and Zimmermann, K. (1992). On the utilization of spatial structures for cognitively plausible and efficient reasoning. In *Procs. of the 1992 IEEE International Conference on Systems, Man, and Cybernetics Proceedings, Chicago.*

Guesgen, H.-W. (1989). Spatial reasoning based on Allen's temporal logic. Technical Report TR-89-049, ICSI, Berkeley, CA.

Hernández, D. (1984). Modulare Softwarebausteine zur Wissensrepräsentation. IMMD (IV) and RRZE Universität Erlangen-Nürnberg.

Hernández, D. (1991). Relative representation of spatial knowledge: The 2-D case. In Mark and Frank (1991), pp. 373–385.

Hernández, D. (1992). *Qualitative Representation of Spatial Knowledge.* Ph.D. thesis, Institut für Informatik, Technische Universität München.

Hernández, D. and Zimmermann, K. (1992). Default reasoning and the qualitative representation of spatial knowledge. In preparation.

Jungert, E. (1988). Extended symbolic projection used in a knowledge structure for spatial reasoning. In *Proceedings of the 4th BPRA Conf. on Pattern Recognition*. Springer-Verlag, Berlin.

de Kleer, J. (1986). An assumption based truth maintenance system. *Artificial Intelligence, 28*, 127–162.

Kuipers, B. J. and Levitt, T. (1988). Navigation and mapping in large-scale space. *AI Magazine, 9*(2), 25–43.

Kumar, V. (1992). Algorithms for constraint satisfaction problems: A survey. *AI Magazine, 13*(1), 32–44.

Levitt, T. and Lawton, D. (1990). Qualitative navigation for mobile robots. *Artificial Intelligence, 44*, 305–360.

Mackworth, A. K. (1987). Constraint satisfaction. In Shapiro (1987).

Mark, D. M. and Frank, A. U. (Eds.). (1991). *Cognitive and Linguistic Aspects of Geographic Space*. NATO Advanced Studies Institute. Kluwer, Dordrecht.

McAllester, D. (1980). An outlook on truth maintenance. Memo 551, MIT Artificial Intelligence Laboratory, Cambridge, MA.

McDermott, D. V. (1983). Contexts and data dependencies: A synthesis. *IEEE Transactions on Pattern Analysis and Machine Intelligence, 5*(3), 237–246.

McDermott, D. V. (1992). Reasoning, spatial. In Shapiro, E. (Ed.), *Encyclopedia of Artificial Intelligence* (Second edition)., pp. 1322–1334. Wiley.

Meseguer, P. (1989). Constraint satisfaction problems: An overview. *AI Communications, 2*(1), 3–17.

Mukerjee, A. and Joe, G. (1990). A qualitative model for space. In 8th-AAAI (1990), pp. 721–727.

Shapiro, E. (Ed.). (1987). *Encyclopedia of Artificial Intelligence*. Wiley.

Smith, T. R. and Park, K. K. (1992). Algebraic approach to spatial reasoning. *International Journal of Geographical Information Systems, 6*(3), 177–192.

Stallman, R. and Sussman, G. J. (1977). Forward reasoning and dependency-directed backtracking. *Artificial Intelligence, 9*(2), 135–196.

Topaloglou, T. (1991). Representation and management issues for large spatial knowledge bases. Technical Report, Dept. of Computer Science, University of Toronto, Toronto, Ontario.

Vilain, M., Kautz, H., and van Beek, P. (1990). Constraint propagation algorithms for temporal reasoning: A revised report. In Weld and de Kleer (1990). Revised version of a paper that appeared in Proceedings of AAAI 1986, pp. 377–382.

Weld, D. S. and de Kleer, J. (Eds.). (1990). *Readings in Qualitative Reasoning about Physical Systems*. Morgan Kaufmann, San Mateo, CA.

Qualitative Triangulation for Spatial Reasoning

Gérard F. Ligozat

LIMSI, Université Paris XI
91403 Orsay Cedex, France

Abstract. This paper presents a systematic way of defining qualitative calculi for spatial reasoning. These calculi, which derive from the concept of qualitative triangulation, allow inference about the relative relationships of punctual objects in two-dimensional space. After introducing the general concept of qualitative triangulation, we discuss the main aspects of some important members of this family of calculi, including the so-called flip-flop calculus, which subsumes the relative calculus in dimension one, and the calculus introduced by Freksa (orientation-based spatial inference). This allows us to present in a general setting the notions of coarse and fine inference, as well as the conceptual neighborhood properties of sets of spatial relations. We also show how these calculi can be used for actual inference, and how switching from a particular calculus to a refinement of it can be used to strengthen the inference.

1 Introduction

There has been a recent upsurge of activity in the domain of qualitative spatial reasoning. This is hardly surprizing, since the ability to reason about space and its properties is such a crucial factor in systems having to decide and act in physical space. However, it sems that principled approaches allowing the spatial component the same degree of independence granted to the temporal component are only beginning to emerge.

Consider two typical examples where qualitative spatial reasoning is needed. The first example, considered by [Ran91], is osmosis between two cells. Here, the relations between spatial entities in terms of containment are the relevant aspects to be modelled. Orientation does not play a significant role in this case. The second example is wayfinding in a large-scale environment. Here, the subject who is trying to elaborate a route may generally be represented as a punctual entity, and the notion of orientation comes to the foreground.

Among the various approaches to qualitative spatial reasoning, we consequently first distinguish between the axiomatic topological approach exemplified by Randell [Ran92], or Egenhofer [EgAl92], where such notions as interior, closure, containment are considered, and approaches where orientation is a fundamental aspect.

We will be concerned with the second kind of approach. As it happens, much work in this domain has been more or less inspired by the approach defined by Allen [All83] for temporal reasoning: relevant relations between spatial entities are considered, and the central part of the reasoning activity is taken up by propagating knowledge about those relations.

Different types of qualitative spatial reasoning systems can be analyzed according to some basic parameters:

• Dimension of the ambient space: 2-D space, 3-D space.
• Absolute vs relative reference.
• Nature of the objects considered : points, rectangles, arbitrary objects.
• Nature of the objects acting as landmarks or reference.

Here, we will be concerned with the 2-D-case and leave entirely aside the case of 3-D space, although the basic notions can probably be carried over to the 3-D case, using tetrahedra instead of triangles. Besides, a general remark can be made about the interest in many situations of considering the 1-dimensional case: it often reveals part of the nature of the more complex 2-D situation in an illuminating way.

The question of the choice of an absolute vs a relative frame of reference has two aspects. The first aspect is that, in many situations, what matters is the relative positions of the objects considered, which makes the choice of an absolute frame of reference, at best, an artifact for description. The second aspect has to do with the types of the objects themselves: if objects are abstracted as points in 2-D space, a single object is not enough to define a frame of reference. Things change if those points are provided with a direction of motion. A way to get round this difficulty is to consider two distinct points in the plane as a point of departure. This idea, which appears to have been first used by Freksa [Fre92], is central to our presentation.

Other objects have been considered, including rectangular objects with a direction of motion [MuJo90], 2-D projections of 3-D objects [Her92], 2-D and 3-D objects [Gus89].

In what follows, we only consider points in a 2-D space.

2 Points in 2-D space

2.1 Motivation

Consider the following hare-shooting story:

Al, Bill, Clint and Doug are hare shooting. Al is walking some distance behind Bill. Clint has left Bill behind and is proceeding to the right.

Clint, turning to Bill: "I think I have spotted a hare on my right!"

Bill: "I can't see anything on my right!"

Doug, who had remained behind Al, on his left, turns to Bill: "I have spotted a hare behind me, on the right!"

On the basis of a suitable analysis of the spatial data contained in this story, is it possible to determine:

- whether one and the same animal can have been spotted by Bill and Doug (coherence)?

- if such is the case, the location of the hare with respect to Al and Bill?

In the following, we describe a family of calculi which allow to answer this kind of questions, on the basis of qualitative directional data.

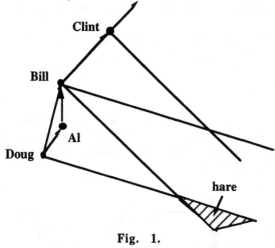

Fig. 1.

2.2 Reasoning with an absolute frame of reference

Here, we consider briefly the first option, which consists in reasoning with a fixed, absolute frame of reference.

Consider 2-dimensional space with a frame of reference consisting of two (non-necessarily orthogonal) axes (Fig. 2). On each axis, a given A point defines three regions which, accordingly with the notations in [Lig91], we label 0 ("before A"), 1 ("at A"), and 2 ("after A"). The 1-D point calculus involved is just the usual point-based calculus considered in temporal reasoning, which is called A_1 calculus in [Lig90], [Lig91]. The corresponding transitivity table is Table 1. In this table, the notation [0,2] denotes the disjunction of relations 0, 1, and 2.

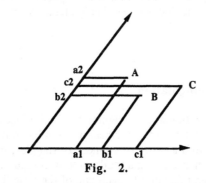

Fig. 2.

	0	1	2
0	0	0	[0,2]
1	0	1	2
2	[0,2]	2	2

Table 1.

Each point is then defined by its two projections, and the corresponding calculus amounts to reasoning separately on the two components. In other words, we get what might be called the $A_1 \times A_1$ -calculus. Table 1 can be used separately on each component.

For example, in the case illustrated in Fig. 2, from the known facts:

B (2,0) A   and   C (2,2) B

we can deduce

C (2, [0,2]) A.

Remark. The nine relations which appear in this absolute calculus correspond to a partition of the plane (defined by a given point) as represented in Fig. 3 (a). Anticipating on notions we define below, we can describe the neighborhood structure of the situation as in Fig. 3 (b).

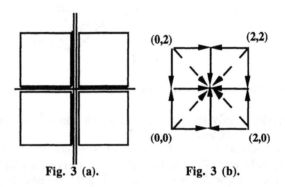

Fig. 3 (a).   Fig. 3 (b).

Example. Consider the hare-shooting story. A reasonable choice consists in taking the current location of Al as an origin, and AB (Al, Bill) as giving the direction of the y-axis. Then Clint is located in region (2,2). However, we should also take into account the fact that the projection of Clint with respect to AB is farther from A than B. This would lead to consider local, rather than global, systems of reference.

2.3   Quantitative triangulation

According to Webster's dictionary, "triangulation is the process of determining the distance between points on the earth's surface, or the relative positions of points, by dividing up a large area into a series of connected triangles, measuring a base line between two points, and then locating a third point by computing both the size of the angles made by lines from this point to each end of the base line and the lengths of these lines".

Suppose Al (A) wishes to determine the position of Clint (C). Al can be helped by Bill (B) and both can proceed as follows: using direction (AB) as a direction of reference, Bill measures the angle u between (AB) and (BC), while Al measures the angle v between (AB) and (AC) (Fig. 4 (a)).

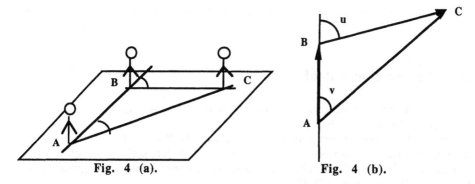

Fig. 4 (a).   Fig. 4 (b).

If the exact measures of u and v are known, as well as the distance between A and B, there is no ambiguity on the precise location of Clint (Fig. 4 (b)). The basic idea of qualitative triangulation is to introduce qualitative knowledge about the values for (u,v), and to try to determine what inferences can be made in this context.

Suppose now that Doug (D) comes into the picture. Suppose we know the location of Doug with respect to Bill and Clint. The question arises of inferring what the location of Doug is with respect to Al and Bill (Fig.5).

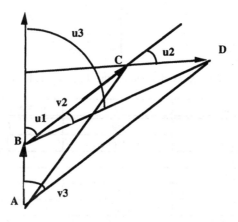

Fig. 5.

The basic idea of qualitative triangulation consists in using qualitative knowledge about the values of the angles u and v, and in propagating this knowledge as new points are considered. As it turns out, this idea of reasoning about partially known angular values gives rise to a whole family of spatial calculi, which include Freksa's calculus as a special member. We first examine a quite particular case.

Before we do so, we can make a general observation:

Fact
Let (u_1, v_1) define the location of C with respect to (AB)
(u_2, v_2) define the location of D with respect to (BC)
(u_3, v_3) define the location of D with respect to (AB).
Then $u_3 = u_1 + v_2$.

We now briefly describe the coarsest version of qualitative triangulation: the "flip-flop" calculus.

3 The flip-flop calculus

Suppose the only measurements our people can make are multiples of 180°: for each angle, the only possible results are:

0°, 180°, 360°, between 0° and 180°, between 180° and 360°.

In other words, each informer can only answer: "right before me", "right behind me", "on my left", "on my right". Then, the plane is partitioned into seven regions which are labeled 0 to 6 (Figure 5, Table 1).

Fig. 6. Regions defined in the flip-flop calculus

u	v	name of region
180	180	0
180	undefined	1
180	0	2
undefined	0	3
0	0	4
< 180	< 180	5
> 180	> 180	6

Table 2. Characterization in terms of angles

The answer to the question of where D is located with respect to (AB) knowing its location with respect to (BC) and the location of C with respect to (AB) is answered by Table 3. A particular case is C=B, where (BC) cannot be used to locate D (C is in the region labeled 3 with respect to (AB)). Consequently, Table 3 contains a blank line for this case. In this table, [a, b] denotes the disjunction of all relations c such that $a \leq c \leq b$ (the numbering refers to Table 2).

Example. Consider the hare-shooting story again. Consider A (Al), B (Bill), C (Clint), D (Doug). From our knowledge of Clint's location, we can only deduce that C with respect to AB is in region 5. The hare (H) is in region 5 with respect to BC. By Table 3, this implies regions 0, 1, 2, 5, or 6 for H with respect to AB. On the other hand, D is in region 6 with respect to AB, and the (other) hare H' in region 6 with respect to BD. By Table 3 again, H' is in regions 0, 1, 2, 5, or 6 with respect to AB. Although this is a very coarse inference, it shows that H and H' could be the same animal, which would be located anywhere but at Bill's location, or right before him.

D / (BC) C / (AB)	0	1	2	3	4	5	6
0	4	3	[0,2]	0	0	6	5
1	4	3	2	1	0	6	5
2	4	3	2	2	[0,2]	6	5
3							
4	[0,2]	3	4	4	4	5	6
5	6	3	5	5	5	[0,2], 5, 6	[4,6]
6	5	3	6	6	6	[4,6]	[0, 2], 5,6

Table 3. The flip-flop calculus

Clearly, the flip-flop calculus involves what happens when all points considered are restricted to stay on the line (AB). This can be considered in its own right by defining a one-dimensional calculus.

4 One-dimensional qualitative triangulation

Consider two distinct points A and B on a line. They define a relative orientation AB, and partition the line into five regions, which can be labeled 0 to 4 , see [Lig91].

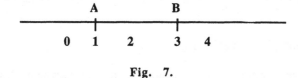

Fig. 7.

Now, a new point C is located by virtue of being in one of the five relations with respect to (A,B). If we introduce another point D , its location with respect to (BC) will be expressed by one of the five labels relative to (B,C). Now the relation of D with respect to the original couple (A,B) is given by Table 4, which is just the one-dimensional part of Table 3 (in the sense that 2-dimensional regions, namely 5 and 6, are left out).

	0	1	2	3	4
0	4	3	[0,2]	0	0
1	4	3	2	1	0
2	4	3	2	2	[0,2]
3		no	meaning		
4	[0,2]	4	4	4	4

Table 4. One-dimensional qualitative triangulation

Notice in particular that C being in relation 0, 1, or 2 with respect to (A,B) implies that the orientation defined by (BC) is opposite to the original (AB).

4.1 Coarse reasoning and fine reasoning in one dimension

What is gained in considering pairs of points on the line, rather than the usual point calculus?

Suppose we only retain from the information about C / (AB) the u-part (knowledge of angle u = (AB,BC)). This means that we do not make the distinction between regions 0,1, and 2 (what comes before B). In the same way, we only keep the v-part of the information about D / (BC), i.e., we merge regions 2, 3, 4. Then from Table 4 we can derive Table 5, by merging the first 3 rows (replacing 0, 1, or 2 by [0,2]), and then merging the last 3 columns of Table 4.

	0	1	[2,4]
[0,2]	4	3	[0,2]
4	[0,2]	3	4

Table 5

A quick examination shows that this table is essentially equivalent to Table 2, in a relative setting. Equivalently, it can be deduced from the general fact mentioned above, in terms of adding angles around B.

This discussion shows that the benefit of using two points instead of one can be measured by the cases where, in Table 4, we can find 0, 1, or 2 instead of just [0,2].

4.2 The topology of relations

An important aspect of qualitative reasoning is the way in which it can model the properties of continuity of its domain. In particular, qualitative spatial reasoning should reflect the natural neighborhood relations between spatial relations. In each of the qualitative triangulation calculi we are examining, the neighborhood structure can be described in a convenient way by an associated incidence graph.

In order to do this, we consider the partition of the plane defined by a given set of relations. In the "flip-flop" case, this partition comprizes seven regions. Each of these regions is homeomorphic (isomorphic as a topological space) to $\mathbf{R}^p$, if p is its dimension. Using the language of cell decomposition, each of these regions is an open cell of dimension 0 (a point), one (a segment without its end-points) or two (the interior of a disk).

We now have a way of defining the notion of (primitive) conceptual neighborhood:

- A relation r_1 corresponding to a cell of dimension p is an edge of a relation r_2 of dimension p+1 if the closure of r_2 contains r_1.
- Two relations are primitive neighbors if one is an edge of the other.
- The relation of neighborhood is the symmetric, transitive closure of the relation of primitive neighborhood.

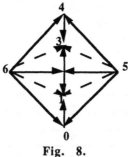

Fig. 8.

The conceptual neighborhood structure in the case of the flip-flop relations is represented in Fig. 8. In this representation, solid arrows represent the edge relation. Dotted arrows are used to represent the transitive closure of the edge relation.

5   Freksa's calculus as qualitative triangulation

In [Fre92], Freksa defines a qualitative calculus with 15 primitive relations of a point C with respect to two distinct points A and B. This calculus is in fact a special case of qualitative triangulation. Indeed, consider the case where for angles u and v, the only possible results are:
- either 0°,   90°,   180°,   270°,
- or one of the intermediate ranges, viz. (0, 90), (90, 80), (180, 270), and (270, 360).

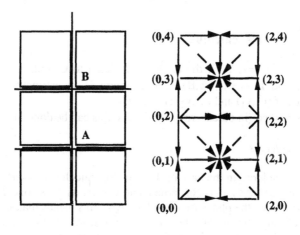

Fig. 9 (a).   Fig. 9 (b).

We get a total of fifteen regions which make up a partition of the plane, as shown in Fig. 9 (a). Since this partition can be considered as obtained from the product of partitioning the line perpendicularly to AB (regions 0, 1, 2), and on the axis defined by AB (regions 0, 1, 2, 3, 4), we can label these regions accordingly (Fig. 9 (b)). Table 6 gives the correspondence between pairs of angles and regions.

The neighborhood structure of the set of relations is represented in Fig. 9 (b), using the same representation as above.

We can forget about the quantitative origins of the regions, and reason about them in a purely symbolic way.

u	v	Name of region
0	0	(1,4)
undefined	0	(1,3)
(0, 90)	(0,90)	(2,4)
90	(0,90)	(2,3)
180	0	(1,2)
(90, 180)	(0,90)	(2,2)
(90, 180)	90	(2,1)
(90, 180)	(90, 180)	(2,0)
180	undefined	(1,1)
180	180	(1,0)
(180, 270)	(180, 270)	(0,0)
270	(180, 270)	(0,1)
(180, 270)	(270, 360)	(0,2)
(270, 360)	270	(0,3)
(270, 360)	(270, 360)	(0,4)

Table 6.

Inference in Freksa's calculus

Let again (u_i, v_i), $i=1, 2, 3$, the angles associated to C with respect to (A,B), D w.r.t. (B,C), and D w.r.t. (A,B). The problem is to estimate (u_3, v_3) when qualitative values of (u_1, v_1) and (u_2, v_2) are known.

Again, as remarked (in other terms) by Freksa, this can be done in two steps:

The coarse grain calculus

As already considered in the case of the flip-flop calculus, this consists in taking only u into account as far as C is concerned, and v as far as D is concerned. In terms of the decomposition of the plane associated to (AB), this means that, once (A,B) has been used for defining a frame of reference, we only worry about B it its 9 adjoining regions. This part of the calculus is simple, because of the fact

$$u_3 = u_1 + v_2$$

This also shows that the nine regions around B defined by (BC) are what matters for this part of the calculus.

As a consequence, we get the first composition table in [Fre92]. According to the ranges of u_1 and v_2 involved, the resulting range for u_3 is 90 or 180°. The encoding chosen by Freksa allows a nice expression of the result in terms of integers modulo 8.

The fine grain calculus

The fine grain calculus consists in making use of the extra information provided by u_2, v_1 to restrict the values of v_3. In this way, some results can be strengthened. The second table in [Fre92] gives part of the resulting table. Here is an example of a case where fine grain calculus improves on the coarse version.

Example 1. To give a first illustration of the way Freksa's calculus works, let us consider the case where C is in region (2,2) with respect to (AB), and D in region (0,2) with respect to (BC) (Fig. 10).

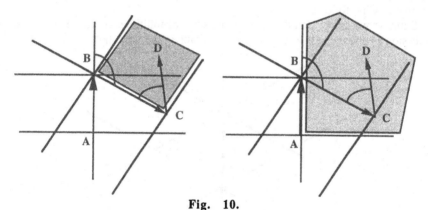

Fig. 10.

As Table 6 shows, relation (2,2) corresponds to $u_1 \in (90, 180)$, $v_1 \in (0, 90)$; relation (0,2) corresponds to $u_2 \in (180, 270)$ and $v_2 \in (270, 360)$.

Coarse reasoning amounts to adding u_1 and v_2: the conclusion is that $u_3 \in (0, 180)$, which defines five neighboring regions: (2, 0), (2,1), (2,2), (2,3), and (2,4). However, consideration of $v_3 = (AB, AD)$ shows that this angle is less than 90 degrees. Hence, fine grain reasoning implies than D is in fact in one of the three shaded regions in Fig. 10: (2,2), (2, 3), and (2, 4).

Example 2. Consider once again the hare-shooting example. Using the same notations as in section 2, we know that C / AB is in (2,4) (which is characterized by $u_1 \in (0, 90)$). Since Clint spots H on his right, but Bill does not see H, we assume that H / BC is the disjunction of (2,2), (2,3), and (2,4), which we can write as (2,[2,4]) (and is characterized by $v_2 \in (0, 90)$). Hence, by coarse reasoning,

H / (AB) corresponds to (2,[0,4]) (characterized by $u_3 \in$ (0, 180)). In this case, in fact, fine reasoning does not improve on the coarse result.

On the other hand, D / (AB) is in (0,0). Since H' / (BD) is in (0,4), fine reasoning implies that H' / (AB) is in ([0,2],0).

Notice that the intersection of (2,[0,4]) and ([0,2],0) is (2,0).

On the basis of those two results, we can answer the two questions raised in section 2:
- The hypothesis that there is a unique hare (H = H') is coherent.
- If such is the case, H / (AB) is in (2,0).

6 Defining other qualitative triangulation calculi

Clearly, this mode of reasoning, which is based on angular knowledge in a certain scale, is just a particular member in a series of possibilities. We only examine an example in what follows.

Suppose we can evaluate angles on the scale of multiples of 60°. Then the same discussion as above leads to partitioning the plane in the way shown in Fig. 11 (a).

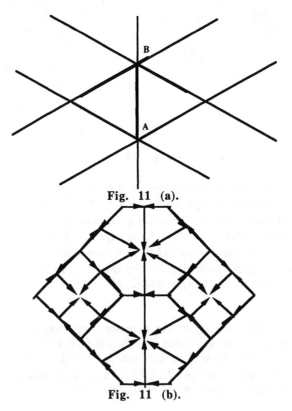

Fig. 11 (a).

Fig. 11 (b).

The topological structure of the resulting set of relations is shown in Fig. 11 (b).

Spatial inference in this context relies on the same mechanisms as before. In particular, coarse reasoning is based on summing the angles. In the same manner as in Freksa's calculus, a transitivity table can be established for fine reasoning, which makes the further constraints defined by v_3 explicit.

Example. Suppose C is in the region corresponding to $(60 < u_1 < 120, 0 < v_1 < 60)$ with respect to (AB), and D in the region corresponding to $(240 < u_2 < 300, \quad 300 < v_2 < 360)$ with respect to (BC).

Then, coarse reasoning shows that D is in the relation corresponding to $(60 < u_3 < 180)$ with respect to (AB). Fine reasoning shows that furthermore we have $0 < v_3 < 60$, restricting D to be in one of threee (instead of five) relations with respect to (AB) (Fig. 12).

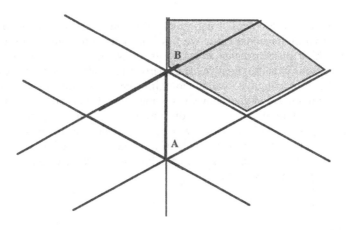

Fig. 12.

7 Refining a qualitative calculus

Refinement corresponds to the idea of getting better knowledge in a situation already considered, and using the new information for sharpening the previous level of inference.

As an example, consider the situation in the example of section 5. Suppose now that we know the following estimates (see Fig. 13):

$$- 90 < u_1 < 135, \quad 45 < v_1 < 90$$
$$- 225 < u_2 < 270, \quad 270 < v_2 < 315$$

This information is consistent with the preceding one, which it sharpens. Now, it allows us to use the calculus based on 45° angles, hence to conclude that D is in fact in the shaded area in Fig. 13.

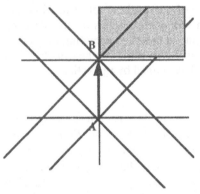

Fig. 13

Conclusion

We have examined the problem of defining qualitative spatial calculi for point-like objects in the two-dimensional case. We have introduced the notion of qualitative triangulation, which arises in a natural way when knowledge of the angles involved in estimating directions is approximative. The coarsest kind of calculus one gets in this way is the flip-flop calculus, which contains the one-dimensional version of qualitative triangulation. We then showed how the calculus introduced by Freksa is a particular case of qualitative triangulation, and gave examples of more complex calculi where a finer level of knowledge is used. One gets in this way a family of calculi which should allow stronger levels of inference than used up to now in this context, with the advantage that a given calculus may be refined to a stronger one when more information is available.

References

[All83]   J. F.Allen, Maintaining Knowledge about Temporal Intervals, *Communications of the ACM* **26**, 11 (1983) 832-843.

[BeLi85] H. Bestougeff and G. Ligozat, Parametrized abstract objects for linguistic information processing, *in*: *Proceedings of the European Chapter of the Association for Computational Linguistics*, Geneva, (1985), 107-115.

[BeLi89] H. Bestougeff and G. Ligozat, *Outils logiques pour le traitement du temps: de la linguistique à l'intelligence artificielle*, Masson, Paris, 1989 .

[EgAl92] M.J. Egenhofer, K. Al-Taha, Reasoning about Gradual Changes of Topological Relationships, *in Frank, A.U., Campari, I. and Formentini, U. (Eds.) Theories and Methods of Spatio-Temporal Reasoning in Geographic Space, Proceedings of the International Conference GIS- From Space to Territory, Pisa, Italy, September* 1992, 196-219.

[Fre91]   C. Freksa, Qualitative Spatial Reasoning, in: *D.M. Mark & A.U. Frank (Eds.), Cognitive and Linguistic Aspects of Geographic Space, Kluwer, Dordrecht, 1991.*

[Fre92]   C. Freksa, Using Orientation Information for Qualitative Spatial Reasoning, *in Frank, A.U., Campari, I. and Formentini, U. (Eds.) Theories and Methods of Spatio-Temporal Reasoning in Geographic Space, Proceedings of the International Conference GIS- From Space to Territory, Pisa, Italy, September* 1992, 162-178.

[Gus89]   H.W. Güsgen, Spatial reasoning based on Allen's temporal Logic, *ICSI TR-89-049, International Computer Science Institute, Berkeley, CA,* 1989.

[He91]   D. Hernandez, Relative Representation of Spatial Knowledge: the 2-D case, in: *D.M. Mark & A.U. Frank (Eds.), Cognitive and Linguistic Aspects of Geographic Space, Kluwer, Dordrecht,* 1991, 373-385.

[Lig90]   G. Ligozat, Weak Representations of Interval Algebras, *Proc. AAAI-90,* 715-720.

[Lig91]   G. Ligozat, Generalized Interval Calculi, *Proc. AAAI-91,* 1991, 234-240.

[MuJo90]A. Mukerjee and G. Joe, A Qualitative Model for Space, *Proc. AAAI-90,* 1990, 721-727.

[Ran91]   D. Randell, Analysing the familiar: a logical representation of space and time, *Third International Workshop on semantics of Time, Space, and Movement, Toulouse,* 1991.

Enhancing Qualitative Spatial Reasoning — Combining Orientation and Distance

Kai Zimmermann

Dept. of Computer Science, Hamburg University,
Bodenstedtstraße 16, 22765 Hamburg, Germany

Abstract. In recent years several qualitative reasoning approaches have been developed in the spatial domain. Although means exists to cope with orientation, position, and topological relations, few of them use the concept of distance. In this paper a new approach is presented that allows to use and to combine knowledge about distances and positions in a qualitative way. It is based on perceptual and cognitive considerations about the capabilities of humans navigating within their environments. The basic spatial reference system, the formalism in which the distance relations are represented, and an implementation that uses multiple domain experts communicating via a black board structure will be described.

1 Introduction

In recent years many approaches to the qualitative representation of space have been developed. While capable of dealing with orientation and topological relations few of them use the very important concept of distance. In this paper we present an approach for dealing with spatial knowledge, that combines orientation and position with the ability to represent and use distances, qualitatively. An implementation of the system is described that uses domain experts communicating via a black board structure.

2 The New Approach: Combining Orientation, Position and Distance

As the basic *spatial reference frame* we use the representation developed by Freksa [1], which is comparable to the approach of Mukerjee & Joe [2], but uses vectors and points instead of rectangular objects, which makes it more suitable for the easy representation of distances. It will be extended by the explicit representation of the occuring edges and their lengths. We will show how length information, i.e. distances, can be transformed into position knowledge and vice versa, and how this combination adds to the result of the inference process.

The next section will give a short introduction to the qualitative representation of position and orientation we use. The following section will explain some properties of the Δ-calculus used to represent the relations between distances qualitatively. The last section will define the edges and distances used within the new approach and give an example of the inference process.

2.1 The Spatial Reference Frame

As the basic spatial representation we use the notation and inference mechanisms developed presented in [1]. It is based on a vector AB and a point C that is located wrt. this vector. The possible locations of C are then described by qualitative abstraction. See Fig. 1 for the reference frame and the chosen iconic representation. This formalism has been extended to allow not only for chaining of directly following references, i.e. from $AB{:}C$ and $BC{:}D$ the relation $AB{:}D$ can be concluded, as in the original paper, but also for the computation of all possible combinations, see Freksa & Zimmermann [3]. E.g. given $AB{:}C$ and $AB{:}D$ you can compute $BC{:}D$, $BD{:}C$, $CD{:}A$, and so on. See Fig. 2 for an example. Other work is coping with the combination of space and time to allow reasoning about motion, e.g. Zimmermann and Freksa [4].

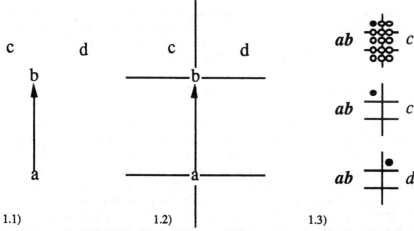

Fig. 1 1) Consider somebody walking from a to b. On his way he observes c in front and to the left of b and d in front and to the right. 2) By introducing the two lines orthogonal w.r.t.ab through a and b and the line through a and b we get an orientation grid with 15 different positions: six areas, seven places on the lines, and two points. 3) The positions of c and d can now be described in terms of these 15 spatial relations which is depicted iconically.

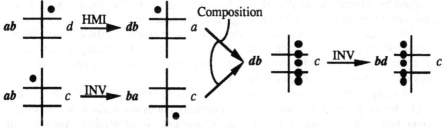

Fig. 2 A sample inference within the chosen representation. Consider somebody walking from A to B and then to D in the above example. One might ask where the point C is to be expected wrt. the second vector BD. The computation of $BD{:}C$ from $AB{:}C$ and $AB{:}D$ yields: C may be anywhere to the left of the line coinciding with vector BD. INV is an operation that inverts the orientation of the vector AB, transforming $AB{:}C$ into $BA{:}C$. HMI is short for *homing-inverse*. Homing is the operation that transforms $AB{:}C$ into $BC{:}A$, i.e. the reference vector is shifted. Thus, homing-inverse yields $CB{:}A$. Composition is an operation that chains two vectors, $AB{:}C$ and $BC{:}D$, and yields $AB{:}D$. For details see [3].

Notice that as well point A and B can be used as the standpoint of the observer, Fig. 3. This allows us to combine motion sequences and panorama views observed at one point, like those used by Kuipers and Levitt [5] and Schlieder [6].

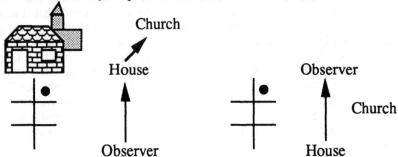

Fig. 3 Both, "A church can be seen behind the house and to the right." and "When I came from the house I passed a church to my right." can be expressed within this representation system. In both cases the house is used as reference object and the church is the entity to be located with respect to the stand point of the observer.

2.2 Representing Point-Like Measures: The Δ-Calculus

Many different approaches have been developed to deal with point-like measures, in this case the length of the involved edges, i.e. distances. Most of the approaches simply use numeric values based upon the mathematical set of real numbers, e.g. points or intervals. Because exact numeric measures accour rarely in human reality several qualitative approaches have been developed. Consequently the qualitative approaches to deal with such measures use relations between points. To overcome the restrictions of ordinary relational algebra which allows only three distinctions (<, =, >) between points, several approaches have been developed, e.g. Order of Magnitude calculus, see Mavrovouniotis and Stephanopoulus [7] or Raiman [8], used in technical domains, where this kind of knowledge can be sufficient to predict the outcome of differential equations. But since most distances in a modeled scene are in the same Order of Magnitude, nothing would be gained. Δ-Calculus has been developed by Zimmermann [9, 10, 11] to represent point-like measures from the spatial domain based on the following cognitive considerations:

1. Observed measures from the world always have a positive dimension. We do not think of depth being a negative height, for example. Especially distances are always positive, since an edge can not have a negative length. Thus there is no need to provide means to model negative values.

2. Most approaches offer too much operations to be cognitive plausible. For example the multiplication of two observed distances does not make sense normally, although there is no problem if you are using numbers. What would be the result of 'The distance from point a to b times the distance from point b to c'? You can only answer this question if you use a numerical representation or escape into a two-dimensional solution leaving you with an area as the result, incomparable to the input values. Direct multiplication is therefore excluded from Δ-Calculus.

3. Relational approaches offer the ability to represent unknown comparison results. But their major disadvantage has been their lack of expressiveness. We try to overcome this problem in that we allow for the representation of the difference

between two point-like measures. This allows us to compute addition and subtraction of point-like measures, in the domain of space this means the chaining in the same or the opposite direction of two lines.

The following figures give short examples on how Δ-calculus is represented and used. Areas, proportions, i.e. the ratio of two measures, and volumes are represented as tupels and triples of point-like measures. For details see the above mentioned papers.

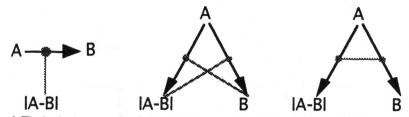

Fig. 4 The basic representation for Δ-calculus meaning: A is by difference |A-B| greater than B. The graphical representation exploiting the symmetry of the smaller measure and the difference measure, depicting the fact that A is greater than b by exact the value of |A-B|. Note that |A-B| is just a symbolic label and no numeric computation occurs. This symbolic difference can now be related to observed measures, again.

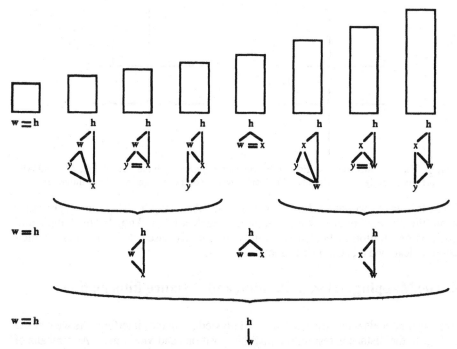

Fig. 5: Proportion classes at different levels of granularity, where w stands for the width and h for the height of the rectangles depicted above the diagrammatic representation of the relation patterns. Each class can be identified by a specific pattern between two measures and the differences that occur when comparing them. The classes form conceptual neighborhoods such as those introduced for temporal relations by Freksa [12].

3 The Distances Used

In the above described reference frame three vectors occur explicitly: The vectors *AB*, *BC* and *CD*. These are now mapped from vectors to edges without direction, since we want to exploit their distances. Additionally we introduce the orthogonal distance of point *C* wrt. vector *AB*, Dx, and the distance of point C to the two orthogonal lines, DyA and DyB. See Fig. 6 for the resulting edges.

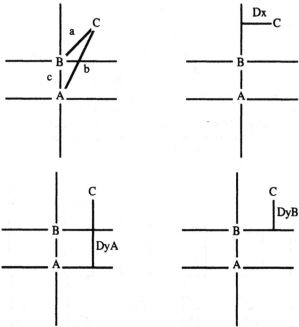

Fig. 6: The introduced edges. Edge a, b, and c coincidence with the vectors BC, AC, and AB, correspondingly. The edges Dx, DyA and DyB decompose edges a and b orthogonally.

From this edges we take a further abstraction: their length. These are represented symbolically and related via Δ-calculus. Each kind of knowledge, length and position/orientation is treated separately by agenda based domain experts which communicate via a black board structure.

4 The Mapping Between Position and Distance Information

This section deals with how the different knowledge sources interfere. As we can see in Fig. 8, the distances restrict the possible positions and vice versa. As a means of communication a black board agenda has been chosen to which each inference component signals new facts.

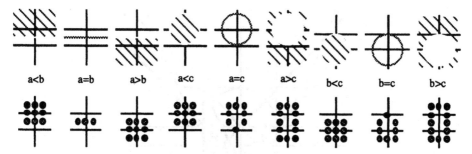

Fig. 8: The mapping from distance knowledge to position knowledge and vice versa. For each possible relation between the length of two edges of the triangle a, b, and c the possible positional relations within the reference frame are given. For the black dots the mapping can be converted meaningfully, i.e. one can map the position into a single relation between the lengths of the edges. For the gray dots every relation between the lengths of the edges are possible.

Note that the different logical combinations of the results of the mapping for each distance relation resemble the combination of the source relations. Thus, from a<c and b>c follows a sharper result because the intersection of the single results can be taken.

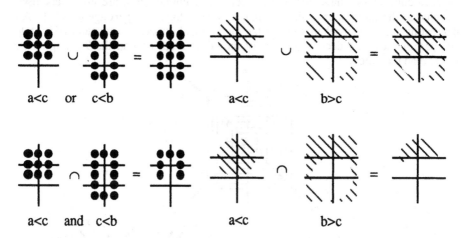

Fig. 9: The combination of more than one assertion. Note, that although within the qualitative spatial representation the shape of the restricted area and its small size can not be represented, this information is still available within the composed knowledge bases for means of visualization, for example.

The following Fig. 10 depicts the restrictions that are introduced by relating not only edges a, b, and c but also Dx, DyA, and DyB via first order Δ-calculus. The exact description of the areas and the corresponding constraints are not given due to the restricted publication space.

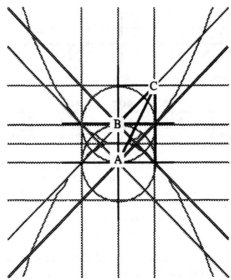

Fig. 10: The resulting areas from relating each of the edges to each other.

No Δ-calculus classes have been used directly, yet, although they are stored and used to enhance the inferences in the specialised length domain expert agenda [11]. An interesting fact is that when you use the hierarchical classes introduced in Fig. 5 to compare the x and y decompositions of the edges areas around the 90° angles are larger than the others.

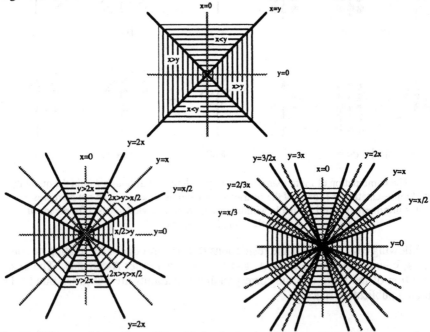

Fig. 11: When combining Dx and Dy with Δ-calculus the areas around the 90 degree angle are larger. The given ratios correspond to the Δ-calculus proportion classes of Fig. 5.

5 Conclusion

In this paper we described a qualitative representation for the spatial domain and a set of operations to deal with it. We then showed how the representation of relations between point-like measures can be enhanced by taking into account the specific spatial domain restrictions. Both representations can be combined efficiently to map distance knowledge onto positions and vice versa.

References

1. C. Freksa: Using orientation information for qualitative spatial reasoning. In: A.U. Frank, I. Campari, U. Formentini (Eds.) *Theories and Methods of Spatio-Temporal Reasoning in Geographic Space*. Springer, New York, 1992, 162-178.

2. A. Mukerjee, G. Joe: A qualitative model for space, *Proc. AAAI-90*, 721-727.

3. C. Freksa, K. Zimmermann: On the utilization of spatial structures for cognitively plausible and efficient reasoning. In: *Proc. IEEE International Conference on Systems, Man, and Cybernetics*. Chicago, 1992.

4. K. Zimmermann, C. Freksa: Enhancing Spatial Reasoning by the Concept of Motion. In A. Sloman et al. (Eds.): *Prospects for Artificial Intelligence*. IOS Press 1993, 140-147.

5. B.J. Kuipers, T.S. Levitt: Navigation and Mapping in Large Scale Space. *AI Magazine* 9 (1988) 25-43.

6. Schlieder, C.: Anordnung und Sichtbarkeit. Eine Charakterisierung unvollständigen räumlichen Wissens. Ph.D. thesis, Hamburg, 1991.

7. M.L. Mavrovouniotis, G. Stephanopoulus: Formal Order of Magnitude Reasoning in Process Engineering. *Computer Chemical Engineering* 12 (1988) 867–880.

8. O. Raiman: Order of Magnitude Reasoning. *Proc. of AAAI-88*, 100–104.

9. K. Zimmermann: *SEqO – Ein System zur Erforschung qualitativer Objekt-repräsentationen*. Report FKI-154-91, Technische Universität München. Diplomarbeit, 1991.

10. K. Zimmermann: A Proposal for Representing Object Sizes. In Simone Pribbenow and Christoph Schlieder (Eds.): *ECAI 92 workshop notes „Spatial Concepts: Connecting Cognitive Theories with Formal Representations."* 1992, Vienna.

11. K. Zimmermann: *Measuring without Measures – The Δ-Calculus*. Report, 1993.

12. C. Freksa: Temporal reasoning based on semi-intervals. *Artificial Intelligence* 54 (1992) 199-227.

Map Semantics

Ian Pratt

Department of Computer Science,
University of Manchester,
Manchester M13 9PL,
United Kingdom

Abstract. Could there be a *semantics of maps*—a theory which would stand in the same relation to maps as (ordinary) semantics stands in to (ordinary) languages? And what would such a theory look like? This paper addresses these questions from the standpoint of the doctrine that a semantics for a language is a theory which gives the truth-conditions for expressions in that language. We show how a semantics for certain simple sorts of maps can be constructed by analogy with minimal model semantics for the predicate calculus under default reasoning. We go on to show, however, that other sorts of maps challenge a fundamental assumption about truth-conditional approaches to semantics. In particular, we see how the figure/ground ambiguity inherent in some maps forces us to re-think the role played by the syntactical organisation underlying an account of truth-conditions.

1 Introduction

Maps have much in common with written natural languages like English. Both media rely on the use of repeatable symbols—marks on paper—whose meaning is a matter of convention. Both media are useful for representing (and for misrepresenting!) the world about us. And both media are useful as aids to memory and reasoning. However, there are also differences. While maps are specialized to represent space, and have difficulty coping with disjunctive or uncertain information, English exhibits no such limitations. The basic vocabulary of English is vastly greater than the set of symbol-types employed on any map. And maps regularly exploit devices alien to English, such as using continuously variable symbols in an analog way (as when the size of a dot represents the population of a city). Most strikingly perhaps, maps differ from languages like English in the similarities they bear to *drawings* and *photographs*. However stylized or conventional or arbitrary our mapping conventions may be, the resemblance between an aerial photograph and a map of the same scene is hardly deniable; the resemblance between an aerial photograph and page of the English describing the scene is nil.[1]

---

[1] Comparisons between maps and languages are not uncommon, of course. See, for example, Robinson and Petchenik [8], Head [5] and [6], Schlichtmann [9] and Andrews [1].

If maps exhibit some similarities to, and yet some differences from, natural languages, what sort of theory might we construct to explain *how* maps represent the world? Could there be a *semantics of maps*—a theory which would stand in the same relation to maps as (ordinary) semantics stands in to (ordinary) languages? And what would such a theory look like? This paper is an attempt to at least begin to answer these questions.

Semantics may be defined as the study of linguistic meaning. Derivatively, we speak of *a semantics* for a language as a set of rules specifying the meanings of expressions in that language. Now, according to one prevalent school of thought within semantics, to give the meaning of an expression (in a language) is to state the conditions under which that expression would be true of false. For example, to know the meaning of the German sentence "Schnee ist weiß" is to know that it is true if and only if snow is white; to know the meaning of the German sentence "Äpfel sind blau" is to know that it would be true if and only apples were blue; and so on. On this view, then, a semantics for a language is a theory which gives the truth-conditions for expressions in that language. It must be admitted that the view of a semantics as a specification of truth-conditions, while widespread and influential, is by no means universally accepted. Nevertheless, it will serve here to anchor our comparison of maps with languages—to focus our attention on theoretically important issues.

Since maps, like natural languages, represent the world, it makes sense to ask whether they do so *correctly* or *incorrectly*. To borrow terms familiar in the context of natural language, it makes sense to ask whether a map is *true* or *false*. This suggests the following formulation of our initial question concerning the possibility of a semantics of maps. Can we provide maps (or classes of maps) with a semantics, in the sense of a theory which specifies the conditions under which a given map makes true or false claims? If so, what will such a map semantics look like? How would it differ from the semantics for other languages?

2 The Semantics of Artificial Languages and the Semantics of Maps

In explaining the view that a semantics for a language is a theory giving the truth-conditions of its expressions, we cited as an example the truth-conditions of two (pretty simple) German sentences. Of course, serious attempts at semantics for natural languages attempt to give the truth-conditions for more considerable (in fact, infinite) fragments of those languages. Unfortunately, semantic accounts of even restricted parts of natural language are enormously complicated, and could not possibly be presented here. Moreover, the semantics of many common features of language are not well understood, and progress has been slow in investigating them.

In order to get an idea for what a formal semantics for maps might look like, then, we can turn for inspiration not to *natural* languages, but instead to simple *artificial* languages, for which a semantics—in the sense of a body of rules specifying truth-conditions—have been given. Of all these artificial languages, the simplest and most widely employed is the predicate calculus. The standard semantics for the predicate calculus is well-known, and hardly need be presented in detail here. However, it will help to remind ourselves of some of its principal features; the contrast with with maps will prove instructive.

The predicate calculus allows us to make statements such as:

man(socrates)	*Socrates is a man*
philosopher(plato)	*Plato is a philosopher*
$\forall x(\text{man}(x) \rightarrow \text{mortal}(x))$	*All men are mortal*
$\exists x(\text{philosopher}(x) \rightarrow \text{wise}(x))$	*Some philosophers are wise*

The standard semantics for this language, due to Tarski, takes the form of rules specifying in the truth-conditions of the above formulae.

The key idea behind this a semantics is that of a *structure*. A structure M consists of a *domain D*, which is just any set of entities, together with an *interpretation function I*, which links the symbols in the language to the elements of the domain. For example, the interpretation of a name—say, "socrates"—is simply an element of D (intuitively, the man socrates); the interpretation of a predicate—say, $man(\_)$—is simply a set of elements of D (intuitively, the set of elements which have the property of being a man). The rules for determining the truth-conditions of quantified formulae are complicated to state, and would involve us here in too much irrelevant detail. But the important point is that such rules determine, for any formula ϕ and any structure M, the truth-value of ϕ in M. Thus, the semantics for the predicate calculus gives the truth-conditions of formulae by specifying the structures in which they come out true.

One last detail: given a *collection*, T, of predicate calculus formulae, we say that T is true in a structure M just in case all of the formulae in T are individually true in M. In that case, it is customary to speak of M as being a *model* of (the formulae in) T. We will return to this detail later.

With that brief excursion into the semantics for the predicate calculus behind us, we now come to examine the prospects for a parallel semantics for maps. To fix our ideas, let us consider the simple map of fig. 1, which we can take to indicate the locations of certain sorts of buildings in a village using the symbols □ (house), † (church) an • (well). Schlichtmann[9] has proposed that these symbols can be analyzed as complexes of 'substantive' and 'locational' components. The substantive component, encoded by features of the icon other than its location, conveys the nature of the represented object (e.g. house, church, well), while the locational component, encoded by the position of the icon on the map, conveys the location of the represented object in the obvious way. Thus, a church-icon is seen as having a structure:

$$\dagger + (x, y) \qquad (1)$$

much as a formula in the predicate calculus is seen as being made up of its names, predicates and other symbols.[2] Of course, the syntactic structure of map

---

[2] For the sake of comprehensibility, I have been deliberately sloppy when it comes to distinguishing positions on the maps from the numerical coordinates corresponding to them.

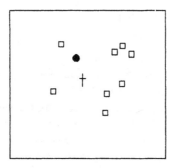

Fig. 1. A simple map with located symbols representing a church, a well and some houses

symbols is likely to be much simpler than the syntactic structure of predicate calculus formulae.

We can construct a semantics for maps such as that of fig. 1 by straightforwardly extending the semantics of the predicate calculus. Again, the key idea is that of a structure M, except that, in addition to the domain D of individuals, M will recognise a space S of *locations*, with M assigning a location to each of the individuals in D. Furthermore, in addition to the *symbolic* interpretation function I taking the map symbols to sets of objects in D, there will be a *spatial* interpretation function μ taking map-coordinates to locations in S. For example, in the map of fig. 1, I would map • to the set of wells, □ into the set of houses and † to the set of churches, while μ would map points on the map surface to locations in space. There would then have to be a truth-rule *something like*:

Let M be a structure with domain D, space S, symbolic interpretation function I, and spatial interpretation function μ. A map-symbol $X +$ (x, y) with substantive component X and locational component (x, y) is *true* according to M just in case there exists an element of D which is in the set $I(X)$ and to which M assigns the location $\mu((x, y))$ in S.

Of course, all this is no more than an initial gesture in the direction of what a formal semantics for such map symbols would look like. For example, we have given no indication of how to incorporate, into a formal semantic framework, graphic variables which vary continuously along one dimension, such as hue, orientation, and shade. Still less clear is how to incorporate, within a formal semantics, the use of multi-dimensional variables, such as texture or shape, to represent substantive (non-locational) information in a non-digital way. Clearly, there is much work of a detailed nature that could be undertaken; but we shall assume that these details can be sorted out, and not concern ourselves with them here.

3 Default Representation

The previous section gave a condition for the truth of an individual symbol pair (1) in a structure, but gave no conditions for the truth of a whole map in a structure. Let us see how we might fill this lacuna.

We return to a remark made above concerning the semantics of the predicate calculus. Given a *collection* T of predicate calculus formulae, we say that T is true in a structure M just in case all of the formulae in T are individually true in M. If we take a map such as that of fig. 1 to be a collection, T, of symbol-pairs such as (1), we might be tempted to say: map T is true in a structure M, just in case all of the individual symbol-pairs in T is true in M. But this, it turns out, would lead to most unintuitive results.

One feature of maps that we have not mentioned so far is that the *absence* of symbols can convey information. Thus, in the map of fig. 1, the absence of a church-icon at a particular location—i.e. the non-occurrence of a symbol-pair of the form (1) in the map—represents the fact that there is *no* church at the corresponding point in space.[3] That is why we cannot adopt the above rule for evaluating the truth of whole maps with respect to a structure: a map should come out as *false* if it fails to mark church at a place where, in fact, there is one. We might say, in view of this observation, that maps can represent information *by default*. How can we modify the kind of formal semantic account suggested above to capture representation by default?

Again, we need look no further than the predicate calculus for inspiration—specifically, at recent work in artificial intelligence on the representation of information by default in this type of formalism.

Consider the following collection of formulae:

$$\forall x((\text{bird}(x) \& \neg\text{abnormal}(x)) \to \text{flies}(x)),$$
$$\text{bird}(\text{tweety}),$$
$$\text{bird}(\text{oscar}),$$
$$\text{abnormal}(\text{oscar}),$$
$$\text{oscar} \neq \text{tweety}$$

with the intuitive interpretation that all birds which are not abnormal fly, that Tweety and Oscar are (distinct) birds, and that Oscar is abnormal. Let us denote this set of formulae by T. Notice that these formulae T say nothing about whether Tweety is abnormal or not; they simply leave the matter open. (Hence they leave open the question of whether Tweety can fly or not.) In terms of the normal semantics for the predicate calculus: there are models of T in which Tweety is abnormal and does not fly, as well as models in which Tweety is not abnormal and does fly.

---

[3] It is worth pointing out that the view that maps represent information *by the absence of symbols* requires some qualification, as Andrews [1], p. 15, makes clear. However, we shall assume that the maps we are concerned with here are to be understood in this way.

But suppose we wanted to *assume* that, in the absence of any reason to the contrary, Tweety is not an abnormal bird—after all, most birds are normal—and so can fly. More generally, we might make the assumption that a thing is *not* abnormal unless it is asserted to be. In making this assumption, we are, in effect, encoding information *by default*. For the mere absence of the assertion that something is abnormal is tantamount to asserting its normality, just as the absence of a church-icon at a point on a map asserts the non-existence of a church at the corresponding location.

How, then, can the truth-conditional semantics be modified to allow the encoding of information by default? A number of possibilities have been proposed in recent years, but by far the most interesting and intuitive is an idea of *minimal model semantics*, which we briefly review here.

We mentioned above that the formulae T have some models in which Tweety is abnormal; this is because T includes no formula saying whether Tweety is abnormal or not. But suppose we confine our attention to those models of T in which the extension of predicate abnormal(x) is *as small as possible*. We can then say that a set of formulae T is true in any structure M, just in case the individual formulae in T are true in M, *and the extension of the predicate* abnormal(x) *in M is minimal*. That way, things get counted as normal unless they have been declared (explicitly or implicitly) to be abnormal.

This approach to semantics involves numerous technical complications, especially when it comes to defining exactly what we mean when we say that a structure is one in which the extension of a predicate is *minimal*. But again we need not concern ourselves with these details here; since they have been satisfactorily worked out elsewhere.[4] What is important is to see how such a semantics can yield a solution to the problem of encoding information by default. A semantics which takes the interpretation of a set of formulae to be the set of models in which as few things are abnormal as possible means that the *absence* of a formula saying that Tweety *is* abnormal effectively encodes the information that Tweety is not abnormal.

Thus, the fact that maps represent information by default need not be a problem when it comes to providing them with a formal semantics. We might say that a map T (considered as a set of symbol-pairs) is true in a structure M, just in case the individual symbol-pairs in T are true in M, *and the extensions of the predicates* church, school, house *etc. in M are minimal*. That way, structures locating objects where the map declares no object to be (by being blank) would be excluded from the map's interpretation, because they would not be minimal.

So far, we have sketched an account of how the semantics of certain simple sorts of maps—specifically, those such as fig. 1 involving located icons—might be presented. And we have seen that the resulting semantics, while different from the semantics of a language like the predicate calculus, will nevertheless be, in broad outline, familiar, and introduces no fundamentally new notions. Our concern in the sequel, however, will be with a quite different aspect of maps

---

[4] For a full technical account of minimal model semantics in AI, see, for example, Brewka [2], ch. 4.

which, as we shall see, challenges a basic assumption about what we expect from a formal semantics. First, however, we need to step back, and take a broader view of the enterprise of semantics.

4 Syntactic Structure, Truth-Conditions and Use

The semantics of the predicate calculus, with or without minimization conventions, takes for granted a particular decomposition of the expressions in that language into constituent symbols: names, predicates, variables, connectives and quantifiers, together with various syntactic devices for combining those symbols into formulae. But it is important to realize that this decomposition is not only important when it comes to explaining the truth-conditions of formulae: it is also important in explaining the operation of any system which *uses* these formulae.

One of the attractive features of formal languages is that their expressions can be manipulated according to precise rules, in particular, according to rules contained in a computer program. That is why such languages are so fundamental in AI: computers can use them for reasoning. Now, when expressions in the predicate calculus are used in reasoning, they are operated on by processes sensitive to their syntactic structure. Consider a theorem-proving program inferring the conclusion mortal(socrates) from the premises man(socrates) and $\forall x(\text{man}(x) \rightarrow \text{mortal}(x))$. What that theorem-prover must do is to detect a certain pattern in the syntactic structure of the premises, and to perform manipulations on that structure to generate the conclusion. In other words, the symbols in the formal language—more precisely, their physical instantiations in a computer—are the raw material for the theorem-proving processes. Any explanation of that process must recognise the relevant symbols as constituents of the formulae being manipulated.

Hence we see that the syntactic structure of expressions in a language can play a double role. First, it is fundamental to the semantics for that language, in that the rules specifying the truth-conditions of formulae make reference to the symbols in those formulae and the way they are combined. Second, it is fundamental to understanding the way formulae in that language are manipulated in reasoning, because reasoning processes which use formulae must be causally sensitive to their syntactic structure.

Now of course, these remarks are not in themselves particularly surprising. Obviously, syntactic structure is important to the semantics of the predicate calculus; and obviously, processes for manipulation of predicate calculus formulae must be causally sensitive to their syntactic structure. However, when we come to consider the case of maps, we will find ourselves confronted by a more complex situation in which there is no unique decomposition into symbols that is appropriate both for specifying truth-conditions and for explaining use. This, we shall argue, constitutes a crucial difference between maps and languages such as the predicate calculus.

The main observation of this section is the following. The imposition of syntactic structure on a representational medium—its decomposition into symbols and their modes of combination—may serve as a vehicle for more than one task. We have identified two such tasks: (i) a specification of the truth-conditions of expressions in that language and (ii) an account of how expressions in that language are processed by those who use them. This distinction will prove crucial when it comes to comparing maps with language.

5 Figure/Ground Ambiguity and Syntactic Ambiguity

In section 2, we sketched some semantic rules for maps involving a collection of located icons, such as that of fig. 1. In the present section, we examine a different kind of map, and pose the question of what a formal semantics for this kind of map might look like.

Consider the maps of fig. 2a)–c), which show a number of lakes (dark) and islands (light). How might we give a formal semantics for maps such as these?

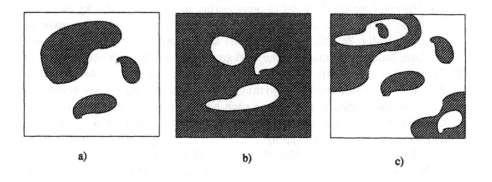

a)   b)   c)

Fig. 2. Three fictional maps showing: a) lakes, b) islands, c) lakes and islands

The first question we need to answer is: *which things are the symbols?* Well, in fig. 2a), it seems natural to take the the areas representing *lakes* to be symbols. (See Schlichtmann [9], pp. 24, who recognises such symbolic elements.) Here, there is no symbol representing the body of land in which the lakes are located; the land is taken to be background, and only the lakes have symbols corresponding to them. Moreover, under normal mapping conventions, the absence of a lake-symbol covering a particular point on the map can be taken to indicate by default that the corresponding point in physical space does not lie in a lake.

In fig. 2b), by contrast, it seems more natural to take the the areas representing *islands* to be symbols. Here, there is no symbol representing the body of water in which the islands are located; the water is taken to be background; only the islands have symbols corresponding to them. Moreover, under normal mapping conventions, the absence of a island-symbol covering a particular point on the map can be taken to indicate by default that the corresponding point in physical does not lie on an island.

But what about fig. 2c)? Here, we have a complex arrangement of islands, lakes, islands within lakes and lakes within islands, so that it is no longer clear which things—the islands or the lakes—are represented by symbols, and which things—the lakes or the islands—are represented by the absence of symbols. Yet a decision has to be made on this matter if we are to give a formal semantics for these maps. Let us consider the options.

The following options exist for analysing the maps such as those of fig. 2 into structures of symbols.

1. The first option is that one category of surface (i.e. land or water) is represented by symbols, while the other is represented by absence of symbols. This option comes in two versions:
   (a) The symbols are the patches representing land, and the background, i.e. those parts of the map covered by no symbols, the patches representing water. In fig. 2c), the land-symbols are not simply connected (i.e. they have holes in them), and the background shows through the holes.
   (b) The symbols are the patches representing water, and the background, i.e. those parts of the map covered by no symbols, the patches representing land. In fig. 2c), the water-symbols are not simply connected (i.e. they have holes in them), and the background shows through the holes.
2. The patches representing water and those representing land are *both* symbols. Again, these water- and land-symbols are not in general simply connected (i.e. they have holes in them), but each hole is exactly filled by a symbol of the other type, so that the two types of the symbols together cover the map completely, without overlapping.
3. The closed curves representing *the outlines of* bodies of water or *the outlines of* bodies of land are symbols. Notice that one such closed curve (say, corresponding to the outer boundary of a lake) make have other closed curves (corresponding to the outer boundaries of islands) within it; these may in turn have yet more closed curves within them, and so on, up to arbitrary depth.

   There is an alternative (and, assuming that the map is drawn on an ordinary piece of paper, technically equivalent) way of viewing this option. Instead of taking the symbols to be closed curves on the map, we can take them to be the simply-connected regions (i.e. regions with no holes in them) inside those closed curves. And we allow that a small lake-symbol may be superimposed on top of part of an island symbol, and vice versa, up to arbitrary depth. On this view, then, the two types of the symbols together cover the map completely, but do not have holes in them and are in general overlaid one on top of the other.

Corresponding to the three options for symbolic decomposition above, we might suggest the following semantics rules for giving the meanings of such maps (we use μ throughout to denote the function taking points on the map to the locations they represent):

1. The semantics for the first option must again be given in two versions:

   (a) Let L be a map symbol (a connected dark region of the map surface).
   The symbol L is *true* in a structure M, just in case, for all points (x, y) on the map, if (x, y) is in L, then the point $\mu((x, y))$ is on land according to M.

   An entire map is true in a structure M if (i) all the symbols L in that map are true in M, and (ii) there is no other structure M' in which all the symbols in the map are true, and such that the set of locations on land in M' is a subset of the set of locations on land in M.

   (In other words, we insist that M is a land-minimal model of the map.)

   (b) Let L be a map symbol (a connected *light* region of the map surface).
   The symbol L is *true* in a structure M, just in case, for all points (x, y) on the map, if (x, y) is in L, then the point $\mu((x, y))$ is on water according to M.

   An entire map is true in a structure M if (i) all the symbols L in that map are true in M, and (ii) there is no other structure M' in which all the symbols in the map are true, and such that the set of locations in water in M' is a subset of the set of locations in water in M.

   (In other words, we insist that M is a water-minimal model of the map.)

2. This time we suppose that the map is a collection symbols-pairs of the form $\langle l, L \rangle$ or $\langle w, L \rangle$ where L is again a connected region of the map surface, and l and w are symbols indicating the shading (w for dark and l for light). Symbol-pairs of the former type (the light patches) denote land, and those of the latter type (the dark patches) denote water.

   The symbol-pair $\langle l, L \rangle$ is *true* in a structure M, just in case, for all points (x, y) on the map, if (x, y) is in L, then the point $\mu((x, y))$ is on land according to M

   The symbol-pair $\langle w, L \rangle$ is *true* in a structure M, if and only if, for all points (x, y) on the map, if (x, y) is in L, then the point $\mu((x, y))$ is on water according to M.

   An entire map is true in a structure M if all the symbol-pairs $\langle l, L \rangle$ and $\langle w, L \rangle$ are true in M.

   (There is no need for any minimal model semantics here, for there is no representation by default. Assuming that the symbols really do cover the map without gaps, the status of any point (land or water) with in the region covered by the map is decided, since every point on the is covered by exactly one symbol.)

3. Here, we take the map to consist of a set of symbol-pairs $\langle l, \delta \rangle$ (closed curves enclosing light patches) and $\langle w, \delta \rangle$ (closed curves enclosing dark patches).

   A symbol-pair $\langle l, \delta \rangle$ is true in a structure M if, for all points (x, y) within δ such that there is no $\langle w, \gamma \rangle$ with γ inside δ and with (x, y) inside γ, $\mu((x, y))$ is on land.

   A symbol-pair $\langle w, \delta \rangle$ is true in a structure M if, for all points (x, y) within δ such that there is no $\langle l, \gamma \rangle$ with γ inside δ and with (x, y) inside γ, $\mu((x, y))$ is on water.

An entire map is true in a structure if all the symbol-pairs $\langle l, \delta \rangle$ and $\langle w, \delta \rangle$ are true in that structure.

(Thus we have a default semantics in which more specific symbols—those enclosing smaller areas—override less specific ones. This kind of semantics is again familiar from AI, and can be put on a formal basis in a number of ways. One such way involves a minimal model approach, where the definition of a 'minimal' model is modified so as to make more specific symbols take priority over less specific ones when it comes to deciding which models are minimal. However, the technical details are far too complex to spell out here.[5])

Thus we see that of the three methods of decomposing the maps of fig. 2 into symbols lead to three different accounts of the semantics of those maps. Yet each is, from the point of view of specifying the *truth-conditions* of those maps, equally adequate, since they all in the end lead to the same truth-conditions. The map itself—the pattern of coloured patches—is neutral on the question of how those patches are organised into symbols and their modes of combination, just as a drawing or photograph is neutral as to how the marks it contains should be perceived. The map itself does not decide what is figure (the symbols) and what is (back)ground. In that sense, the map can be said to be syntactically ambiguous: there is more than one option when it comes to imposing a syntactic structure on the map in order to provide it with a semantics.

Nothing like this sort of syntactic ambiguity is observed for artificial languages such as the predicate calculus. With such languages, we *begin* with the individual symbols and their modes of combination: there can be no question of the sort of figure/ground ambiguity that can arise with maps such as those of fig. 2.

The main observation of this section the following. When giving a semantics for maps such as those of fig. 2, a number of options present themselves, depending on how the map is to be decomposed into symbols, or, as we put it, depending on what syntactic structure we impose on the map. There is more than one syntactic structure in terms of which adequate semantic rules can be given. The map itself is neutral with respect to these different syntactic structures. Nothing like this phenomenon is observed for artificial languages such as the predicate calculus.

6 Figure/Ground Ambiguity and the Use of Maps

In section 4, we observed that the imposition of syntactic structure on a representational medium may serve as a vehicle for two tasks: (i) a specification of the truth-conditions of expressions in that medium and (ii) an account of how expressions in that medium are processed by those who understand them. In section 5, we observed that an account of the truth-conditions of maps such as those of fig. 2 could be given equally well in three different ways, with each way corresponding to the imposition of a different syntactic structure on the map.

---

[5] See, for example, Brewka [2], section 4.3

These observations lead to the following question. Given that the different methods of syntactic decomposition of maps can all be made the basis of adequate truth-conditional semantics, are these three decomposition methods all equally suited to an account of map-*use*? For example, when humans (or computers) read maps, which aspects or features of the maps are the items on which the relevant psychological processes (or programs) involved operate?

The answer is, of course: *it depends on what the map is being used for.* Consider the map of fig. 3, and let us suppose that we are using the map to plan a route between two lakeshore points, X and Y. We distinguish two cases: first, the journey is to be undertaken in a *boat*, and the route proceeds over water, with the islands regarded as obstacles; second, the journey is to be undertaken in a *jeep*, and the route proceeds over land, with the lakes regarded as obstacles.

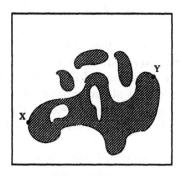

Fig. 3. A route-planning task with a map marking lakes and islands

The task of planning routes through spaces strewn with obstacles has been extensively studied in robotics and computational geometry. (For a surveys, see the papers in Kapur and Mundy [7].) Most such algorithms assume that the scene has been segmented in such a way that obstacles are identified as units: that is, they assume that there is a list of data-structures available, each member of which contains information about the outline of some obstacle. The algorithms then process these obstacles, extracting critical line-segments and combining these to generate a route.[6] Here, then, the data-structures which the route-planning algorithm processes correspond to the outer boundaries of the obstacles; there is (in most cases) no symbol that represents any area of free space.[7] Thus when doing route-planning, it is the *obstacles* (or their outside boundaries) that are the figure, and the free space in which movement is possible

---

[6] For an exception, see the algorithm in Chubb [4], who discusses route-planning using a bitmap.

[7] Actually, this last statement is a simplification, since some route-planning algorithms *do* construct explicit representations of free space, for example Brooks [3]. However,

the (back)ground. Clearly, planning journeys by boat and planning journeys by jeep require different views of what counts as figure and what counts as ground.

This is not to say that the there is no way of decomposing the map into data-structures which would be useful for both over-water and over-land route-planning: in fact, data-structures representing land-water boundary curves would be well-suited here.[8] However, the above example does illustrate the general point that the way to organise a map into symbols depends critically on what the reader wants the map *for* (on a given occasion). The map itself is neutral on this question of distinguishing figure from (back)ground; it is the map *reader* who is responsible for this. Put another way: the map itself is neutral on the question of how it is decomposed into symbols—islands or lakes, buildings or courtyards, fields or field-boundaries; it is the map *reader* who must organise what he perceives into units that serve his current reasoning purposes.[9]

7 Can there be a Formal Semantics for Maps?

In artificial languages, expressions have a canonical syntactic decomposition. And the symbols of the language—the ultimate components of any expression—must be recognised *both* by an account of the truth-conditions of expressions in the language, *and* by any account of how expressions in that language (or data-structures which mirror them) might be used in thought or communication. We illustrated this dual role of syntactic structure in the case of the predicate calculus.

To *some* extent, the same is true of maps. Some aspects of representation in maps, such as the use of icons to represent items such as houses or churches or wells, exhibit the kind of obvious and canonical syntactic structure characteristic of artificial languages (even though it is typically much simpler). The account we sketched of the truth-conditions of maps such as that of fig. 1 certainly relied on this symbolic decomposition. It seems reasonable to assume that an account of the way map readers understand such maps must do so as well: anyone who understands the map of fig. 1 *must* perceive the squares, circles, etc. as repeatable figures, located at various points on a blank (back)ground.

Other aspects of representation in maps, however, are different. We considered, as an example, the differentiation of land from water by lines or coloured regions. (We might just as well have considered the marking of buildings as black patches on a white background, or the depiction of farmland divided by walls and hedges and ditches using linear features.) Such maps admit of many different kinds of organisation into component structures, each one requiring different

---

even these algorithms decompose free space in a way which is sensitive to the obstacles in it, and require explicit representations of the obstacles in order to effect a suitable decomposition.

[8] For a concrete proposal as to the organisation of geographic objects in a computer system, see Worboys [10].

[9] The role of the reader in determining the syntactic structure of maps is discussed in detail in Head [5] and [6]; we need not go into greater detail on this point here.

sorts of semantic rules to give the intended truth-conditions. Yet it is a crucial feature of the map that it remains resolutely neutral as to which of these syntactic decompositions should be imposed on it. For, when we consider how the map is *used*, we must recognise that different reasoning tasks require different perceptual organisations of the map into significant components; and there is apparently no basic level of significance common to all of these processes, and which could support a sensible account of the map's truth-conditions.

It is this organisational ambiguity, then, so essential to maps yet so completely eschewed by artificial languages, that constitutes an essential difference between them. Can there be a formal semantics for maps? Well, yes and no. Yes: by imposing a syntactic structure on a map, we can provide rules which give its truth-conditions. But whichever syntactic structure we impose, we will end up recognising as symbols units which play no part in the explanation of how maps are *used* for some reasoning tasks. For it is, we suggest, part of the very essence of maps—what sets them apart from artificial languages like the predicate calculus—that they exhibit this sort of ambiguity.

Thus, the attempt to extend the ideas of formal semantics to maps leads us to new insights about what we take a semantics to be, and what we expect of it. In particular, we saw how, in artificial languages for which a formal semantics has been supplied, it is taken for granted that the syntactic structure presupposed by the semantic rules also plays a role in any understanding of how expressions in those languages are processed. We have argued that this assumption of the dual role of syntactic structure may have to be abandoned in the case of maps. Whatever we decide about the the best way to give an account of the truth-conditions of maps—and we have suggested that there will in general be more than one option—the separation of semantic from processing concerns will have to be recognised.

References

1. J.H. Andrews: Map and Language/ A metaphor extended. Cartographica vol. 27(1), 1–19 (1990)
2. Gerhard Brewka: Nonmonotonic Reasoning: Logical Foundations of Commonsense. Cambridge: Cambridge University Press, 1991
3. R. Brooks. Solving the Find-Path Problem by Good Representation of Free Space. Proceedings, AAAI. Los Altos, CA: Morgan Kaufmann 1982, pp. 381–386
4. D. Chubb: An introduction and analysis of a straight line path algorithm for use in binary domains. In A. Kak and Su-shing Chen (eds.): Spatial Reasoning and Multi-Sensor Fusion: Proceedings of the 1987 workshop, Los Altos, CA: Morgan Kaufmann, 1987, pp. 220–229
5. C. Grant Head: The map as natural language: a paradigm for understanding. In C. Board (ed.): New Insights in Cartographic Communication. Cartographica Monograph 31, University of Toronto 1984, pp. 1–32
6. C. Grant Head: Mapping as language or semiotic system: review and comment. In David M. Mark and Andrew U. Frank (eds.): Cognitive and Linguistic Aspects of Geographic Space. NATO ASI series D: Behavioural and Social Sciences vol. 63. Dordrecht: Kluwer 1991, pp. 237–262

7. D. Kapur and J.L. Mundy (eds.): Artificial Intelligence 37(1–3), Special volume on geometric reasoning, (1988)
8. Arthur H. Robinson and Barbara Bartz Petchenik: The nature of maps. London and Chicago: University of Chicago Press, 1976
9. Hansgeorg Schlichtmann: Characteristic traits of the semiotic system 'Map Symbolism'. The Cartographic Journal 22, 23–30 (1985)
10. Michael F. Worboys: A generic model for planar geographical objects. International Journal of Geographical Information Systems, 6(5), 353–372 (1992).

DEVELOPMENT OF A CARTOGRAPHIC LANGUAGE

J. Raul Ramirez, Ph.D.

Center for Mapping, The Ohio State University
1216 Kinnear Road, Columbus, Ohio 43212

Abstract. The development of a cartographic language for topographic maps is presented. Map components are identified as part of the analytical study of maps and based on these components the elements of the language, alphabet and *grammar*, are identified. Then, the alphabetic symbols together with the components of the grammar: operations, rules and writing mechanism, are discussed in detail.

1 Background

In recent years, there has been a great deal of interest in automating the generation of digital cartographic products (in vector format) by users of spatially referenced digital data. Among the many data sources used to generate these digital products, analog maps are the most readily available.

Analog maps are produced by expert cartographers on some kind of permanent material, through a combination of manual, mechanical, photographic and, lately, computer-aided means. One of the most useful types of analog maps is the reference or topographic map.

Topographic maps accurately portray the surface of the earth and "the spatial association of a selection of diverse geographical phenomena" (Robinson et al., 1984). They are widely used in engineering, city planning, natural resources and other fields requiring spatially referenced elevation and planimetric information.

For many centuries, the production of topographic maps was a laborious, manual operation. Usually, it began with the collection of terrain information by ground survey methods and later (in this century) by remote sensing. This information was used to generate a graphic representation or map. As indicated by Bertin (1983) "graphic representation is the transcription, into the graphic sign-system of 'information' known through the intermediary of any sign-system". The graphic representation was called the "base map" and was transformed into the final map by expert cartographers.

Most topographic maps are produced in a national map series by a government agency under very strict production guidelines. But even each one of these manually made maps has unique characteristics that reflect the tastes and personalities of the individuals responsible for its production. Examples of analog topographic maps are the 1:24,000 quadrangles of the U.S. Geological Survey.

Traditionally, map production has been considered more a craft or art than a science and this is partially reflected in the definition of cartography found in the Multilingual Dictionary of Technical Terms in Cartography (ICA, 1973):

> The art, science and technology of making maps, together with their
> study as scientific documents and works of art...

In the past, a cartographer was a combination of artist and scientist who took great pride in the final appearance of the map. According to Imhof (1982):"Artistic talent, aesthetic sensitivity, sense of proportion, of harmony, of form and color, and of graphical interplay are indispensable to the creation of a beautiful as well as to a clear, expressive map". The development of the skills necessary to make maps started when an individual began as an apprentice under the direction of an expert cartographer. Years later, after many hours of practice and tedious work, the apprentice would be considered proficient enough to be a cartographer.

With the advent of computers, new kinds of cartographers and cartographic products were born. The new cartographers are scientists interested in a precise portrait of the terrain, its analysis and interpretation, and in finding efficient ways to generate the spatially referenced data badly needed when using Geographic Information Systems (GIS) and other new products and technology. One of the new cartographic products, the digital map, is displayed on a graphic computer screen and has a transient nature. Digital maps are called "temporary" maps by Riffe (1970). These cartographic products exist as graphic entities for short periods of time. Even though they may look like analog products, there are major differences between them. Some of these differences are conceptual and some are operational, and their discussion is beyond the scope of this paper.

Regardless of their nature, maps are remarkable tools capable of carrying a large amount of information in visual form. For a given application, they are similar and are made in similar fashion all over the world. For example, topographic map users will be able to understand any topographic map made in any place of the world, regardless of their national language or the agency that made the map. They may not be able to recognize the specific meaning of all the marks, but will be able to understand most of the information carried by the map.

From the viewpoint of automation, there is not yet a solution to generate computer-compatible vector maps from analog maps. Even though high resolution scanners are commonplace today and raster images of maps can be generated efficiently, very little progress has been made in automating the generation of vector maps. A fundamental reason for the lack of progress seems to be the limited understanding of maps.

At the Center for Mapping of The Ohio State University, we have been conducting an analytical study of topographic maps in order to improve our understanding and have found that any topographic map has up to *eleven components*: cartographic projection, contour interval specification, credit and notes, direction arrow, heading, positional diagram, positional reference frame, quality of data sources, representational signs, scale representation and surface of reference (Ramirez, 1988). These components can be expressed in a cartographic language. The cartographic language is composed of an alphabet and a grammar. The cartographic alphabet has four graphic signs and the cartographic grammar has three elements: operations, rules and a writing mechanism. Natural and numerical signs and their grammars are part of the cartographic language but will not be discussed here. All together, they allow the computerized manipulation of maps in a consistent and systematic manner and constitute the fundamental tools for the development of automatic map conversion and expert cartographic systems. A description follows.

2 Map Components

Any topographic map has some or all of the components shown in the second column of Table 1.

The meaning of these components follows:

-*Cartographic projection* is the component that identifies the map projection used in the map.

-*Contour interval specification* is the component that carries the prescribed elevation difference between successive contour lines for a specific map sheet.

-*Credit and notes* is the map component which includes information such as: map disclaimers, warnings, publisher's name, copyright's owner, map sheet number, number of copies printed, publication date, production method, revision date, and so forth.

-*Direction arrow* is the component that indicates in a map the primary reference direction relative to the earth or any celestial body (for example, the direction of true North).

-The *heading* component is taken here following Bertin (1983). It is composed of the title block and the legend. The *title block* carries information that identifies the represented area and the map producer. The *legend* is the link between the cartographic language and the natural language of the map producer. It is equivalent to a bilingual dictionary. It uses graphic signs and natural language (the language of the producer) to define the equivalencies between the components of both languages.

-The *positional diagram* is the map component that shows the position of a map sheet with respect to the whole area of interest.

-The *positional reference frame* is the component that connects the terrain to the displayed area on a map.

Table No. 1
Information Grouping of Components of a Topographic Map

Information	Map Components
Complementary	Credit and Notes Direction Arrow Heading (Title Block) Positional Diagram
External	Cartographic Projection Quality of Data Sources Surface of Reference
Local	Contour Interval Specification Positional Reference Frame Scale Representation
Internal	Heading (Legend) Representational Signs

-The *quality of data sources* provides the basic information from which the quality of a map can be evaluated.

-*Representational signs* are those signs located in the *display area*, which is the area delimited by the positional reference frame of a map. Three kinds of representational signs are found: graphic, natural language and numerical language signs.

-*Scale representation*, in very simple terms, is the reduction ratio from the surface represented to the map representation. It can be expressed graphically or numerically.

-*Surface of reference* is the surface used as the datum for coordinate computation.

These components have been identified by studying topographic maps of many countries and organizations. They include all the information carried on topographic maps. They may be designated by different names by other authors or they may be subdivided or compressed in a different number of components but in any case, they comprise *all* the information contained in a topographic map.

In general, a map only needs to have *representational signs* to be recognized as such, but most topographic maps have several of those components and national map series usually have all of them. Figure 1 is a very simple example of a topographic map with all eleven components and Figure 2 is a schematic representation of the map of Figure 1. Figure 2 is the result of the analytical study of the map of Figure 1.

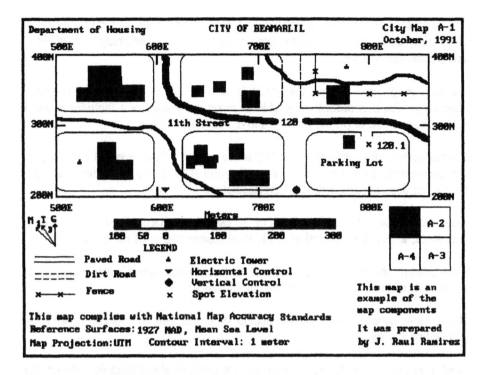

Figure 1. A Topographic Map with Eleven Components

Ramirez (1988, 1992) found map components can be grouped based on the type of information they carry (see Table No. 1). The first column of this table shows information *common* to several map components (it does not include all the types of cartographic information). Information is understood here in the context of Umberto Eco's *semiotics* (Eco, 1976). Ramirez introduced a set of definitions to differentiate the information carried by the components of a topographic map and they are the base of the above grouping. These definitions follow.

Actual cartographic information (ACI) is the information carried by a map at a given moment. *Total cartographic information (TCI)* is the maximum information a topographic map can carry. Total cartographic information is composed of the *essential cartographic information, complementary cartographic information and skeletal information.*

Essential cartographic information (ECI) is the portion of the total cartographic information which allows a reader to perceive the horizontal and vertical positions of the surface represented in their true relationship, in agreement with the producer's conception. Essential cartographic information is composed of external, local and internal information.

Complementary cartographic information (CCI) is the portion of the total cartographic information that can be removed from a map without disturbing the essential cartographic information.

Skeletal cartographic information (SCI) is the portion of the total cartographic information which is needed to be able to recognize a map as a multidimensional graphic representation (in the most primitive fashion).

External cartographic information (EXI) is the portion of the essential cartographic information that, if removed, affects the perception of the global representation of the terrain, but does not affect its local representation.

Local information (LOI) is the portion of the essential cartographic information that, if removed, affects the perception of the local representation of the terrain.

Internal information (INI) is the portion of the essential cartographic information that, if removed, makes the terrain representation disappear.

Global representation is the representation of the terrain in connection with the surrounding reality.

Local representation is the representation of the terrain isolated from the surrounding reality.

Formulas 2.1 and 2.2 relate these different types of information.

$$TCI = ECI + CCI + SCI \qquad (2.1)$$

$$ECI = EXI + LOI + INI \qquad (2.2)$$

For a topographic map with all components, formula (2.3) applies,

$$ACI = TCI \qquad (2.3)$$

For a map without all components, formula (2.4) holds,

$$ACI < TCI \qquad (2.4)$$

These topographic map components and their grouping (based on the information they carry) play an important role in the development of the cartographic language, specifically in the determination of the cartographic alphabet as shown in the next sections.

3 The Cartographic Language

As indicated earlier, the cartographic language is composed of the alphabet and the grammar. The alphabet is the set of primitive signs from which all cartographic elements can be generated. It is equivalent to the alphabet of any natural language (for example, *a, b, c, d,...,* and so forth, for the English language). The grammar is the

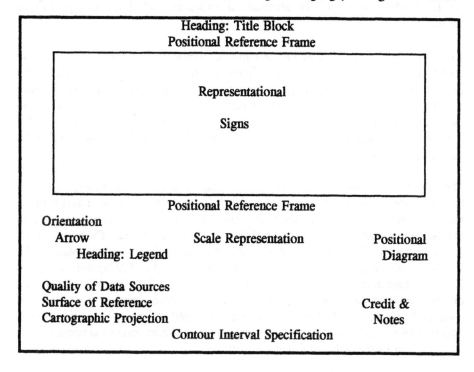

Figure 2. Schematic Topographic Map Components Representation

set of operations, rules and writing mechanisms that allows (and constrains) the generation of cartographic elements from the cartographic alphabet. Cartographic elements are the terrain features represented on a topographic map (for example, the *outline of a building*). In the next paragraphs, the cartographic alphabet and grammar are discussed in some detail.

3.1 The Cartographic Alphabet

To determine the cartographic alphabet, Ramirez (1988) followed Chomsky's conception of *structural language*. In agreement with Chomsky (1972), the central notion of the theory of linguistics is the *linguistic level*. Each linguistic level has a fixed alphabet (*primes*) and the elements of a level are formed by an operation called *concatenation* (^). The elements of a level become the alphabet for the next higher level.

A language has a structure of levels, ranging from the most primitive level where the primes are the letters of the alphabet, to the most advanced level comprised of the most complex linguistic expressions of the language. The development of a natural structural language starts by defining the level of morphemes whose primes are the letters of the alphabet. The next level contains words whose primes are morphemes and so forth, until the transformational level is reached. The transformational level is the level where any type of grammatical expression of that language can be constructed.

Ramirez (1988) assumed that the cartographic language is composed of a set of *cartographic levels* equivalent to the linguistic levels. Each level has its *alphabet (primes)* and a set of *operations* to form the elements of that level. At the most primitive level, the primes are the *cartographic alphabet*, equivalent to the letters of the alphabet of any natural language; in the most advanced level, the elements of that level are the *cartographic elements* and therefore, the expression of that level is the *cartographic product* or *map*.

The conceptual method to determine the cartographic alphabet is as follows: from the most advanced level, the primes are identified. Those primes are the elements of the previous level. In the previous level, the primes are identified. This is repeated until the most elemental level is reached and the primes are identified. These primes are the cartographic alphabet. In order to identify the different cartographic levels, Ramirez used principles of information theory, following Eco (1976) and established 5 cartographic levels (see Table 2 and Formulas 2.1 to 2.4). Each level carries a different type of information. By removing information, starting from the most complex level (the complete topographic map), a simplified version of the topographic map is obtained each time. At the end, at the most primitive level or simplified map, the primes or cartographic alphabet are identified.

Table No. 2
Cartographic Levels

Level	Information
5	SCI + INI + LOI + EXI + CCI
4	SCI + INI + LOI + EXI
3	SCI + INI + LOI
2	SCI + INI
1	SCI

In terms of Chomsky's structural language, Ramirez's method is equivalent to taking a natural language and identifying the *grammatical transformations* (its primes), then, studying only the grammatical transformations as the elements of interest. The *phrase structures* (the primes of the grammatical transformations) are then identified. When the phrase structures are taken as the elements of interest, the *words* (the primes of the phrase structure level) are identified. Words are then studied to identify their primes, called *morphemes*. Finally, morphemes are studied to identify the *letters of the alphabet* (its primes), the most basic component of any natural language.

To determine the cartographic alphabet, Ramirez (1988) studied the skeletal representation of different map series and individual topographic maps. Figure 3 is an example of the skeletal representation of the map shown in Figure 1 and is similar to the skeletal representations studied by Ramirez. A total of 143 different cartographic elements were found and studied from these maps. These 143 cartographic elements do not cover *all cartographic elements*, but are representative of *all kinds* of elements.

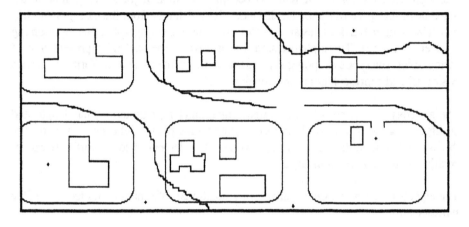

Figure 3. Skeletal Representation of a Map

As a result of this study, he found there are four basic graphic signs in the cartographic alphabet. They are shown in Figure 4. These signs have been used for many years by cartographers in the production of analog maps, and they are used today in the production of computer-compatible vector maps.

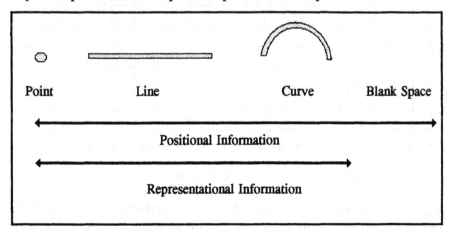

Figure 4. The Cartographic Alphabet

These four signs are the basic primitives from which more complex signs (such as *polygon*, *area*, etc.) or cartographic elements (*man-made* and *natural* features) can be constructed. A brief discussion of these signs follows.

Point is the sign that occupies no area and has no length. From this view point, the alphabet sign *point* is different from the cartographic point which occupies an area (for example, the cartographic point representing an *individual tree*). As a matter of fact, the alphabet sign *point* is the skeletal representation of cartographic points. It carries positional and representational information.

The alphabet sign *line* has length and occupies no area. It joins two points on the plane or in the space (the *shortest* distance). It is different from the cartographic line which has length and occupies an area (for example, the cartographic line representing a street in a map). The alphabet sign *line* is part of the skeletal representation of cartographic lines, areas, polygons, cartographic elements and so forth. It carries positional and representational information.

The alphabet sign *curve* has a non-linear functional representation, has length and occupies no area. It joins, at least three points on the plane or the space. It is part of the skeletal representation of areas, cartographic elements and so forth. It carries positional and representational information.

The alphabet sign *blank space* carries only positional information and has no visible representation.

Figure 5 illustrates the idea of positional and representational information as well as the use of the *blank space* sign.

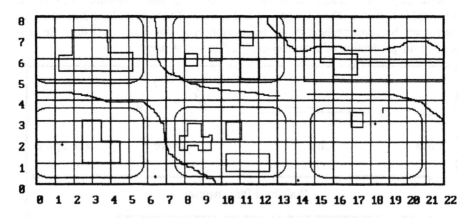

Figure 5. Positional and Representational Information

In a graphic representation, each location has positional information. In Figure 5, this is indicated by the mesh of the coordinate grid. For example, the upper-left corner of the display has no graphic symbols (just blank space) but has positional information

(0,7). From this point of view, it is not an empty space because it carries positional information. We cannot remove this information from the graphic representation without altering or even destroying the graphic representation. On the other hand, the lower-left corner of the map has blank space (which carries positional information) and a graphic symbol (composed of the alphabet signs *curve* and *line*) which carries representational and positional information (some type of cartographic element is located there).

The alphabet signs described above are used in the generation of topographic map representations. The rules and regulations used to achieve that are called the cartographic grammar and will be discussed next.

3.2 Cartographic Grammar

As mentioned earlier, the cartographic grammar is composed of *operations, rules* and a *writing mechanism*. A description of each follows.

Cartographic Operations. There are four cartographic operations: *concatenation (^), image construction (&), coordinate transformation (@)* and *addition (+)*.

Concatenation (^) is the operation which allows the connection of two alphabetic signs to create a more complex sign, two complex signs, or a complex sign and an alphabetic sign to create even more complex signs.

These complex or more complex signs are called *skeletal representation (SK)* and they are the most elemental representation of a topographic map. At this level (see Figure 3), the map user may be able to recognize none or just a few of the cartographic elements represented (due to the limited information in the skeletal representation). Formula 3.1 is the general expression of the concatenation operation.

$$SK = A \,\hat{}\, B \,\hat{}\, C...$$ (3.1)

where, A, B and C are cartographic signs or complex signs.

Ramirez (1988, 1992) found a set of definitions and rules that control the use of the concatenation operation. An example of a definition is given next.

> *When a cartographic feature is formed by a single cartographic, natural or numerical alphabet character, the concatenation operation takes place with the unit operator (I). The unit operator is defined in such a way that $I\,\hat{}\,I=I$, $A\,\hat{}\,I=A$, where A is any cartographic, natural or numerical alphabet character. The unit operator does not need to be explicitly indicated.*

Figure 6 illustrates the use of the concatenation operation.

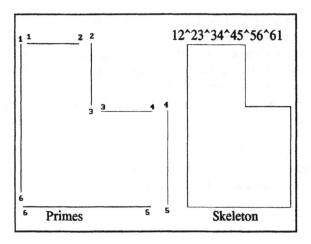

Figure 6. Example of the Concatenation Operation

Image construction (&) is the operation that adds to the skeletal representation some or all Bertin's (1983) visual variables: *size (SI), value (VA), pattern (PA), color (CO), orientation (OR) and shape (SH)* to create a cartographic element at the original scale (CEg). Formula 3.2 is the general expression of this operation.

$$CEg = SK \& SI \& VA \& PA \& CO \& OR \& SH \qquad (3.2)$$

The variable *size* uses the change in the dimensions of a graphic sign to impart to it a specific meaning. The width of cartographic lines and the magnitude of cartographic points are examples of this variable. The variable *value* expresses the different degrees of grays used in the representation of a graphic sign. The variable *pattern* represents the design used in the construction of a graphic sign. Line types, line symbols, cross-hatching and area patterning are examples of this variable.

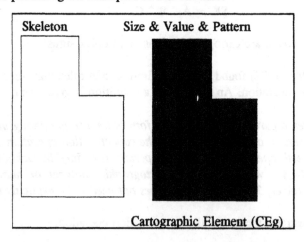

Figure 7. Example of the Image Construction Operation

The variable color represents the use of hues in graphic signs to attach a specific meaning to them. For example, in a topographic map, the color blue is used to indicate water. The variable *orientation* uses the alignment (with respect to a given direction) of graphic signs as a way of assigning a particular meaning to them. The variable *shape* uses the outline of a graphic sign to represent a specific terrain feature.

Figure 7 illustrates the use of the image construction operation on the skeletal representation of Figure 6.

Coordinate transformation (@) is the operation that takes the cartographic element (at the original size) and modifies it to reduce it to the map size, location and orientation. Three transformations can be applied to a cartographic element: scaling, translation and rotation. *Scaling (Sc)* is the process by which the size of the cartographic element is changed from the original size to the map size. *Translation (Tr)* is the process by which the origin of the local coordinate system is shifted. Local coordinate system is the one based on map units. *Rotation (Ro)* is the process by which the orientation of the reference axes of the local coordinate system is changed. Formula 3.3 is the formal definition of this operation and Figure 8 illustrates the coordinate transformation operation as applied to the cartographic element of Figure 7.

$$CE = CEg @ Sc @ Tr @ Ro \tag{3.3}$$

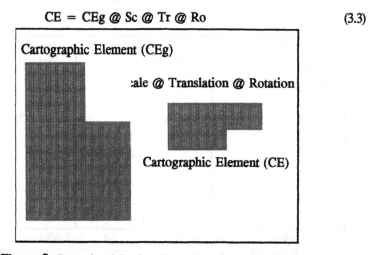

Figure 8. Example of the Coordinate Transformation Operation

Addition (+) is the algebraic operation which allows insertion/removal of cartographic elements (or a portion of them) to/from a topographic map. From the point of view of this operation a map is considered, at the beginning, as a blank space filled only with positional information. Then, this space (SP_i) is modified by the addition operation by adding (or removing) cartographic elements carrying locational information resulting in a new version of the space (SP_{i+1}). Formula 3.4 is the formal definition of this operation and Figure 9 is an example of its use.

$$SP_{i+1} = SP_i + CE \tag{3.4}$$

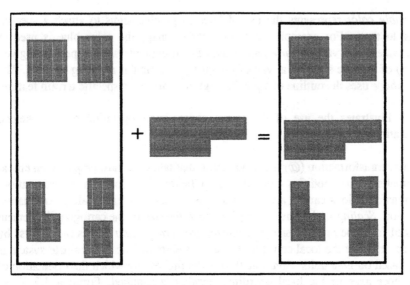

Figure 9. Example of the Addition Operation

Figure 10 illustrates the sequential use of cartographic operations from the terrain to the final map.

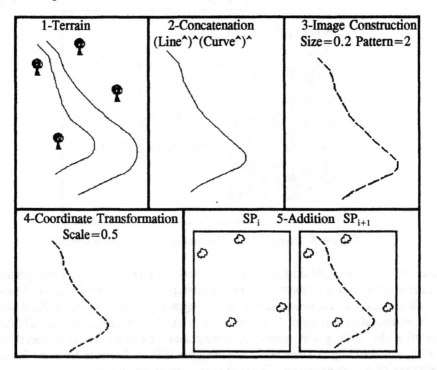

Figure 10. The Use of Cartographic Operations

Cartographic Rules. They are the regulations for constructing cartographic elements. There are three different sources of cartographic rules: (i) map planning and design; (ii) element's priority, and; (iii) cartographic element representation.

Map planning and design are the processes of selecting all the general and particular characteristics of a map or map series, such as: map components, components outline, scale, projection, surface of reference, units, specific graphic symbols to be used, characteristics of those symbols and so forth.

Element priority is the order of placement of cartographic elements on a map. Features with a higher priority are placed on the map before features with a lower priority. This priority is related to each particular application and generally, is related to the level of importance and permanency of terrain features.

Cartographic element representation is how terrain features are represented on a map, and the inter-relation of these representations. For example, on most topographic maps, topography is represented by contour lines.

Cartographic rules can be grouped as *general, layer related* and *priority* rules. *General rules* apply to topographic maps and topographic map series as a whole. For example, the following rule applies to the 7.5-minute series (1:24,000 scale) of the United States Geological Survey:

> *Every 7.5-minute map should have a legend and a title block as part of the map heading. Title block must include the following information: Quadrangle and State or States' names, county name, map series ID, and agency. The legend must show the road classification and route signs and must be placed on the lower-right margin of the map.*

Layer related rules apply to each family of cartographic elements, in particular, and to inter-family relations. Table 3 shows the nine major families of cartographic elements presented in topographic maps, in agreement with the Mapping Science Committee (1990).

Table No. 3

Cartographic Elements Major Families

Boundaries
Geodetic Control
Hydrology
Hypsography
Miscellaneous Cultural Features
Non-Vegetative Features
Public Land Survey System
Transportation System
Vegetation

There are rules for each one of these families of elements. For example, the following rule applies to hydrology:

Two natural flowing water features cannot cross each other.

The following rule applies to hypsography:

Contour lines of the same type should not cross each other.

The following inter-family rule applies to hydrology and hypsography:

A standing water body cannot be crossed by contours.

There are more than *fifty* general, layer-related and inter-family cartographic rules at this time, and as the analytical study of maps progresses, more rules are expected to be found.

Priority rules are map purpose dependent. Two major criteria are used in their establishment: *importance and permanency*. The most important cartographic features are those that will remain unchanged on a topographic map in those situations where some features must be altered (for example, in overcrowded areas). If two features are of equal importance, then the most permanent is the one that will remain unchanged. Permanency is related to the longevity of terrain features. The longer a terrain feature stays without undergoing any change, the more permanent its cartographic representation. Table 4 shows a proposed priority for the cartographic features of Table 3.

Table No. 4
Features Priority

1	Geodetic Control
2	Transportation System
3	Miscellaneous Cultural Features
4	Public Land Survey System
5	Boundaries
6	Hydrology
7	Vegetation
8	Non-Vegetative Features
9	Hypsography

In this table, the most important family of features is 1 and the least important is 9. In each family of features there is a sub-classification of the corresponding features. For example, in the *Transportation System*, highways have a higher priority than county roads (because highways are considered more important). A priority classification for each family of features for the U.S. Geological Survey 1:24,000 quadrangle series has been developed at the Center for Mapping (Ramirez and Lee, 1991).

Writing Mechanism. The writing mechanism of the cartographic language is the *Universal Mapping Command (UMC)* (Ramirez, 1991). The general expression of a UMC is:

$$\text{UMC} = +(@(K(((\{cA^\hat{}\}^\hat{}[\{c_1B\}]^\hat{}...)(<U><V><U_1><V_1>)))) \quad (3.5)$$

A UMC is a formula-like expression that allows analytic representation of any graphic element of a topographic map. UMC's make use of the cartographic alphabet, cartographic operations plus some additional operators (K, c, c_1) to express cartographic elements. They are discussed next.

K is the *cartographic element restriction* which is one or more constraints imposed over the whole cartographic element. The general expression of K is:

$$K = <\text{element restriction } \textit{if condition (optional)}> \quad (3.6)$$
$$\text{Condition} = <\text{condition}>$$

For example, if the ground features to be represented are houses and need to be represented as closed shapes, then their cartographic representations can be restricted in such a way that all of them are closed shapes. In such a case $K = <Closed\ Shape>$. K is used to impose not only graphic representation constraints, but also data management constraints. For example, to have all houses with an area less than 200 m^2 represented in the same information layer i, K will be written as $K = <Layer\ i$ *if Area lt 200* $m^2>$. Both restrictions can be expressed as $K = <Closed\ Shape><Layer\ i\ if\ Area\ lt\ 200\ m^2>$.

The c and c_1 are the alphabet character restrictions. *Alphabet character restrictions* are constraints imposed over individual and/or consecutive alphabet characters in a cartographic element. The general expression of alphabet character restrictions is:

$$c = <\text{alphabet restriction } \textit{if condition (optional)}> \quad (3.7)$$
$$\text{Condition} = <\text{condition}>$$

For example, if the ground features to be represented are houses (as in the previous example) and need to be represented with square intersections (if the computed angle differs from ninety degrees by less than some value, such as 2^0), then c or c_1 can be written as:

$$c = <Square\ Intersection\ if\ Angle\ less\ than\ 2^0>$$
$$\text{Angle} = \text{Absolute Value} <\text{Intersection Angle - } 90^0>$$

In the general UMC expression (Formula 3.5), A and B are alphabet characters from the cartographic alphabet or text or numerical expressions; U and U_1 are space dimensions (coordinate values such as X, Y, and Z); V and V_l are image construction operators.

Figure 11 is an example of the UMC to write a given type of building in a map and Figure 12 is an example of UMC for the type of road of Figure 10.

It must be noticed that everything in a UMC, except the space dimensions (U), are known before a map of a map series is made. This is a very important characteristic that can be exploited from the viewpoint of automation. Another important characteristic of a UMC is the fact that they guarantee consistency, even if manual computer-aided methods are used to produce topographic maps. This happens because once the UMC's are set, all the activity of the operators is limited to collecting the locations of the ground features to be represented.

There is a UMC per each type of terrain feature to be represented on a topographic map. For example, if the houses of the area to be mapped need to be differentiated as residential and business, there will be two UMC's for the houses: one to represent houses-residential and one to represent houses-business.

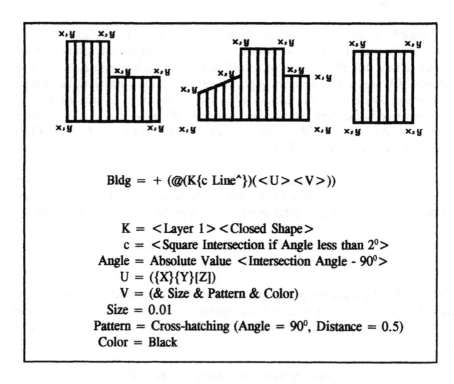

$$Bldg = + (@(K\{c\ Line^\})(<U><V>))$$

$$K = <Layer\ 1><Closed\ Shape>$$
$$c = <Square\ Intersection\ if\ Angle\ less\ than\ 2^0>$$
$$Angle = Absolute\ Value\ <Intersection\ Angle - 90^0>$$
$$U = (\{X\}\{Y\}[Z])$$
$$V = (\&\ Size\ \&\ Pattern\ \&\ Color)$$
$$Size = 0.01$$
$$Pattern = Cross\text{-}hatching\ (Angle = 90^0,\ Distance = 0.5)$$
$$Color = Black$$

Figure 11. Example of Universal Mapping Command

3.3 Assembling the Cartographic Language

A formal definition of the cartographic language is presented here.

Cartographic language is defined as,

> A set (finite or infinite) of symbols, each finite in dimension and constructed out of a finite set of signs (the alphabet), in agreement with a set of rules (the grammar) by means of which the cartographer communicates his/her conception of reality through a cartographic record such as a map.

The cartographic language is composed by the three sets of signs: natural language signs (the alphabet of the language of the map producer), numerical signs (1, 2, 3, and so forth) and graphic signs (the cartographic alphabet). The natural language signs generate expression in the natural language called *names* and *labels*; the numerical signs generate *numerical names and labels*; the graphic symbols are the shapes or cartographic elements constructed from the cartographic alphabet.

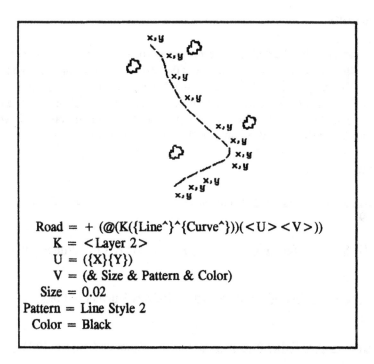

Road = + (@(K({Line^}^{Curve^}))(<U><V>))
K = <Layer 2>
U = ({X}{Y})
V = (& Size & Pattern & Color)
Size = 0.02
Pattern = Line Style 2
Color = Black

Figure 12. UMC for the Road Type of Figure 10

The set of rules for constructing natural language and numerical names and labels are the same rules of the grammars of the natural language and numerical language plus cartographic rules from the cartographic language. The rules for constructing graphic symbols (cartographic elements) come from the cartographic language.

These rules are used with the Universal Mapping Commands and the operations of concatenation, image construction, addition and coordinate transformation to generate cartographic elements and ultimately, the topographic map.

Once a particular cartographic series or cartographic product is planned and components selected and designed, the cartographic grammar is fully defined. This is the major difference between the cartographic language and the natural languages. In the cartographic language, grammar is not universal but depends on the cartographic product or series. In a natural language, the grammar is universal. The cartographic grammar is unique and fully defined for a given cartographic product or cartographic product series.

After the cartographic grammar is defined, we use the cartographic alphabet, cartographic operations and universal mapping commands to write a cartographic product. All the data of universal mapping commands (except the space dimensions U) are known after planning is completed. Once the space dimensions are known, the process of writing a cartographic product can be automated.

4 Conclusions

A complete cartographic language for topographic maps has been presented. This language constitutes the basis for automation of the production of computer-compatible vector maps. This language has been used in the development of a cartographic model for the 1:24,000 quadrangle series of the United States Geological Survey (Ramirez and Lee, 1991), and is used in the solution for generating computer-compatible vector maps from analog documents developed at the Center for Mapping.

References

Bertin, J. Semiology of Graphics, Madison: The University of Wisconsin Press, 1983, pp. 4-19.

Chomsky, N. Syntactic Structures, The Hague: Mouton & Co., Printers, 1972, pp. 18-25.

Eco, U. A Theory of Semiotics, Bloomington: Indiana University Press, 1976, pp. 33-47.

Imhof, E. Cartographic Relief Presentation, Berlin: Walter de Gruyter, 1982, p. 359.

International Cartographic Association Multilingual Dictionary of Technical Terms in Cartography, ICA, 1973, p. 1.

Mapping Science Committee Spatial Data Needs: the Future of the National Mapping Program, National Research Council, 1990, p. 15.

Rife, P. "Conventional Map, Temporary Map, or Nomap? International Yearbook of Cartography, 1970.

Ramirez, J.R. A Map Representation Theory for the Evaluation of Digital Exchange Formats, Columbus: The Ohio State University, Department of Geodetic Science and Surveying, Report No. 389, 1988, pp. 32-75.

----- "Computer-Aided Mapping Systems: The Next Generation, PE&RS, Vol. 57, No.1, January 1991, pp. 85-88.

----- Modern Cartographic Theory: An Introduction, First Draft (unpublished), 1992, pp.120-211.

Ramirez, J.R, Lee, D. The Development of a Cartographic Model for Consistent and Efficient Map Production, The Ohio State University, Center For Mapping, Final Report, U.S. Geological Survey Grant No. 14-08-0002-G1884, 1991, pp. 71-98.

Robinson, A.H, Sale, R.D., Morrison, J.L., Muehrcke, P.C Elements of Cartography, Madison: The University of Wisconsin Press, 1984, p. 7.

Spatial queries and data models*

Leila De Floriani - Paola Marzano

Dipartimento di Informatica e Scienze dell'Informazione
Università di Genova
Viale Benedetto XV, 3 - 16132 Genova - ITALY

Enrico Puppo

Istituto per la Matematica Applicata
Consiglio Nazionale delle Ricerche
Via De Marini, 6 (Torre di Francia) - 16149 Genova - ITALY

Abstract

We present a unified framework for classifying and answering spatial queries relevant to a Geographic Information System. We classify spatial queries into topological, set-theoretic, and metric queries, on the basis of the kind of relationships between the query object and entities in the search space involved. For answering such queries, we propose an approach that combines an object-based description of spatial entities, provided by a topological model, with a partition of the space embedding such entities, given by a spatial index. In particular, we propose a new unified topological model, called the Plane Euclidean Graph (PEG), that is capable of describing point, line, and region data, and that incorporates relational operators on such entities. We briefly describe major techniques, rooted in computational geometry, for solving interference queries and overlays on such a data model. Finally, we describe the use of a superimposed spatial index for speeding up searches and answering queries involving distances.

1   Introduction

One main concern for a spatial database supporting a Geographical Information System is the ability of handling spatial data and of integrating them with other kinds of information (e.g., tabular data). This requirement involves some fundamental topics, like the definition and classification of spatial relationships

---

*This work has been supported by a research grant of the Italian National Research Council.

and queries, the development of efficient data models and structures for representing spatial data, and the design of efficient algorithms for solving geometric problems involved in answering spatial queries.

The above topics correspond to complementary research streams, that often evolved independently. The classification of spatial relations starts from classical topology [30], and has been object of research in the context of GIS during the last years [7,8,18,9,33]. The development of spatial models has been deeply studied both in geometric modeling (e.g., [25,21,28,27]), in computational geometry (e.g., [13,24,14,5]), and, specifically, in GIS (e.g., [10,15,6,32,22,33,23]). Finally, the efficient solution of geometric problems is mainly investigated in computational geometry (e.g., [24]).

At the state-of-the-art, several geographic information systems have been developed that are able to answer "some" spatial queries on geographic data. Anyway, though some work has been done towards an integration of spatial relationships, data models, and geometric algorithms (e.g., [7,22,16,29,33]), the concordance between a complete and integrated formalization and an efficient implementation is still an open issue.

The basic aim of this paper is to investigate the formalization of spatial queries, and to propose a combined object-based and space-based representation of spatial entities, that is able to support suitable algorithmic techniques for the solution of geometric problems involved in answering spatial queries.

Informally, spatial queries can be defined as queries about spatial relations (e.g.: intersection, containment, boundary, adjacency, proximity, etc.) of entities geometrically defined and located in space. Thus, a first problem arising when analyzing queries and operations relevant to a given application, is to define the spatial entities and their possible relationships in the context of the application considered. In Geographic Information Systems, we essentially deal with plane maps composed of three basic spatial entities (points, lines, and regions), which have a well-defined geometric characterization. Here, we propose a formal framework for characterizing spatial entities and relations, and we give a classification of spatial relations within such a framework. The classification we adopt is similar, though not coincident, to classifications proposed by other authors [7,33], and it is aimed to the identification of three classes that, besides being characterized by the different underlying geometric concepts, reflect the different computational problems met in answering the corresponding queries.

Topological relations encode the connecting structure of a map and can be further classified into adjacency, boundary, and co-boundary relations among the basic spatial entities. Set-theoretic relations are based on concepts like inclusion, intersection, coincidence, element-of, and encode spatial interferences between entities, even from different maps. Metric relations involve the concept of distance and encode spatial proximity.

Once spatial relations have been defined, the definition of queries induced by them is straightforward: a spatial query is based on a spatial entity q, called the *query object*, a set of entities S of a map, called the *query space*, and a spatial relation $\Re$. The answer to such a query must report all entities in S that are in relation $\Re$ with q. The algorithmic technique adopted for answering a spatial query is highly dependent of the relation inducing the query and of the data model used to encode spatial entities.

Under an *object-based* approach, the representation of spatial entities is independent of their position in space. Map data can be represented in a model where each entity is described explicitly and exactly (a point by its location, a line segment by its equation and endpoints, a region by its border). When topological relations between entities are stored as well, we obtain a *topological model* that provides a complete structural representation of a map.

In this paper, we propose a unified data model able to describe point, line and region data, either isolated or topologically related, called a *Plane Euclidean Graph (PEG)*. A PEG provides a combined geometric and topological representation of a map, which is invariant under affine transformations, and it incorporates relational operators on such entities.

We show how topological queries and set-theoretic queries about objects within a single map can be efficiently answered through the relational operators of the PEG. Other interference queries, involving objects from different maps, can be solved through efficient and effective algorithmic techniques rooted in computational geometry. We give a brief description of such techniques, and we also show how they can be used for answering multiple queries and computing map overlays.

The use of a topological model, however, is not sufficient for answering queries involving proximity constraints. Moreover, interference queries and overlays can be sped up by localizing interference computations. This is possible if data are represented in the context of a *space-based* structure. In order to complete our model with one such structure, we superimpose a spatial index on the PEG, which acts as a pruning device in spatial searches.

Spatial indexing structures have been developed by several authors and they are extensively used in GISs [11,28]. Spatial indices are often superimposed on unstructured sets of spatial entities, without an explicit encoding of their mutual relations [11,17]. Our proposal is the use of a hybrid model, i.e., a model combining the topological representation provided by the PEG, with a superimposed spatial index, which provides a partition of the space embedding the PEG. A hybrid model supports both an object-based and a space-based approach in representing objects and in answering queries in the context of a GIS.

2 Spatial entities and relations

In this work, we deal with plane *maps*. Entities composing a map are embedded in the plane. Thus, each entity must belong to one of three distinct classes, depending of its dimension:

- 0-dimensional entities form the class **P** of *points*;

- 1-dimensional entities form the class **L** of *lines*;

- 2-dimensional entities form the class **R** of *regions*.

Given a generic class **C**, we denote with dim(**C**) the dimension of entities of **C** (dim(**P**)=0, dim(**L**)=1, and dim(**R**)=2). Elements of each class have a well-defined structural and geometric characterization, that allows a complete and concise description of each entity:

- A *point* is characterized by its coordinates.

- A *line* is characterized by its endpoints, that define its *boundary*, and by its geometry. We have three different types of lines:

  - straight-line segment: there is no need for geometry, since endpoints provide a complete description;
  - polygonal chain: the geometry is represented by a sequence of points in the plane;
  - curve: the geometry is represented by an equation.

  In the following, we will restrict our consideration to the first two types, assuming that curves can be approximated by polygonal chains.

- A *region* is completely characterized by its boundary. The boundary is composed of a set of closed and simple polygonal chains:

  - a chain defines the *outer boundary* of a region;
  - other chains may define *inner boundaries*, separating a region from other regions completely contained into it.

  Notice that a closed chain may either consist of a single line, or be composed of a (closed) sequence of lines; in this latter case, the boundary of a region will contain both such lines and their endpoints. By Jordan theorem, a region can be defined as the portion of plane inside the outer boundary and outside all inner boundaries.

The concept of boundary, defined above, will be used in the following, together with its symmetric concept of co-boundary: the *co-boundary* of an entity O is the set of all entities that contain O in their boundary. Given an entity O, we denote by ∂O the boundary of O, and by γO the co-boundary of O. Notice that the boundary of an object is composed of entities of lower dimension, while its co-boundary is composed of entities of higher dimension. Thus, points have an empty boundary, and regions have an empty co-boundary (in 2D). We also define the *immediate boundary* and *co-boundary*, denoted by $\partial_1 O$ and by $\gamma_1 O$, respectively, as the subsets of ∂O and γO that contain only entities of one dimension lower or one dimension higher than O, respectively. The immediate boundary of regions and the immediate co-boundary of points are composed only of lines.

For now, we define a *map* as a triple $M = (P, L, R)$, where $P \subset \mathbf{P}$, $L \subset \mathbf{L}$, and $R \subset \mathbf{R}$. Notice that, for the consistency of the map, L shall contain at least all lines defining the boundary of entities in R, and, similarly, P shall contain at least all points defining the boundary of entities in L. Notice that a map M does not necessarily contain *only* points and lines that are part of the boundary of some other lines and regions of the map, respectively: isolated points and lines can be present, that denote *features* of the map. For instance, if M is the map of a countryside, relevant regions could denote fields, woods, pastures, etc.; on the other hand, some lines will denote roads and streams, and some points will denote houses, huts, springs, etc., that do not necessarily coincide with points and lines on the border of regions.

In the context of a single map, we do not allow general intersection between spatial entities: we break lines into chains by inserting points in the map wherever a line intersects another line. Such intersection points are introduced not only for convenience, but they usually have some semantics associated with: where a river intersects a road there will be a bridge; where a road intersects the border of a farm there will be a gate.

A special case of a map is the *network*: in a network only lines representing communication channels (e.g., rivers, roads, railways, pipes), and only branches and crossing points of such lines are important, while regions are irrelevant. Thus, a network can be regarded as a map $N = (P, L, \emptyset)$.

In the following Subsections, we will try to formalize and classify relationships between spatial entities that can be relevant to the definition of spatial queries. We propose a classification into three major groups, according to a topological, set-theoretic, or metric point of view. Of course, our classification is not exhaustive (e.g., we do not consider here order relations [7]). Nevertheless, we introduce a framework that can help to formalize other relations on spatial entities, and to define queries based on such relations.

A *relation* $\Re$ is defined between two classes C_1 and C_2; in general, we will denote by $\overset{\Re}{\sim}$ an unordered relation, and by $\overset{\Re}{\rightsquigarrow}$ an ordered relation. Thus, given an ordered relation $\Re$ on $\mathbf{C}_1 \times \mathbf{C}_2$, and two entities $a \in \mathbf{C}_1$ and $b \in \mathbf{C}_2$, we will denote with $a \overset{\Re}{\leadsto} b$ the predicate "a is in relation $\Re$ with b".

2.1 Topological relations

The definition of a spatial entity introduced above induces topological relations between entities of the three classes, namely, relations between an entity and lower dimensional entities of its boundary, or, symmetrically, between an entity and higher dimensional entities of its co-boundary, and relations between entities of the same class that are adjacent through some entity of a different class. Nine *topological relations* can be defined by combining $\mathbf{P}$, $\mathbf{L}$ and $\mathbf{R}$ in pairs. As topological relations are defined through concepts like boundary, co-boundary, and adjacency, we actually consider them in the context of a single map $M = (P, L, R)$, that is consistent with the above concepts[1].

Topological relations can be classified into three different groups, depending of the relative dimensions of the classes involved; each group is composed in turn of three relations.

- *Adjacency relations* are characterized by $\dim(C_1) = \dim(C_2)$.

  PP: let $a, b \in P$, then $a \overset{PP}{\sim} b$ iff $\gamma_1 a \cap \gamma_1 b \neq \emptyset$ (i.e., iff there exist a line $l \in L$ such that a and b are endpoints of l);

  LL: let $a, b \in L$, then $a \overset{LL}{\sim} b$ iff $\partial a \cap \partial b \neq \emptyset$ (i.e., a and b have an endpoint in common);

  RR: let $a, b \in L$, then $a \overset{RR}{\sim} b$ iff $\partial_1 a \cap \partial_1 b \neq \emptyset$ (i.e., there exist a line $l \in L$ separating a from b).

- *Boundary relations* are characterized by $\dim(C_1) > \dim(C_2)$. The following definition holds for the three relations LP, RP, and RL, by proper instantiation of sets C_1 and C_2:

  $C_1 C_2$: let $a \in C_1$ and $b \in C_2$, then $a \overset{C_1 C_2}{\leadsto} b$ iff $b \in \partial a$.

- *Co-boundary relations* are characterized by $\dim(C_1) < \dim(C_2)$, and they are symmetric with respect to the previous ones. The following definition holds for the three relations PL, PR, and LR, by proper instantiation of sets C_1 and C_2:

  $C_1 C_2$: let $a \in C_1$ and $b \in C_2$, then $a \overset{C_1 C_2}{\leadsto} b$ iff $b \in \gamma a$.

---

[1] In the following, we use interchangeably a set C and its corresponding class $\mathbf{C}$, whenever no ambiguity arises.

In Section 4, we will introduce a model and a data structure that allow an efficient encoding and retrieval of information on topological relations involving a given entity of a map. Such relations will be useful in order to answer queries that involve computing the boundary or the co-boundary of objects, or queries that involve navigation of the map by following adjacencies.

2.2 Set-theoretic relations

So far, we have considered spatial entities as topological objects. Under a different point of view, points can be considered as basic elements, and lines and regions can be regarded as infinite and continuous sets of points. This approach allows us to define *set-theoretic relations*, that are based on concepts like coincidence, element-of, inclusion, and intersection, denoted with the usual symbols $\equiv$, $\in$, $\subset$, and $\cap$, respectively. Notice that, in order to compare spatial objects through set operators, they do not need to be part of a given map. Thus, the following relations will be defined in general on the three spatial classes rather than on subsets of such classes composing a map.

We consider four groups of relations, one for each set operator, and, for each group, we consider all possible ordered pairs of classes $C_1 C_2$ that lead to a meaningful relation. The following definitions are just a straightforward translation of set operators into our formalism.

- *Coincidence relations* are unordered and require $C_1 \equiv C_2 \equiv C$; thus, they can be defined on pairs **PP**, **LL**, and **RR**. A relation is trivially defined as follows:

  $\equiv_C$: let $a, b \in C$, then $a \overset{\equiv C}{\rightsquigarrow} b$ iff $a \equiv b$.

- *Element-of relations* are ordered and can be defined on pairs **PL** and **PR**, and, symmetrically, for pairs **LP** and **RP**. The following general definitions hold both for $C \equiv L$ and for $C \equiv R$:

  $\in_C$: let $a \in P$ and $b \in C$, then $a \overset{\in C}{\rightsquigarrow} b$ iff $a \in b$.

  $\ni_C$: let $a \in C$ and $b \in P$, then $a \overset{\ni C}{\rightsquigarrow} b$ iff $b \in a$.

  Notice that relation $\in_R$ is fundamental for defining of one of the most important and common spatial queries, known as point location.

- *Containment relations* are ordered and are defined for pair **LR** and, symmetrically, for pair **RL** as follows:

  $\subset$: let $a \in L$ and $b \in R$, then $a \overset{\subset}{\rightsquigarrow} b$ iff $a \subset b$.

  $\supset$: let $a \in R$ and $b \in L$, then $a \overset{\supset}{\rightsquigarrow} b$ iff $b \subset a$.

- *Intersection relations* are defined on all pairs obtained by combining classes **L** and **R** and have the following general definition:

$$\cap_{C_1, C_2}: \text{let } a \in C_1 \text{ and } b \in C_2, \text{ then } a \overset{\cap C_1 C_2}{\sim} b \text{ iff } a \cap b \neq \emptyset.$$

Notice that, as for topological relations, each possible ordered pair of classes appear in at least one set-theoretic relation. Moreover, pairs **LL** and **RR** appear in both coincidence and intersection relations, and pairs **RL** and **LR** appear in both containment and intersection relations. This is obvious, since coincidence and containment are special cases of set intersection.

2.3 Metric relations

Some important spatial queries, like proximity queries, involve distances between spatial entities. Given a metric d on the plane (usually the Euclidean metric), the distance $d(p, q)$ between two points p and q is extended to other entities X and Y (either lines or regions) as follows:

$$d(p, X) = \min_{q \in X} d(p, q);$$

$$d(X, Y) = \min_{p \in X} d(p, Y).$$

Thus, the distance between a pair of entities is always defined, independently of their classes. Moreover, given an entity O and a group of entities G, if we consider the distance between O and each entity $P \in G$, we can define metric properties of elements in G with respect to O, like having minimum or maximum distance from O, or lying within a given range from O.

In the following, we define some relations based on metric concepts, that are relevant to the definition of spatial queries. Notice that metric relations are not absolute relations between a pair of entities, but they are relative relations defined with respect to either a reference set (min/max relations), or to a reference value (range relations).

- *Min/max relations* are ordered relations defined between a pair of entities $a \in C_1$ and $b \in C_2$, and with respect to a reference set $B \subset C_2$.

  $\min_B$: let $a \in C_1$, $b \in C_2$ and $B \subset C_2$, then $a \overset{\min_B}{\leadsto} b$ iff $\forall b' \in B$, $d(a, b) \leq d(a, b')$.

  $\max_B$: let $a \in C_1$, $b \in C_2$ and $B \subset C_2$, then $a \overset{\max_B}{\leadsto} b$ iff $\forall b' \in B$, $d(a, b) \geq d(a, b')$.

  $\min_{B,k}$: let $a \in C_1$, $b \in C_2$ and $B \subset C_2$, then $a \overset{\min_{B,k}}{\leadsto} b$ iff $\#\{b' \in B \mid d(a, b) > d(a, b')\} < k$ (i.e., there are less than k entities of B that are closer to a than b is).

- *Range relations* are unordered relations defined between a pair of entities $a \in \mathbf{C}_1$ and $b \in \mathbf{C}_2$, and with respect to a reference value r.

$<r$: let $a \in \mathbf{C}_1$, $b \in \mathbf{C}_2$ and $r \in \mathbb{R}$, then $a \overset{<r}{\sim} b$ iff $d(a,b) < r$.

$>r$: let $a \in \mathbf{C}_1$, $b \in \mathbf{C}_2$ and $r \in \mathbb{R}$, then $a \overset{>r}{\sim} b$ iff $d(a,b) > r$.

Relations $\leq r$ and $\geq r$ can be defined analogously. Notice, anyway, that any range relation can be transformed into an equivalent containment (or element-of) relation as follows: let a, b and r be as before, and let us define

$$G_{a,r} = \{p \in \mathbb{E}^2 \,|\, d(p,a) < r\}$$

called the *enlarged a*. Then, we have:

$a \overset{<r}{\sim} b$ iff $b \overset{\subseteq}{\leadsto} G_{a,r}$;

$a \overset{>r}{\sim} b$ iff $\neg(b \overset{\subseteq}{\leadsto} G_{a,r})$.

Relations $< r$ and $\leq r$ can be similarly transformed. Enlarged sets can be difficult to compute for regions. Nevertheless, the above transformation is straightforward if a is a segment or a point: in the latter case, for instance, $G_{a,r}$ is a circle of radius r centered at a.

3 Spatial queries

After defining the relations on spatial entities, we define spatial queries. In our approach, a query is based on a *query object* q, that is an entity from a given class $\mathbf{C}_q$, and a *query space* S, that is a subset of another class $\mathbf{C}_S$. Usually, the query space will be one of the three sets P, L and R defining a *query map* M.

A generic query asks for all entities in the query space that are in a given relation with the query object. More formally, given q and S as above, and a relation $\Re$ defined on $\mathbf{C}_q \times \mathbf{C}_S$, a generic query is denoted by a triple $\prec q, S, \Re \succ$, and it is defined as follows[2]:

$$\text{return all } s \in S \text{ that satisfy } q \overset{\Re}{\leadsto} s$$

We will also say that $A = \{s \in S \mid q \overset{\Re}{\leadsto} s\}$ is an *answer* to query $\prec q, S, \Re \succ$, and we will denote $\prec q, S, \Re \succ \longmapsto A$.

All spatial queries we consider are defined by proper instantiations of classes $\mathbf{C}_q$ and $\mathbf{C}_S$, and of relation $\Re$ in the generic definition given above. From the classification of relations we have given in the previous Section, we obtain three

---

[2] This definition is for a generic ordered query; the definition for an unordered query is analogous by substituting $\leadsto$ with $\sim$.

groups of spatial queries; each query directly corresponds to a spatial relation. In the following, we do not give the tedious list of all queries, but we rather discuss the three classes of queries, by focusing on the problems related to their answering, and by giving some relevant examples.

We will denote as before with P, L, and R sets of points, lines, and regions, respectively, and we will denote with p, l, and r a generic point, line, and region, respectively.

3.1 Topological queries

As pointed out in Section 2.1, topological relations are defined on pairs of sets that are topologically consistent; one way to guarantee such consistency is to consider sets from a single map $M = (P, L, R)$. Thus, for a given query, the query space S will be either P, or L, or R, or a suitable subset of one of them, while the query object q will be an entity of M (i.e., $q \in P \cup L \cup R$). We list as examples two region-based and one point-based topological queries, that are often used in geographic applications, together with their informal expression:

$\prec r, R, \overset{RR}{\sim} \succ$: "return all regions of M that are neighbors of r";

$\prec r, L, \overset{RL}{\leadsto} \succ$: "return the border of region r";

$\prec p, P, \overset{PP}{\sim} \succ$: "return all points that are immediately adjacent to point p".

Notice that the point-based query given above is especially useful for traversing networks, like road maps.

Topological queries can be answered efficiently if map M is represented through a model that encodes topological relations between spatial entities, and if such model is implemented through a data structure that allows efficient retrieval of such relations. In Section 4, we will present and discuss one such model and data structure. We will see that topological queries are in general "easier" than other queries, i.e., the algorithms for retrieving them have a low time complexity.

3.2 Set-theoretic queries

Queries based on set-theoretic relations can have different meanings, and different solutions, when either q belongs to the same map of space S or not.

A set-theoretic query about an entity of the map is generically related with the localization of point or lineal features, and will be solved by using relational operators of the model that we will introduce in Section 4. In other words, if the query object and the query space come from the same map, there exist a structure that "contextualizes" the query object in the space where the answer must be retrieved, and the relational operators incorporated in the model allow an immediate answer (see Section 4.1).

In this Section we will focus on queries induced by set-theoretic relations, when the query object does not belong to the query map. In this case, the contextualization of q must be performed through suitable algorithms that are able to retrieve the entities in S that spatially interfere with q, once q is "plunged" into the query map. For this reason we will also refer to this latter group of queries as *interference queries*.

Coincidence queries are similar to queries in standard databases, as the coordinates of points provide search keys. A coincidence query for points can be easily and efficiently answered if the points in the query space are maintained sorted (e.g., in lexicographic order) in a balanced tree. A coincidence query for lines can be answered analogously, by considering first coincidences of endpoints, and then coincidence of geometries. A coincidence query for regions is slightly more complicated, but it can still be answered in a similar way by scanning the region boundary.

The other queries need suitable algorithms for geometric search, that either work on unstructured data, or are able to exploit the structure of the search space in order to retrieve an answer efficiently. Here, we only list two important queries, corresponding to two basic interference problems: *point location* and *segment intersection*.

$\prec p, R, \overset{\epsilon \mathbf{R}}{\leadsto} \succ$: "return the region of M containing point p";

$\prec l, L, \overset{\cap \mathbf{LL}}{\leadsto} \succ$: "return all lines of M intersecting line l".

In Section 5, we will consider the basic techniques offered by computational geometry to search efficiently planar subdivisions (like maps are), and to intersect sets of segments. We will show that all interference queries can be answered through such techniques, combined with the exploration of the topological structure encoding the query map.

3.3 Metric queries

Metric queries are based on metric relations. Thus, they are concerned with the distance between the query object and the entities in the query space. The topological structure of the query map is not much helpful in finding answers to such queries, since the spatial proximity of entities is not necessarily related to their topological adjacency/incidence relations.

In some cases, we can transform metric queries into other non-metric queries: as we anticipated in Section 2.3, range queries can be transformed into interference queries by using enlarged sets. A typical range query is the following:

$\prec p, P, \overset{\leq r}{\sim} \succ$: "return all points of P that lie closer than a distance r from point p".

Such a query can be substituted with the equivalent query:

$\prec G_{p,r}, P, \overset{3}{\leadsto} \succ$: "return all points of P inside region $G_{p,r}$",

where region $G_{p,r}$ is the enlarged set of p, i.e., a circle of radius r centered at p.

Unfortunately, there is no way to cheat metric problems that arise with min/max queries, which are, however, very important in the context of spatial databases. A typical example of min query is the following:

$\prec p, P, \overset{\min P}{\leadsto} \succ$: "return the point in P that lies closest to point p".

Auxiliary topological structures superimposed on the query map, like Voronoi diagrams, can help answering the above query; on the other hand, such structures are useful only in some cases, while they require a considerable amount of storage [12].

In order to answer min/max queries efficiently, we would like to have spatial information encoded into models that are more space-based than object-based. On the other hand, space-based structures do not allow an explicit encoding of spatial entities, and, especially, they do not offer any support for encoding topological relations, that are the basis of a large part of spatial queries. Thus, if we want to maintain a topological object-centered model for spatial entities, while guaranteeing efficient spatial search through space-based structures, we must integrate the two approaches.

A solution consists of using *spatial indexes*, like *grids* or *quadtree-based structures*, as superstructures on the topological model encoding the maps. As spatial indexes allow efficient spatial search, they are not only essential for answering metric queries, but they can also help speeding-up the treatment of other spatial queries, and of operations like map overlay, that we will consider later. Finally, spatial indexes help to organize information for efficient disk storage. In Section 6, we will discuss the use of spatial indexes, and their integration with the topological model presented in Section 4.

3.4 Multiple queries

It often happens that a query induced by the same relation $\Re$ is made on the same query space S for many query objects from a set Q, that we call the *query set*: a typical example is when we want to classify a whole set of entities from a given map with respect to another map. A *multiple query* is defined by considering all queries about entities of Q as a whole, as follows:

$$\forall q \in Q, \text{ return all } s \in S \text{ that satisfy } q \stackrel{\Re}{\rightsquigarrow} s.$$

We will denote the multiple query defined above by $\prec Q, S, \Re \succ$.

A multiple query can be treated either as a "set of queries", or as a "query about a set". The former approach is trivial, and the corresponding solution is to answer independently to each query about an object $q \in Q$. The latter approach considers using some global technique that, given Q, S, and $\Re$, is able to answer to $\prec Q, S, \Re \succ$ at a reduced computational cost, with respecto to the time needed when considering each entity independently.

The trivial approach is convenient whenever the cost of answering a single query is of the same order of the size of its output: the cost of answering a multiple query cannot be smaller than the size of the global output, that is equal to the sum of all sizes of outputs of single queries, i.e., to the cost of answering all single queries. Thus, we cannot hope to achieve better results with the global approach. For instance, all topological queries can be answered efficiently in this way.

In other cases, especially for interference queries, a single query has a cost whose order is higher than the output size, being a function $f(n)$ (either logarithmic or linear) of the size n of S. Solving single queries independently gives a total cost of $m f(n)$, where m is the cardinality of Q. We will see in Section 5 how, depending on the relation $\Re$ inducing the query, we can use suitable algorithms to achieve a better performance under the global approach. A typical example of such queries is multiple line intersection, e.g., intersection between two networks.

A global approach based on sorting can be used to answer coincidence queries: if objects in Q and S are maintained sorted according to some key (e.g., lexicographically, for points), a multiple coincidence query can be answered in time $O(\min(m, n))$ by scanning the sorted lists in parallel, while the trivial approach would require at least $O(m \log(n))$ time.

Also multiple metric queries, that could be answered by using spatial indexes, can be faced efficiently with a pseudo-global approach: spatial buckets often correspond to disk pages; if entities in the query set are sorted according to the spatial bucket containing each entity, single queries involving search in the same spatial region can be grouped, and disk access is thus minimized.

3.5   Overlays

Most geographic databases organize data in several *thematic maps*: information on entities contained in a given portion of space are spread over different maps, depending on their semantics. An *overlay* of maps consists in taking two or more such maps, and producing a new map, that is a fusion of a part or all the information contained in the input maps.

More formally, given two maps $M' = (P', L', R')$ and $M'' = (P'', L'', R'')$, their overlay $M' \oplus M''$ will be a map consisting of:

- all points of $P' \cup P''$ plus all intersections between lines of L' and L'';

- all lines obtained by breaking the lines of $L' \cup L''$ at intersection points;

- all regions obtained by fragmenting the regions of $R' \cup R''$ with the network of lines above, i.e., all maximal portions of space that have the property of being completely contained into one region of R' and into one region of R''.

Such new entities will be topologically related among them, and the new map could be included in the database if a suitable model is built that represents it.

Windowing, i.e., the extraction of a portion of map contained inside a given (usually rectangular) window is, at least geometrically, a special case of an overlay, where one of the two maps is trivially the window itself.

While the aim of the spatial queries is to report entities (contained in the database), the aim of overlays is to build new entities. It is not our purpose here to make a thorough discussion of all aspects and problems related with overlay operations. Nevertheless, we have introduced overlays in this Section, because the geometric problems involved in their computation (essentially, region and segment intersection, and point classification), are the same met when facing multiple interference queries. Thus, algorithms for answering multiple interference queries are also useful to solve overlays (see Section 5).

4   A topological model for maps

As already mentioned in the introduction, the problem of representing spatial entities, both in 2D and in 3D, together with their relations in a comprehensive and global model has been faced by several authors, in the fields of geometric modeling, computational geometry, and GIS.

In the 2D case, most literature in geometric modeling and computational geometry is concerned with the representation of *planar subdivisions*, i.e., partitions of a planar domain into simply connected regions, induced by a *Plane Straight Line Graph (PSLG)* [24,13,31]. A PSLG is a connected Euclidean graph with non-crossing straight-line edges and no dangling chains, i.e., such that every vertex has at least degree two. A PSLG Σ partitions a portion of the Euclidean plane into a set of simply connected regions, bounded by edges of Σ and containing no edge or vertex of Σ in their interior. Thus, points and lines in planar subdivisions are introduced only as parts of the boundaries of regions, while no isolated points or open chains are allowed.

In geographical information theory, some authors have proposed models for 2D maps based on *simplicial complexes*, that are special cases of PSLGs with triangular facets [10,6,33]. Some other authors have pointed out the need for models that overcome the limits of PSLGs by encoding also point and lineal features without introducing dummy entities [22,23]; this trend is present on a more general basis in recent research in geometric modeling [26,27].

The model we propose here extends a planar subdivision by allowing the graph inducing the subdivision to be any planar graph with piecewise linear edges, and possibly containing isolated points and/or dangling chains. The result is similar in principle to the *single valued vector maps* proposed in [22], but here we focus on the explicit encoding of all spatial entities, and mainly on the embedding of relational operators in the model. These requirements lead to the development of an efficient data structure for representing the model, and of efficient accessing algorithms that encode relational operators.

4.1   The Plane Euclidean Graph

The *Plane Euclidean Graph (PEG)* is defined by an Euclidean graph $G = (V, E, F)$, where:

- V, called the set of *vertices*, is a set of points in the plane;

- E, called the set of *edges*, is a set of polygonal chains having their endpoints in V, and such that any two edges of E never cross (i.e., they never intersect, except at their endpoints). A *polygonal chain* of E is defined as a sequence of points $e = (v, p_1, \ldots, p_k, w)$, where $v, w \in V$ and $p_1, \ldots, p_k \notin V$; points $p_1, \ldots, p_k$ are called *joints* of e.

- F, called the set of *faces*, is a set of maximal regions f bounded by chains of E, such that for every two points P and Q inside f, there exists a curve on the plane that joins P to Q without intersecting any edge of E.

In practice, a *PEG* is induced by any Euclidean graph with piecewise linear edges, with the only restriction of being a plane graph. The planarity is necessary to guarantee that topological relations between entities in the PEG are well defined.

Given a map $M = (P, L, R)$, defined as in Section 2, we can represent M through a PEG $G_M = (V, E, F)$ where $V \equiv P$, $E \equiv L$, and $F \equiv R$; this notation is intended to make a distinction between spatial entities and their representations embedded in the topological model. In a PEG:

- A vertex of degree zero represents a point feature.

- A vertex of degree one represents an endpoint of a lineal feature.

- A vertex of degree two represents a junction point, either on the border between two regions or on a lineal feature: as edges are polygonal chains, junction points should not be introduced (unless they represent point features lying on edges) in order to make the topology of the model as simpler as possible. Notice that a closed chain that does not contain relevant points can be represented, without introducing vertices, as a sequence of joints, where the first and the last joints are coincident.

- A vertex of degree more than two represents either a branch point or a cross point. Besides being necessary in order to ensure the planarity of the PEG, branch and cross vertices are relevant to the semantics of the map.

- An edge completely contained inside a face represents a lineal feature.

- An edge that separate two regions represents a border edge.

Entities in a PEG G are topologically related, like the spatial entities they represent, according to the definitions given in Section 2.1. Moreover, entities representing point and lineal features of the map are related to edges and faces of the model through element-of and containment relations given in Section 2.2.

We have defined *relational operators* on the PEG, capable of returning objects topologically related with a given entity of the model. Relational operators directly reflect the three classes of adjacency, boundary, and co-boundary relations we introduced in Section 2.1. Let $G = (V, E, F)$ be a PEG, and let $v \in V$, $e \in E$, and $f \in F$ generic elements of G. We define on G nine topological operators as follows:

- $VV(v)$: returns all vertices of V that are adjacent to v in G;

- $VE(v)$: returns all edges of E that are part of the co-boundary of v in G;

- $VF(v)$: returns all faces of F that are part of the co-boundary of v in G;

- $EV(e)$: returns the two vertices of V that are endpoints of e in G;

- $EE(e)$: returns all edges of E that are adjacent to e in G;

- $EF(e)$: returns the two faces of F that form the co-boundary of e in G;

- $FV(f)$: returns all vertices of V that are part of the boundary of f in G;

- $FE(f)$: returns all edges of E that are part of the boundary of f in G;

- $FF(f)$: returns all faces of F that are adjacent to f in G.

Notice that the above operators directly answer topological queries we defined in Section 3.1. Moreover, we define the following set-theoretic operators, that encode relations involving point and lineal features af the map. Such operators directly answer to set-theoretic queries when the query object is part of the query map (the intersection operator is not given because intersections are not allowed within a PEG):

- $\in(v)$: returns the face of F containing point feature v;

- $\subset(e)$: returns the face of F containing lineal feature e;

- $\ni(f)$: returns all point features of V belonging to face f in G;

- $\supset(f)$: returns all lineal features of E contained into face f in G.

In Section 5, we will consider geometric algorithms for solving interference queries that manipulate PEGs. As such algorithms always deal with straight-line segments, and regions bounded by such kind of lines, we introduce here the *expanded graph* of a PEG. The idea is simply to divide each edge (polygonal chain) of the PEG into the straight-line segments forming it, and to add these segments to the set of edges, and their endpoints (corresponding to joints) to the set of vertices. Formally, given a PEG G, we define its *expanded graph* $G^* = (V^*, E^*, F)$ as follows:

- $V^* = V \cup \{p \mid p \text{ is a joint of } e \text{ for some } e \in E\}$;

- $E^* = \{(p, q) \mid \exists e \in E \text{ such that } p \text{ and } q \text{ are either vertices or joints with consecutive positions on } e\}$.

Notice that the faces of a PEG and of its expanded graph are coincident.

4.2 A data structure for the PEG

In order to encode a PEG efficiently, we need to define a data structure that fulfills the following requirements:

1. all entities of a graph $G = (V, E, F)$ are explicitly encoded;

2. a part of the relations used by the relational operators are explicitly encoded;

3. all other relations can be retrieved efficiently, i.e., all relational operators can be computed in a time that is linear in their output size.

4. the storage space is as small as possible.

The data structure we propose for representing a *PEG* is a modification of the well known DCEL data structure [24], extensively used for encoding planar subdivisions. The data structure is edge-oriented, as it encodes explicitly relations between each edge and its either adjacent, or boundary, or co-boundary entities. A specification of the data structure follows:

- For each vertex:

  - its two coordinates;
  - if the vertex is isolated, a pointer to the face containing it, otherwise, a pointer to one of the edges incident into it;

- For each edge:

  - a pointer to its geometry (a chain of joints);
  - two pointers to its endpoints (vertices);
  - two pointers to its incident faces (in case the edge is a lineal feature, the two faces will be coincident);
  - four pointers to the first adjacent edges met by rotating counterclockwise and clockwise about its endpoints, respectively.

- For each face:

  - a pointer to a list of edges, containing one edge for each connected component of its boundary;
  - a pointer to a list of edges, containing one edge for each connected component of its contained edges (lineal features);
  - a pointer to a list of isolated points contained in the face.

The list of vertices can be maintained sorted (e.g., lexicographically), and stored into a balanced tree; also, similar lists of edges and vertices could be maintained.

It is easy to show that such data structure fulfills the requirements stated above. A complete analysis of its space complexity and of the time complexity of relational operators implemented on it is omitted here, for brevity.

This data structure is intended for main memory; disk storage, instead, requires partitioning information among different disk pages. It is not immediately clear, and it is not our subject here, how to distribute entities such that links between them can be retrieved efficiently. Spatial indexes we discuss in Section 6 can be used to group together into buckets entities that are spatially close one another.

5   Algorithms for interference queries

In this Section, we briefly outline how geometric problems that arise in answering interference queries can be solved by using efficient algorithmic techniques developed in computational geometry.

Interference queries induced by element-of, containment and intersection relations can be answered efficiently by solving the point location problem, mentioned in Section 3.2. Let G be a PEG representing a map, let G^* be its expanded graph, and let n be the number of non-isolated vertices of G^*. Then, the location of a query point p in G would require $O(n)$ time by using a "brute force" approach (exhaustive search). A more interesting approach based on a preprocessing of the query map is the so-called *slab method* [24]. The expanded PEG G^* is intersected with $n + 1$ horizontal strips obtained by drawing a horizontal straight-line through each of its n vertices. The intersection of G^* with each slab defines a set of non-intersecting segments, thus partitioning G^* into trapezoids. Slabs are sorted by y-coordinate (with $O(n \log n)$ preprocessing time), and the slab containing the query point p can be retrieved in $O(\log n)$ time by applying a binary search. Similarly, trapezoids within a slab are sorted along the x-axis, and the trapezoid (or the segment) containing p can be retrieved in $O(\log n)$ time by applying another binary search within the slab. The trapezoidal diagram can be computed in $O(n^2)$ preprocessing time, but its main drawback is an $O(n^2)$ storage cost.

Other methods for point location have been proposed in the literature, that require linear storage cost. Among them, we can mention the *chain method* [20], that is based on the decomposition of a G^* into a set of monotone chains; point location can be performed in $O(\log^2 n)$ time, using $O(n)$ storage and with $O(n \log n)$ preprocessing time. Another method, which also achieves $O(\log n)$ query time, is the *triangulation refinement method* [19], that requires $O(n)$ storage and $O(n \log n)$ preprocessing time.

The segment intersection problem defined in Section 3.2 would require again $O(n)$ processing time, if solved through exhaustive search. The performance can be improved by first applying a point location algorithm to the endpoints of the query segment l ($O(\log n)$ time), and then traversing G^* while testing the intersections of l only with the edges bounding (or contained into) the faces crossed by l.

A different approach can be used for solving multiple interference queries, involving multiple line intersection. This approach is based on a sweep-line technique, and it can also be used successfully for computing overlays [1]. Let $G_1^* = (V_1^*, E_1^*, F_1)$ and $G_2^* = (V_2^*, E_2^*, F_2)$ be two expanded PEGs with n and m vertices, respectively. The endpoints of the edges of E_1^* and E_2^* are first sorted by x-coordinate (with $O((n + m) \log(n + m))$ preprocessing time).

Then, a vertical sweep-line is moved over the $n + m$ segments from left to right, stopping at each endpoint and at each intersection point between two lines. For each such event, an ordered collection of the line segments intersected by the sweep-line is maintained, called the sweep-line status. At each event, the algorithm performs one of the following operations, depending of the kind of event:

- the left endpoint of a new segment s is met: the new segment s enters the sweep-line status, that is updated accordingly;

- the right endpoint of a segment s is met: the segments leaves the sweep-line status, that is updated accordingly;

- an intersection point between two segments $s_1 \in E_1$ and $s_2 \in E_2$ is met: s_1 and s_2 are exchanged in the sweep-line status.

At each event, all the pairs of segments that become adjacent in the status are checked for possible intersections. If an intersection is found, it is added in the sorted list of events. The event list and the sweep-line status are maintained into two dynamic structures, a priority queue and a balanced tree, respectively. This leads to a global time complexity of $O((k + n + m) \log(n + m))$, where k is the number of intersections.

More recently, an optimal algorithm has been proposed [4], which achieves an $O((n + m) \log(n + m) + k)$ time complexity. Although the crucial tool is still the sweep-line, not only the status but the entire scene to the right of the sweep-line is maintained. In order to reach the optimal computational cost, refined tools (such as segment tree, topological sweep and cost amortization) are used. Unlike the suboptimal algorithm presented above, which uses $O(n + m)$ storage, this technique uses $O(n + k)$ space. Therefore, it remains an open question whether optimal time and space performances are simultaneously attainable.

6 Spatial indexes

The topological model we have defined in Section 4 gives an object-based representation of maps, that offers useful support to the solution of several problems related to spatial queries and overlays. On the other hand, as already pointed out, for some purposes, it would be convenient to organize spatial data according to a space-based scheme, where objects having close locations in space are grouped together. Major reasons for adopting a space-based scheme are the following:

1. answering metric queries efficiently;

2. speeding-up interference queries and overlays by localization of interference computation;

3. achieving efficient disk storage and retrieval.

It is not our purpose to discuss here the third point, that would deserve a separate and thorough study. Our aim is to obtain a *hybrid model*, that is able to support both an object-based and a space-based approach to the representation of spatial entities, by superimposing a spatial index to the topological model. Approaches based on hybrid structures have been proposed also in three dimensions [2,28,3].

A *spatial index* is essentially a space partitioning technique that allows locating efficiently any spatial entity in the portion of space it occupies. The simplest and most popular example of a spatial index is the *regular rectangular grid*, that partitions a rectangular domain using equally-spaced rectangular cells. A point can be located in constant time, while a line or a region can be located in a time that is linear in the number of cells it crosses. A hierarchical approach to spatial indexing is the *quadtree*, where space is recursively subdivided into quadrants. In this case, the quadrant containing a point can be found in a time that is linear in the height of the quadtree, i.e., logarithmic in the maximum number of its quadrants. In the following we will refer to both a cell of a grid and a quadrant of a quadtree with the generic term of *bucket*.

The main difference between using buckets in the context of a hybrid model and following a raster approach in encoding spatial data, is that here buckets are not atomic entities: a bucket is rather a container of (parts of) spatial entities, that are fully described in the topological model. Suitable links must be set between the spatial index and the topological model, in order to support efficient retrieval of entities contained or crossing a given bucket.

If a regular grid is used, there is a priori no bound on the maximum number of entities per bucket: with a coarse grid, a large number of entities per bucket is expected; conversely, with a fine grid, a large number of almost empty buckets is expected. Quadtree based structures, like PM-quadtrees, PMR-quadtrees, R-trees, and R_+-trees [28], allow a control over the maximum number of spatial entities per bucket by refining the spatial index only where the density of entities is higher. PM-quadtrees are well-suited as spatial indices for maps, since they allow exact representation of point, line, and region data; PMR-quadtrees are especially well-suited for networks, since they are designed to represent line data.

R-trees and R_+-trees are based on a binary partitioning technique, they allow the representation of generic geometric objects, and they are especially suited for mapping the spatial index on disk, because their structure reflects the B-tree structure of disk pages.

Independently of the spatial index used, queries and overlays are solved by localizing computations and searches into buckets, provided that efficient retrieval of spatial entities in a bucket is supported. In the following, we discuss the major techniques adopted. For brevity, we consider the main classes of problems, and we give, for each class, a relevant example.

Metric queries are solved by searching first the bucket(s) containing the query object, and then searching its adjacent buckets, if necessary. Let p be a query point, P be a query space of points, and $\prec p, P, \overset{\min P}{\leadsto} \succ$ be a min query for p with respect to P.

1. The bucket B containing p is found, and the set P_B of points of P lying in B is retrieved.

2. An exhaustive search is performed on P_B to find the point $p' \in P_B$ that is closest to p.

3. If either P_B is empty or the circle centered at p and through p' intersects adjacent buckets that have not been visited yet, points 2 and 3 are repeated on such adjacent buckets.

4. Otherwise p' is the answer.

Such a technique is explained in details in [17] for metric search based on PMR-quadtrees.

Interference queries are improved simply by localizing computation into buckets crossing or containing the query object. Let l be a query line, L be a query space of lines, and $\prec l, L, \overset{\cap L L}{\leadsto} \succ$ be an intersection query for l on L.

1. All buckets $B_1, \ldots, B_k$ intersected by l are found, and all corresponding subsets $L_{B_1}, \ldots, L_{B_k}$ of L are retrieved.

2. An intersection algorithm is applied for each subset independently.

Multiple interference queries can be solved similarly by first selecting all buckets crossing the query objects, and then applying the suitable geometric algorithm inside each bucket. If the number of query objects is large enough, i.e., if most bucket are expected to be involved, it is better to apply the following technique described for overlays.

Overlays are improved by scanning all buckets and solving the problem locally inside each bucket, as above. Let M' and M'' be two maps; we want to compute $M' \oplus M''$. For each bucket B in the space covered by the two maps do:

1. Retrieve the entities M'_B and M''_B of the two maps crossing bucket B.

2. Solve $M'_B \oplus M''_B$.

Experiences of efficient overlays on large maps using regular grids are discussed in [11].

7   Concluding remarks

We have introduced a formal framework for two-dimensional spatial entities and we have classified three major groups of spatial relations on such entities: topological, set-theoretic, and metric relations. On the basis of spatial relations, we have formally defined spatial queries, and we have discussed geometric problems related to their solution.

We have presented a topological model for representing spatial entities forming maps and their topological relations, and we have discussed how such model supports topological queries. We have reviewed major geometric algorithms for interference queries and overlays. Finally, we have discussed how metric queries can be supported, and interference queries and overlays can be improved by using a spatial index. The integration of a topological model with a spatial index provides a hybrid model that supports both an object-based and a space-based description of two-dimensional geometric entities.

Spatial databases should be built on models that support effective and efficient representations of spatial data and their relations. Moreover, they should incorporate suitable algorithms that allow efficient information retrieval for spatial queries. Our work is a step towards the design of one such database for geographic maps.

Our framework for the formal definition of spatial entities, relations, and queries can be extended to the three-dimensional case for applications like solid object representation in a CAD system [3]. Also, suitable topological models, spatial indexing structures, and geometric algorithms are known in the literature, that can help solving representation and retrieval problems in the three-dimensional case by following our approach.

The efficient encoding of spatial entities on disk is a very important subject in spatial database design, that we plan to tackle in the future in order to implement our approach. Also, the integration of spatial and tabular data in an environment that supports both the structure and the attributes of spatial entities, together with standard alphanumeric information, should be faced.

References

[1] Bentley, J.L, Ottmann, T.A., "Algorithms for reporting and counting geometric intersections", *IEEE Transactions on Computers*, 28, pp. 643-647, 1979.

[2] Brunet, P., Navazo, I., "Solid representation and operation using extended octrees", *ACM Transaction on Graphics*, 8, 1989.

[3] Bruzzone, E., De Floriani, L., Pellegrinelli, M., "A hierarchical spatial index for cell complexes", in *Proceedings 3rd International Symposium on Large Spatial Data Bases*, Singapore, June 1993, (in print).

[4] Chazelle, B., Edelsbrunner, H., "Optimal solution for intersecting line segments", in *Proceedings 29th IEEE Symp. on Foundation of Computer Science*, October 1988.

[5] Dobkin, D.P., Laszlo, M.J., "Primitives for the manipulation of three-dimensional subdivisions", *Algorithmica*, 5(4), pp.3-32, 1989.

[6] Egenhofer, M., Frank, A.U., Jackson, J.P., "A topological model for spatial databases", *Lecture Notes in Computer Science*, N.409, pp.271-286, 1989.

[7] Egenhofer, M., "A formal definition of binary topological relationships", *Lecture Notes in Computer Science*, N.367, pp.457-473, 1989.

[8] Egenhofer, M., Herring, J., "A mathematical framework for the definition of topological relationships", in *Proceedings 4th International Symposium on Spatial Data Handling*, pp.803-813, Zurich, Switzerland, July 1990.

[9] Egenhofer, M., Franzosa, R., "Point-set topological spatial relations", *International Journal of Geographical Information Systems*, 5(2), pp.161-174, 1991.

[10] Frank, A., Kuhn, W., "Cell graph: a provable correct method for the storage of geometry", *Proceedings 2nd International Symposium on Spatial data Handling*, seattle, WA, 1986.

[11] Franklin, W.R., et al., "Uniform grids: a technique for intersection detection on serial and parallel machines", in *Proceedings Auto Carto 9*, Baltimore, MD, April 2-7, 1989, pp. 100-109.

[12] Gold, C.M., "The meaning of "neighbour"", *Lecture Notes in Computer Science*, N.639, Springer-Verlag, 1992, pp. 220-235.

[13] Guibas, L., Stolfi, J., "Primitives for the manipulation of general subdivisions and the computation of Voronoi diagrams", *ACM Transactions on Graphics*, 4(2), pp.75-123, 1985.

[14] Günther, O., *Efficient Structures for Geometric Data Management* (LNCS 337), Spriger-Verlag, 1987.

[15] Herring, J., "TIGRIS: topologically integrated GIS", *Proceedings Autocarto 8*, ASPRS/ACSM, Baltimore, MD, pp.282-291, March 1987.

[16] Herring, J., Egenhofer, M.J., Frank, A.U., "Using category theory to model GIS applications", in *Proceedings 4th International Symposium on Spatial Data Handling*, pp.820-829, Zurich, Switzerland, July 1990.

[17] Hoel, E.G., Samet, H., "Efficient processing of spatial queries in line segment database", *Lecture Notes in Computer Science*, N.525, Springer-Verlag, 1991.

[18] Kainz, W., "Spatial relationships - Topology versus order", in *Proceedings 4th International Symposium on Spatial Data Handling*, pp.814-819, Zurich, Switzerland, July 1990.

[19] Kirkpatrick, D.G., "Optimal search in planar subdivision", *SIAM Journal of Computing*, 12(1), pp. 28-33, 1983.

[20] Lee, D.T., Preparata, F.P., "Location of a point in a planar subdivision and its applications", *SIAM Journal on Computing*, 6(3), pp. 594-606, 1977.

[21] M. Mäntylä, *An Introduction to Solid Modeling*, Computer Science Press, Rockville, MD, 1987.

[22] Molenaar, M., "Single valued vector maps - a concept in GIS", *Geo-Informationssysteme*, Vol.2, No.1, 1989.

[23] Pigot, S., "A topological model for a 3D spatial information system", *Proceedings 5th International Symposium on Spatial Data Handling*, Charleston, SC, August 3-7, 1992.

[24] Preparata, F.P., Shamos, M.I., *Computational Geometry: an Introduction*, Springer-Verlag, 1985.

[25] Requicha, A.A.G., Voelcker, H.B., "Solid modeling: a historical summary and contemporary assessment", *IEEE Computer Graphics and Applications*, 2, 2, pp.9-24, 1982.

[26] Rossignac, J.R., O'Connor, M.A., "SGC: a dimensional-independent model for pointsets with internal structures and incomplete boundaries", *Geometric Modeling for Product Engineering*, Wosny, M.J., Turner, J.U., and Preiss, K., Eds., Elsevier Science Publishers B.V. (North Holland), pp. 145-180, 1990.

[27] Rossignac, J.R., "Through the cracks of the solid modeling milestone", *Eurographics 91 State of the Art Report on Solid Modeling*, pp. 23-109, 1991.

[28] H. Samet, *The Design and Analysis of Spatial Data Structures*, Addison-Wesley, Reading, MA, 1990.

[29] Smith, T.R., Park, K.K., "Algebraic approach to spatial reasoning", *International Journal of Geographical Information Systems*, 6(3), pp.177-192, 1992.

[30] Whitney, H., *Geometric Integration Theory*, Princeton University Press, 1957.

[31] Woo, T.C., "A combinatorial analysis of boundary data structure schemata", IEEE Computer Graphics and Applications, 5, 3, pp.19-27, 1985.

[32] Worboys, M.F., Hearnshaw, H.M., Maguire, D.J., "Object-oriented data modelling for spatial databases", *International Journal of Geographical Information Systems*, 4(4), pp.369-383, 1990.

[33] Worboys, M.F., "A generic model for planar geographic objects", *International Journal of Geographical Information Systems*, 6(5), pp.353-372, 1992.

Topological Querying of Multiple Map Layers

Sylvia de Hoop[1*], Peter van Oosterom[2], and Martien Molenaar[1]

[1] Centre for Geographic Information Processing,
Wageningen Agricultural University,
P.O. Box 339, 6700 AH Wageningen, The Netherlands.
Email: dehoop@ds13.lmk.wau.nl, phone +31 8370 82643, fax +31 8370 84643.
[2] TNO Physics and Electronics Laboratory,
P.O. Box 96864, 2509 JG The Hague, The Netherlands.
Email: oosterom@fel.tno.nl, phone +31 70 3264221, fax +31 70 3280961.

Abstract. This paper first recaptures why multiple map layers are required in Geographic Information Systems. The two main motivations are: flexibility in data modeling, and efficient processing of data. In order to make the map layer discussions clearer, we introduce two different types of map layers: a structure layer, and a thematic layer. Though the concept of a structure layer is defined in a general sense, to illustrate its practicability the organization of data in a structure layer is initially represented according to the formal data structure for *single-valued* vector maps as proposed by Molenaar. In order to develop a data model for a multi-layered system, the concept of structure layers, specified for the fds, is extended for *multi-valued* vector maps. It turns out that the data can be modeled in various ways. After that the topic of topological querying of multiple map layers is introduced with a few examples. Map overlay plays a central role in this process. But map overlay is a computationally expensive operation, and therefore several alternative optimization techniques are described for answering the queries efficiently. An important goal of the described multiple map layer query language is that it is a realistic approach. That is, the resulting implementation can be used in an interactive environment with real data sets: with at least several megabytes of geographic data. This is reflected by the case study presented in this paper.

1 Introduction

Most Geographic Information Systems (GISs) use map layers to organize the geographic features. This organization may serve several purposes, among which user convenience and software restrictions [2]. User convenience refers to the possibility to make a thematic division of data, that is separate map layers which represent specific themes. Software restrictions may require that features of a different geometric type may not be mingled in one layer. For example, the combination of point features and area features in one layer may not be allowed. Modern applications demand the vertical integration of a large number

---

* Employed by Siemens Nixdorf Informationsystems, Munich, Germany

of separate thematic descriptions for a single region. This requires consideration of the interaction or overlay of multiple layers of data [6]. For raster models, map overlay is straightforward. Thematic information attached to the individual grid cells is combined when two cells coincide [11, 19]. For vector models, map overlay is a more time-consuming process. This paper concerns vector models. More demanding applications involving GIS require more flexible data models for spatial information. Many of these data models accommodate the explicit representation of features, e.g. the ATKIS model [1], the DIGEST structure [10], the DLG-E data model [15], and the formal data structure (fds) for single-valued vector maps [18]. These feature-oriented data models comprise a basic set of topological elements upon which a set of features is superimposed. This paper attempts to provide a conceptual framework for vertical integration of thematic data based on a common topological structure (the fds).

In Section 2, the definitions of two types of map layers are given. This is done to avoid the confusion related to the concept of map layers. In the subsequent section, the concept of structure layers is elaborated on and the relationships with the formal data structure are given. The different data modeling variants are discussed in Section 4; it also gives their implementation in the Postgres DBMS. The topological querying of multiple map layers is treated in Section 5. A case study on 'topographic data sets' is presented in Section 6. The paper is concluded with some remarks and an indication of future work.

2 Multiple Map Layers

A map layer should not be interpreted as a paper map, but denotes (an abstraction of) a geographic data set, i.e., a set of related geographic features. Each map layer describes a certain aspect of the modeled real world. In order to avoid confusion, a more detailed definition of map layer is required. We distinguish two types of map layers:

- The *structure layer:* a set of logically related features which are all based on the same topological data structure. Structure layers are independent of each other. Therefore, new layers describing a different set of features can always be added independently of the existing layers. Note that the features in a single structure layer do not necessarily belong to the same theme. That is, a structure layer may describe several themes. For example, as a road may be the boundary of a forest, the themes transportation (roads) and land use (forest) both have to be in the same topological data structure (structure layer).
- The *thematic layer:* a set of features that belong to the same theme, including all their thematic attributes or condensed (projected) on some thematic attributes. A few examples of the latter are: the construction year of houses, the water quality of lakes, the population of a country. If the features are part of the same thematic layer, then they are all members of the same structure layer too. Note that within one thematic layer the class (or the geometric

type) of the individual features may be different: both rivers and lakes are in the hydrography layer. With cartographic presentation techniques it is possible to display several thematic layers at the same time.

Thematic layers do not exist explicitly, but they are extracted from structure layers when required. As a structure layer can be considered separate of the other structure layers, it enables a modular approach to geographic data modeling. This is the first reason for using structure layers: it is a *natural* technique to organize the data. The other type of map layers is also a good modeling tool, but with more emphasis on the conceptual view instead of the physical structure (see [17]).

The second reason for using structure layers is that it is more *efficient* in terms of data storage and data manipulation. This is especially true in the case of topologically structured maps. If no topology is required (*spaghetti structure* [21]), then storing all layers together is easy. Only searching a feature that belongs to a certain theme may be time-consuming, because it has to be found among all features instead of among the features of its own theme. However, if topology is required, then combining all structure layers in one large *merged* structure layer becomes a problem when the geometry of the features is represented as a planar graph. That is, at every intersection of two lines there must be a node and both lines must be split at this node. In the merged structure layer, the area and line features are divided into many geometric segments. This fact has also been reported in [3, 4]. The strongly geometric segmented features have two severe drawbacks:

- inefficient storage: each 'segment' (piece of area or line feature) has to be represented
- inefficient manipulation: a feature has to be reconstructed from the segments, for example, in order to calculate the perimeter of an area feature

For the data modeling and efficiency reasons described previously, structure layers should not be merged. However, this makes it impossible to directly solve topological queries that involve features that belong to different layers.

For example, one structure layer describes the hydrography of Europe, and another map layer describes the countries (Figure 1). The only connection between the two separate layers is that the metric information is given in the same coordinate system. A few example queries that must be answered are:

- Through which countries does the river 'Rhine' flow?
- Which rivers flow through more than two countries?

Of course other structure layers can be defined for Europe, e.g. roads, digital elevation models, vegetation. Without loss of generality, the example is restricted to two layers. But to solve queries that deal with multiple independent structure layers, a map overlay has to be performed first. In the case of the example, a new structure layer 'hydrography & countries' is created from the two individual layers. This new layer can be constructed, because topology can be derived from

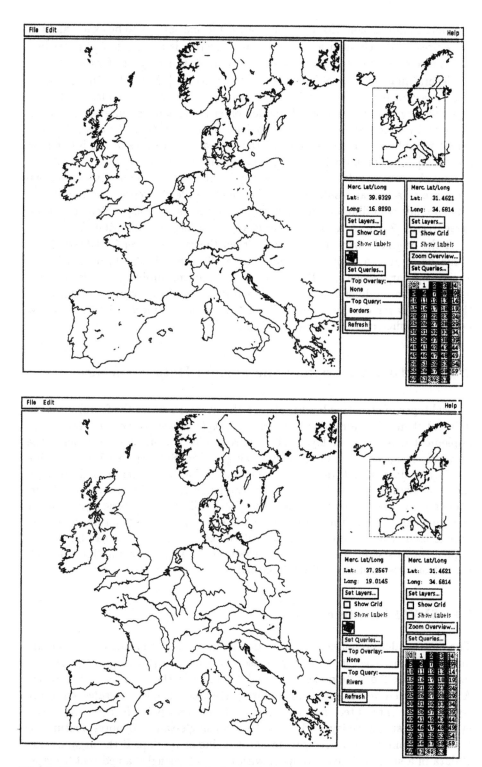

Fig. 1. Two map layers in Europe: hydrography and countries

the metric information. However, the metric information has to be exact, and not finite precision such as in a computer. Map overlay is a complicated procedure and is also expensive to execute from the computational point of view.

Several map overlay algorithms are described in [12, 13, 14, 16]. The first step is performed at the metric level: compute intersections (nodes) and separate the segments of a line which remain on either side of the intersection node. Followed by a reconstruction of the topology in the next step. The new geometric primitives inherit (or are related to) the thematic attributes of the input map layers [24]. The computationally efficient algorithms take advantage of local processing. A problem related to map overlay is the introduction of sliver polygons, these are small areas resulting from the slightly different representations of the same boundary in different map layers [12]. Solutions for this problem have been presented in [5, 25]. A formal algebraic approach has been described in [11].

If the original layers are preserved, then the merged layer is redundant, as it can be produced from the original layers. Therefore it should be considered as a temporary structure. This also avoids updating problems: updating of a feature has to be done in only the original layers, and the complex updating in the merged layer can be avoided. With the aid of this merged layer, it is now possible to efficiently answer topological queries.

3 Structure Layers and the Formal Data Structure

Several types of structure layers exist. In the situation where the features (of the different thematic layers) do not overlap, a *single-valued vector map* (svvm) can implement this structure layer. The 'hydrography & countries' example of the previous section could be implemented using a svvm. Subsection 3.1 will describe the formal data structure (fds) for the svvm. However, if features do overlap then the svvm is not sufficient for the implementation of the structure layer. An example of such a situation is storing both soil type (thematic layer number one) and land cover (thematic layer number two) in one structure layer 'soil type & land cover'. However, a *multi-valued vector map* (mvvm) can be used to implement this structure layer; see Subsection 3.2. In Section 4 an intermediate variant on these options will be described.

3.1 SVVM Structure Layer

The data model presented in this section is based on the formal data structure for single-valued vector maps as proposed by Molenaar [18] (Figure 2). The arrows in Figure 2 denote many-to-one relationships in the direction of the arrow. A single-valued vector map can be interpreted as a single *integrated* structure layer. Where integrated indicates that area, line and point features may exist in the same structure layer. But, as the term single-valued denotes, features of *the same* geometric feature type do not coincide or overlap. If the organization of data in a structure layer complies with the fds for single-valued vector maps,

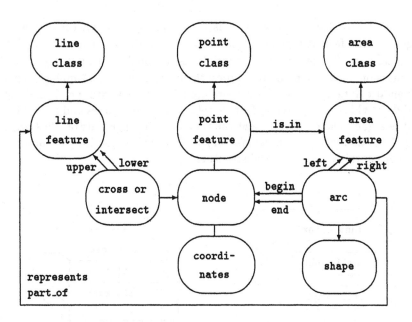

Fig. 2. The fds for a Single-Valued Vector Map

then this structure layer will be called a *svvm structure layer*. A svvm structure layer may represent several themes, e.g. 'hydrography & countries'. It is
a design issue whether the two themes are represented by either two structure
layers (svvm's) or one structure layer (also a svvm). The second solution enables
efficient implementation of topological queries (related to two map layers), by
using the framework of the fds for svvm.

Two main semantic levels can be distinguished in the fds: a geometric level
comprising metric and topology information of the geometric primitives, and
a level of features where the features are described by thematic information
and (indirectly) by the geometric primitives. A level of features encloses aggregation hierarchies and classification (generalization) hierarchies [20]. Spatially
connected features may be put together to form complex features. In turn complex features may be put together to form more complex features, and so on.
This paper considers only the lowest level in the aggregation hierarchy in which
the features have a direct link to the geometry. The classification hierarchies are
reduced to the most refined level of feature classifications (Figure 2). A feature
can only belong to one feature class; a feature class can only contain features
of one geometric feature type, i.e. either point features, or line features, or area
features.

As Figure 2 illustrates, terrain features contain thematic data and geometric
data. The former are represented by the feature classes. The latter are represented by nodes and arcs. The geometric data can be further subdivided into
metric data and topological data. In the fds, the metric information of line

features and area features can be constructed partly through the topological linkages, i.e. the features are connected with the metric data partly through the topological data. To design an appropriate data model for a multi-layered GIS our focus is on the three main data types: thematic, metric and topological data.

Thematic data are directly feature-related, in contrast with metric and topological data which are indirectly feature-related, through nodes and arcs. Therefore, it is a good choice to represent the thematic data per feature. This organization of the thematic data is satisfactory when the GIS describes one or several themes or contains one or several structure layers.

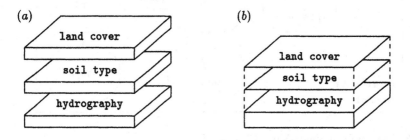

Fig. 3. Structure Layers: (a) multiple SVVM structure layers; (b) a MVVM structure layer

3.2 MVVM Structure Layer

In the situation where area features of different thematic layers do overlap, it is not possible to use the svvm for the implementation of a common structure layer. An example of two such themes are 'land cover' and 'soil type', which are both partitions of the 2D-space. Possible classifications for 'land cover' are industry, forest, residential, agriculture, water, etc. Similarly, possible classifications for 'soil type' are: sand, clay, fen, etc. Example queries involving both themes are:

- Give all agricultural areas with the soil type 'sand'.
- Give all clay regions covered by 'industry' and 'agriculture'.

The representation of several themes in a GIS offers three main options for the organization of metric and topological data in a data model. Before these options are described in the next section, the concept of structure layers is explained in more detail. Beside svvm structure layers, a *multi-valued vector map* (mvvm) can be used to implement a structure layer (Figure 3). The differences between the two types of structure layers are fundamental.

A mvvm structure layer represents several themes, but distinct from a svvm structure layer which may represent several themes. The features of a specific theme in a mvvm structure layer may coincide or overlap with the features of

146

another theme in the same layer. Therefore, the features in a mvvm structure
layer are in general at the geometric level more segmented than in a svvm layer.
A multi-valued structure layer is the map overlay result of two or more structure
layers. In the example presented in the beginning of this subsection, one can use
one mvvm for the representation of the integrated structure layer 'land cover
& soil type'. This would enable the efficient implementation of the topological
queries related to both themes.

More specifically, in comparison with the fds for a single-valued vector map
(see Figure 2), the following applies to the fds for a multi-valued vector map as
well: a node may represent one or more coinciding point features; an arc may
represent (part of) one or more (partly) coinciding line features; with n merged
structure layers, an arc has n coinciding or overlapping area features at its right
hand side and n coinciding or overlapping area features at its left hand side;
a point feature may lie in one or more area features. A line feature consists
of one or more arcs. An area feature consists of at least three arcs and may
additionally comprise several point features in its interior. With this in mind,
the last three extensions for multi-valued vector maps denote many-to-many
relationships. Note that the adjustments for multi-valued vector maps affect
only the geometric level of the fds, including the links between the geometry
and the features, and not the level of features. Thus, in a multi-valued vector
map the original features subsist.

4   Data modeling variants

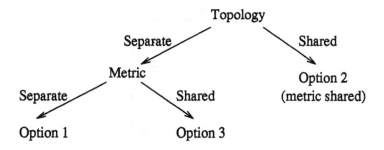

Fig. 4. The three different data models

We proceed with the three different options to organize thematic, metric, and
topological data consistent with the fds when dealing with several descriptions
of the same region. As the thematic data have a direct link to the (related)
features, the emphasis of the data modeling variants is on the (de)composition of
the *metric* data and of the *topological* data. The three options are (see Figure 4):

1. for each individual thematic layer, represent both metric and topological
   data per svvm structure layer separately;

2. for two (or more) thematic layers, represent both metric and topological data in one structure layer (if possible using a svvm, e.g. 'hydrography & countries' otherwise using a mvvm, e.g. 'land cover & soil type');

3. for two (or more) thematic layers, an intermediate structure layer with topological data organized according to a svvm structure layer, and metric data organized according to a mvvm structure layer.

These options are similar to the proposals in [8]. The first two options represent a set of svvm structure layers, and a mvvm structure layer covering all descriptions (see Figure 3), respectively. The third option combines the thematic and topological data as in a svvm structure layer, while the metric information of all descriptions, including the coordinates of all intersection nodes (Figure 5), is gathered as in a mvvm structure layer. The term layer will be quoted in the following, to indicate that it only represents part of a geographic data set, e.g. thematic data and topological data, or thematic data and metric data. In Figure 5 the two top 'layers' comprise thematic and topological data of features that exclude each other spatially as in svvm structure layers. Further, the bottom 'layer' comprises the metric information of the two top 'layers', including the coordinates of the intersection nodes, in accordance with a mvvm structure layer.

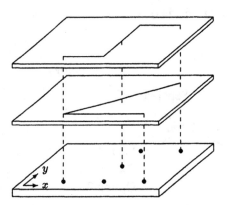

Fig. 5. The metric data, intersections included, centralized

A fourth option would be to combine the thematic and metric data as in a svvm layer, and gather the topological information of all descriptions as in a mvvm layer. But in this case, the derived variant is not consistent with the fds. For efficient topological querying another variant on the three options can be devised; combine the thematic, metric and topological data per structure layer, single-valued as well as multi-valued, and add the topological information, encompassing all descriptions (Figure 6). In Figure 6, the two top layers denote two svvm structure layers, while the bottom 'layer' comprises the topological information of the planar graph obtained after map overlay.

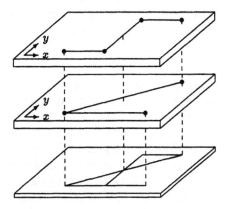

Fig. 6. The topological information, as result of map overlay, added

The above options illustrate the strength of the feature-oriented approach of the fds. When a combination of svvm structure layers is often used then one may consider the permanent integration of these structure layers. With regard to the fds, the integration of structure layers may result in two different types of layers. First, the result may be a mvvm structure layer, which has already been described. Second, the result of the integration of the thematic data of different layers may be a svvm structure layer, which means that feature classes are joined and new features are constructed and assigned to the new feature classes. The latter can be left aside, as the feature-oriented approach assures the explicit preservation of the information content of the original layers after applying map overlay.

Combinations of the options may occur in GISs. We now discuss the pros and cons of the afore-mentioned three options, and the alternative aimed at efficient topological querying. Considerations regarding redundancy and efficiency underlie the choice for a variant.

re 1. set of svvm structure layers The arrangement of non-interrelated svvm structure layers has two severe drawbacks: the same metric data may be stored more than once (redundancy), which may result in an inconsistent dataset, and topological relationships between features in different layers can only be tracked by applying map overlay which is a computationally expensive operation. Map overlay does not always have to be employed to the complete geographic region described in the database, but can in many cases be restricted to the region or features of interest. The benefit of this modeling approach is that features can be accessed directly, and topological queries restricted to one structure layer can be answered quickly. For example, as motorways may run along the border of nature reserves, both motorways and nature reserves may have the same arcs in common (see Figure 2). Therefore, to avoid redundancy these motorways and nature reserves can be stored in one svvm structure layer. The storage of these motorways and nature reserves in separate svvm structure layers requires map overlay when topological relationships between the motorways and nature reserves are analyzed.

re **2. a mvvm structure layer** The gathering of both metric and topological data in one strongly segmented mvvm structure layer has the disadvantage that features have to be reconstructed from small geometric elements each time that they are accessed. Each time geographic information is added to the GIS, map overlay has to be performed. The more layers a GIS accommodates, the smaller the geometric elements will be and the larger the total volume of the data. Accuracy becomes an important factor in this modeling variant. But the advantage is that no additional calculations have to be performed when querying the different themes. It is straightforward to derive topological information of interrelated layers. The creation of a mvvm structure layer may have a temporary character. Applying a mvvm structure layer is especially useful when (topological) querying of interrelated layers is often requested, and the features in the different layers do not exclude each other spatially. For example, the themes 'vegetation' and 'soil type' may be combined in one mvvm structure layer 'vegetation & soil type'. As the (interrelated) topology information is stored explicitly, topological queries related to vegetation, soil type, or both, can be answered immediately.

re **3. an intermediate structure layer** To combine thematic and topological data as in a svvm structure layer, and all metric data (including the intersections) as in a mvvm structure layer has the disadvantage that topologically interrelated features from different layers can only be traced by comparison or applying map overlay additionally. The benefits of this solution are that features do not have to be reconstructed from small geometric elements and that it avoids the redundancy of metric data, as the coordinates pairs are represented only once. For example, the themes 'vegetation' and 'soil type' may also be represented in an intermediate structure layer. The query: 'Give all sand regions covered by heather' cannot be answered at once, but may be answered by selecting the regions from the geographic database which comprise heather or sand. Additional matching of these regions will then provide the answer. As a relationship between vegetation and soil type can be assumed, this third option may reduce the metric data considerably. Note that the themes can also be represented in two separate svvm structure layers.

re **alternative** A set of svvm structure layers extended with topological information encompassing all layers has the advantage that topological information of interrelated layers can be retrieved quickly. Furthermore, the individual features can be accessed quickly. Like the second and third option, this variant also involves the pre-calculation of the intersections of the features in all layers. The added topological data contains only the information of the topological planar graph, i.e. this layer does not contain metric data. Optionally the co-ordinates of the intersection nodes may be preserved. Each node and each arc in the graph are assigned an identifier. Except for the nodes which stem from the intersection of different layers, the nodes and arcs in the topological graph connect the topological data in the combined layer with the metric data in the separate structure layers. The nodes and arcs in the topological graph may contain many references to features in different layers. In addition, arcs may be the border of many area features, i.e. they may contain one or more area features at their left hand side, and one or more features at their right hand side. The number of topological references depends on the number of merged structure layers. Note that this possibility still requires geometric calculation at query time, when the coordinates of the intersection nodes are discarded from the topological graph. This variant involves the redundant storage of topological and metric data. Rules may be defined to guard the consistency of the data. Redundancy can be partly avoided by gathering the metric data as proposed in the second and third options. For example, one svvm structure layer describes 'land cover' in The Netherlands, another svvm structure layer describes 'soil type', and a third svvm structure layer describes 'administrative boundaries'. Map overlay is applied to these layers, which yields a geographic data set (a mvvm structure layer) comprising thematic, metric, and topological data of the original layers, including the metric data of the intersections. The thematic and metric data are removed from this data set. The coordinates of the intersection nodes may be preserved. The topological data remain and are added to the original layers. With the additional topological data set, queries related to multiple structure layers, e.g. 'Give all industrial areas on sand in the province South Holland', can be answered efficiently.

4.1 Relations in Postgres

From the above discussion, we conclude that, an optimal organization of data in a multi-layered GIS will largely depend on the intended application. Therefore, every solution will be a compromise. The following proposal for a data model for a multi-layered GIS, seems quite promising for topological querying a GIS and is explained in the remainder of this section. The data model will be mapped onto the relational model with object-oriented features, as defined in the database management system Postgres [23].

The relations are given in the Appendix. The following basic data types are used: int4, char, bool and point, with with obvious meanings. Variable length arrays of these data types are indicated by square brackets; e.g. int4[]. The set of relations is general and may be used to implement all three data models described in the previous subsection. Four groups of relations (or classes) can be identified: layers, feature classes, topological classes, and metric information. Together with a set of rules, these relations form the complete implementation.

One of the most common queries is to retrieve all features that belong to a given thematic layer (in a certain region). Though this is not a truly topological query, it can be solved with the relations given above using the **thematic_layer** class and the **tlayer** attribute of the area, line, and point features. More complicated topological queries will be described in the next section. The consistency and integrity rules that belong to these relations are:

1. all primary keys must be unique (per definition); note that they are used in the role of object id's;
2. all foreign keys must exist in their own table;
3. in **point_feature** (both single and multi), the **area_feature** must belong to the same structure layer as the point feature itself (has to be checked through the thematic layer);
4. in **arc** (both single and multi), the **area_features** and **line_features** must all belong to the same structure layer (single or multi-valued, respectively);
5. in case of a mvvm structure layer, the references in the classes **point_- feature_multi** (for attribute **areas**) and **arc_multi** (for attributes **left_areas**, **right_areas**, and **lines**) must all be to a different thematic layer per array.

These rules can be enforced by translating them into the Postgres rule system. In this way one can guarantee the consistency and integrity of the data model.

5 Topological querying

In this section, some examples of topological queries related to multiple map layers are worked out in Postgres [23]. Open systems, such as Postgres, offer extensibility, i.e. data types, operators, functions, and index structures can be added to the system. The derivation of a set of binary topological relationships in svvm structure layers, and extended for mvvm structure layers, has been given in [9]. It was proposed to implement the binary topological relationships as predicates. As the parameters of a predicate can be either a constant or a variable, each predicate can represent four types of topological queries. The predicates are given intuitively meaningful names. The feature-oriented approach, followed in this paper, allows it to address the features directly, i.e. an end-user can state queries referring to the features instead of the underlying topological elements. For example, the topological query 'Through which countries does the river Rhine flow?' can be formulated in Postgres as follows:

```
retrieve (area_feature.area_class)
where    Intersect(area_feature.area_class,"Rhine")
```

Note that the string **Rhine** is used as a symbolic identifier of a line feature. This query can encompass separate structure layers. In this example the selected area features are cut in two by the line feature. But depending on the interpretation of the query stated in natural language, several other predicates may be employed as well, e.g. **In**, **Permeate**, and **Meet** (see [9]). For practical use, it will in general be more suitable to group similar relationships between arbitrary features into one generic predicate. In this case, additional conditions provided by the end-user will give the specific relationship requested, e.g. in order to distinguish whether two line features touch or intersect. This may lead to a hierarchical organization of the topological relationships (predicates), as e.g. is given in [22].

A similar example of a topological query, given in Section 2, is: 'Which river flows through more than two countries?' By using the aggregate function count this query may be formulated in Postgres as:

```
/* Select all countries which are intersected by rivers
   and the rivers concerned.
*/
retrieve into hr1(l.line_class, a.area_class)
from     a in area_feature, l in line_feature
where    Intersect(a.area_class, l.line_class)

/* Select every river that also intersects a second country
*/
retrieve unique (hr1a.line_class)
from     hr1a,hr1b in hr1
where    hr1a.line_class   = hr1b.line_class
and      hr1a.area_class  != hr1b.area_class
```

Topological queries may also be applied to check the semantic integrity of the data [8], for instance to sort out whether a set of windmills is located on land among all separate structure layers. Note that the topological relationship is_in between point features and area features is explicitly recorded in the fds (see Sections 3, 4.1). The query may be formulated as: 'Give all windmills which may be erroneously located in lakes'. In Postgres, this query for a svvm structure layer is:

```
retrieve (point_feature_single.point_class)
where    point_feature_single.area = area_feature.aid
and      area_feature.area_class = "lake"
```

6   Theory and Practice: Topographic Data Sets

In this section, a case study is presented. The previous sections contain the theory on map layers and topology, and have shown that it is possible to model the geographic data in various ways. In practice one has to decide in which way the data will be modeled. The case study in this section gives the arguments which were used in the design of a new data model for topographic base data sets of the Topographic Service in The Netherlands. The data sets concerned are related to 1:25,000 and 1:50,000 map scales and may be exchanged by using DIGEST (DIgital Geographic Exchange STandard) [10]. The data model underlying DIGEST is very similar to the fds.

Six thematic layers are distinguished by the Topographic Service: transportation, hydrography, vegetation, buildings, administrative boundaries, and height information. One of the most important questions is: Which themes share a topological data structure, i.e. belong to the same structure layer? One of the first principles is that redundant data storage should be avoided. E.g., as a road may coincide with the border of a forest, they can (partly) be described by the same geometric line primitives and should therefore be in the same topological data structure. But, the integration of different themes in one structure layer may result in too many new nodes caused by the fact that unrelated features of different (related) themes intersect. Features should not be split up unnecessarily; see Section 2. However, a feature that is strongly segmented at the geometric level can still be treated as one feature. This feature-oriented approach is very useful to avoid repetition of the same thematic attributes.

It turns out that most themes are related through the geometry, so they have to be in the same structure layer. However, at the feature level this is not always the case, and 'unrelated' individual features from different map layers may have to be intersected. For example, this occurs in the case of a small road through a forest. In this situation we do not want the forest to be split by the small road into two separate forests. Nodes at the intersection of the road and the boundary of the forest are introduced. For the time being, it was *decided* that nearly all topographic features are in the same svvm structure layer. Except for the administrative boundaries (between municipalities, provinces, etc.) and the height information (height contours and spot heights), which have their own svvm structure layers.

In summary, each structure layer consists of the geometric primitives points (nodes), lines (arcs), and polygons (areas). These geometric primitives are used to create features; e.g. a line feature may consist of several arcs (each running from one node to another node). Also, different types of features may be combined in creating complex features; e.g. rivers (line features) and lakes (area features) may form together a complex water feature. The thematic information is attached to these geographic features. This way of modeling requires a powerful data model, such as described for the fds and implemented by the underlying data model for DIGEST.

It should be noted that the situation in France is quite similar to the situation in The Netherlands: in BDTopo there are two layers, one for topography and one for contour lines [7]. The same division can be noticed in Germany, where the topographic landscape model in ATKIS is divided in a 2D digital situation model and a 3D digital terrain model. The former covers topographic descriptions, like vegetation, transportation, and water(ways). The latter contains height and geomorphological information [1].

7 Conclusion

This paper addresses the confusion about the GIS-concept map layers by introducing two types of map layers: structure layer and thematic layer. These have different meanings, but they are often confused with each other in discussions about map layers. For the implementation of a structure layer, a svvm can be used if the features do not overlap, otherwise a mvvm can be used. Multiple thematic layers can be stored in such a structure layer. Alternatively, multiple structure layers can be used if they must be treated independently of each other. The considerations for these design decisions are presented in this paper: fragmentation of geometric primitives, redundant data storage, efficient implementation of multi-layer topological queries, flexibility in data modeling, and ease of model extensibility.

A set of Postgres relations, which implement the described data models, is given. Some multi-layer queries are given, but more research in this area is required. This can be done in combination with the case of 'topographic data sets'. Another reseach area is the formal description of the mvvm, similar to the formal description of the svvm.

Acknowledgments

Many valuable comments and suggestions on an preliminary version of this paper were made by Paul Strooper. We would also like to thank the Postgres Research Group (University of California at Berkeley) for making their system available.

References

1. Arbeitsgemeinschaft der Vermessungsverwaltungen der Länder der Bundesrepublik Deutschland (AdV). *Amtliches Topographic-Kartographisches Informationssystem (ATKIS)*, 1988. (in German).
2. Stan Aronoff. *Geographic Information Systems: A Management Perspective*. WDL Publications, Ottawa, Canada, 1989.
3. K. Bennis, B. David, I. Morize-Quilio, J.M. Thévenin, and Y. Viémont. GéoGraph: A topological storage model for extensible GIS. In *Auto-Carto 10, Baltimore*, pages 349–367, March 1991.
4. K. Bennis, B. David, I. Quilio, and Y. Viémont. GéoTropics database support alternatives for geographic applications. In *4th International Symposium on Spatial Data Handling, Zürich, Switzerland*, pages 599–610, July 1990.
5. Nicholas R. Chrisman. Epsilon filtering – a technique for automated scale changing. In *Technical Papers of the 43rd Annual Meeting of the American Congress on Surveying and Mapping*, pages 322–331, March 1983.
6. Ian K. Crain. Extremely Large Spatial Information Systems a Quantitative Perspective. In *4th International Symposium on Spatial Data Handling, Zürich, Switzerland*, pages 632–641, July 1990.
7. Benoît David. Personnel communication through email on the topic: Map layers, June 1992.
8. Benoît David and Yann Viemont. Data structure alternatives for very large spatial databases. Technical Report ?, Institut Géographique National – France, 1991.
9. Sylvia de Hoop and Peter van Oosterom. The storage and manipulation of topology in Postgres. In *EGIS'92, Munich, Germany*, pages 1324–1336, March 1992.

10. DGIWG. DIGEST – digital geographic information – exchange standards – edition 1.1. Technical report, Defence Mapping Agency, USA, Digital Geographic Information Working Group, October 1992.
11. Claus Dorenbeck and Max J. Egenhofer. Algebraic optimization of combined overlay operations. In *Auto-Carto 10, Baltimore*, pages 296–312, March 1991.
12. James Dougenik. Whirlpool: A geometric processor for polygon coverage data. In Robert T. Aagenburg, editor, *Auto-Carto IV, Volume II*, November 1979.
13. Andrew U. Frank. Overlay processing in spatial information systems. In *Auto-Carto 8*, pages 16–31, 1987.
14. Wm. Randolph Franklin and Peter Y.F. Wu. A polygon overlay system in PROLOG. In *Auto-Carto 8*, pages 97–106, 1987.
15. Stephen C. Guptill and Robin G. Fegeas. Feature based spatial data models – The choice for global databases in the 1990's. In Helen Mounsey and Roger Tomlinson, editors, *Building Databases for Global Science*, pages 279–295. Taylor & Francis, 1988.
16. Hans-Peter Kriegel, Thomas Brinkhoff, and Ralf Schneider. The combination of spatial access methods and computational geometry in geographic database systems. In *Advances in Spatial Databases, 2nd Symposium, SSD'91, Zürich, Switzerland*, pages 3–21, August 1991.
17. Werner Kuhn. Are Displays Maps or Views? In *Auto-Carto 10, Baltimore*, pages 261–274, March 1991.
18. Martien Molenaar. Single valued vector maps – A concept in Geographic Information Systems. *Geo-Informationssysteme*, 2(1):18–26, 1989.
19. Martien Molenaar. Informatie theoretische aspecten van GIS. In *Vastgoedinformatie in de jaren negentig*, pages 67–80, June 1990. (in Dutch).
20. Martien Molenaar. Object-hierarchies. Why is data standardisation so difficult? In Kockelhoren et al., editor, *Kadaster in Perspectief*, pages 73–86. Dienst van het Kadaster en de Openbare Registers, Apeldoorn, 1991.
21. Donna J. Peuquet. A conceptual framework and comparison of spatial data models. *Cartographica*, 21(4):66–113, 1984.
22. Simon Pigot. Topological Models for 3D Spatial Information Systems. In *Auto-Carto 10, Baltimore*, pages 368–392, 1991.
23. Michael Stonebraker, Lawrence A. Rowe, and Michael Hirohama. The implementation of Postgres. *IEEE Transactions on Knowledge and Data Engineering*, 2(1):125–142, March 1990.
24. Jan W. van Roessel. Attribute propagation and line segment classification in planesweep overlay. In *4th International Symposium on Spatial Data Handling, Zürich*, pages 127–140, Columbus, OH, July 1990. International Geographical Union IGU.
25. Guangyu Zhang and John Tulip. An algorithm for the avoidance of sliver polygons and clusters of points in spatial overlays. In *4th International Symposium on Spatial Data Handling, Zürich, Switzerland*, pages 141–150, July 1990.

Appendix: Data Model in Postgres

```
/* PART 1: LAYERS */
  structure_layer (
      slayer_id = int4,      /* primary key */
      single_valued = bool, /* single or multi-valued layer */
      description = char[]
  )
  thematic_layer (
      tlayer_id = int4,      /* primary key */
      slayer_id = int4,      /* foreign key structure_layer */
      description = char[]
  )
/* PART 2: FEATURE CLASSES */
  area_feature (
      aid = int4,            /* primary key */
      tlayer = int4,         /* foreign key thematic_layer */
      area_class = char[]    /* thematic info */
  )
  line_feature (
      lid = int4,            /* primary key */
      tlayer = int4,         /* foreign key thematic layer */
      line_class = char[]    /* thematic info */
  )
/* PART 3A: TOPOLOGICAL CLASSES, SINGLE VALUED */
  point_feature_single (    /* also: FEATURE CLASS */
      pid = int4,            /* primary key */
      tlayer = int4,         /* foreign key thematic layer */
      point_class = char[]   /* thematic info */
      node_id = int4,        /* foreign key node */
      area = int4            /* foreign key area_feature */
  )
  arc_single (              /* primary key: from_node & to_node */
      from_node = int4,      /* foreign key node */
      to_node = int4,        /* foreign key node */
      left_area = int4,      /* foreign key area_feature */
      right_area = int4,     /* foreign key area_feature */
      line = int4            /* foreign key line_feature */
  )
/* PART 3B: TOPOLOGICAL CLASSES, MULTI VALUED */
  point_feature_multi (     /* also: FEATURE CLASS */
      pid = int4,            /* primary key */
      tlayer = int4,         /* foreign key thematic layer */
      point_class = char[]   /* thematic info */
      node_id = int4,        /* foreign key node */
      areas = int4[]         /* foreign keys area_features */
```

```
  )
  arc_multi (                 /* primary key: from_node & to_node */
      from_node = int4,       /* foreign key node */
      to_node = int4,         /* foreign key node */
      left_areas = int4[],    /* foreign keys area_features */
      right_areas = int4[],   /* foreign keys area_features */
      lines = int4[]          /* foreign keys line_features */
  )
/* PART 4: METRIC INFORMATION */
  node (
      node_id = int4,         /* primary key */
      location = point        /* x,y-coordinates */
  )
```

Towards a conceptual data model for the analysis of spatio-temporal processes: the example of the search for optimal grazing strategies

Jean-Paul Cheylan* and Sylvie Lardon**

* GDR CASSINI and GIP RECLUS and, 17 rue Abbé de l'Epée, 34000 Montpellier, France.
** GDR CASSINI and INRA-SAD, B.P. 27, 31326 Auzeville Cedex, France.

Abstract. This paper addresses a number of conceptual and practical problems of integrating space and time within a GIS environment encountered in an ongoing study of the strategies adopted by a shepherd in managing a flock of sheep at different scales of space and time on an alpine pasture in the southern French Alps. The aim of the study, which is based on the analysis of the timetables used to exploit the vegetatative resource within a set of time-space constraints, is to understand the way that the shepherd partitions space into compartments and the way these are used at different times.The pertinent conceptual issues are reviewed, the problems of formalising this type of problem for analysis are discussed, and alternative ways of structuring the data for GIS application are proposed.By way of conclusion, a preliminary attempt is made to summarise critical issues emerging from the study in the handling of time in spatial analysis and comments are briefly made on directions for future spatio-temoral research into grazing strategies.

1 Introduction

This paper, which reports on research in progress, is concerned with the representation of events as they occur in time and space using GIS. The strategies adopted by a shepherd in managing a flock of sheep on an alpine pasture are used as the framework for discussing methodological problems and their proposed solution. The analysis is based on the content of detailed dairies which suggest that the shepherd employs strategies at different spatial and temporal scales to ensure that the needs of the sheep are satisfied without causing a deterioration of the resources on which they depend.

The emphasis is on the organisational characteristics of the shepherd's time-space behaviour which may be considered to be an "expert system". Although the case study is specific it is representative of a more general class of problems involving the description of dynamic and deformable features which at the same time exhibit a coherence and strategy dictated by a set of spatial constraints. It illustrates the way that complex timetables are conceived and developed as management tools in the temporal and spatial domains.

The paper is divided into three parts. Given the importance of the temporal dimension for this study, the first part reviews the ways in which time has been conceptualised.

The discussion is deliberately brief and concentrates only on concepts that are relevant to this particular study. Reference can be made to the abundant literature on the subject in geography as well as in cognate disciplines for treatment of more general methodological issues. This is followed by a description of the strategies adopted by the shepherd in managing a flock of sheep and a discussion of the methodological problems of handling time in spatial analysis within a GIS framework. In conclusion, a tentative classification is presented based on the empirical work which attempts to formalise the dimensions of the problem of integrating time and space in geographical analysis.

1.1 Conceptualising time

Time is a fundamental component in the collection and recording of information, both as a factor in its own right and as a methological constraint. Determining ways of recording and representing it as a basis for analysis is a fundamental and complex task. A critical aspect of the problem of handling time in spatial enquiry is that of the ephemeral nature of spatial organisation itself, and hence the difficulty of identifying features which themselves vary with time. This is the case of patterns of behaviour of the flock of sheep in this study. Time is itself also an object of analysis: the explanation of an observed spatial structure is frequently undertaken with reference to it origins and genesis. The processes involved are what generate the "archaeology of landscape".

Definitions and data bases. It is imperative at the outset to define the terms that are used explicitly or implicitly in this study. Following the work of Allen (1983) these are:

Facts: Eternal and explicetly bounded truths which represent states of a process, the collection of which can be either discrete or pseudo-continuous.

Events: These modify the state of the world. It is only their end result that matters - that they have been realised or achieved. They may correspond to the instants in, or intervals of, time that define both spatial and temporal facts.

Processes: Courses of action which modify the state of the world. Their operation, activity, and the mechanisms involved are the object of study.

Transitions and mutations: These characterise the temporal succession of two specific states.

Timetables: The collection of intervals of validity and transitions between them. Spatial timetables are those in which the objects are spatially referenced.

Causation: The coupling together of facts, events, and processes.

Strategies: Future causation determined by an actor (eg the shepherd). Analogous to the concept of planning - the dynamic elaboration (adaptive or otherwise) of a suite of actions in time based on the specification of goals, milestones, and desirable outcomes.

Much has been written on temporal data bases (see for example Yeh and Viemont 1992, Snodgrass, 1990). The principal types commonly identified include *static data bases* - the contents of which do not necessarily represent the real state of the world;

{*static with memory* - often referred to as "rollback" or transactional data bases; *historical* - in which historical states are recorded as relations which define the validity of observations; and *temporal* - based on transaction time and periods of validity. This type of data base, in which it is not possible to directly access successive states, can be augmented by including time operators (eg when, beginning, preceding, recovery, etc.) and validity clauses thus making it possible to address temporal queries (Snodgrass, 1987). The data base constructed from information in the shepherd's dairies used in this study is of this type.}

1.2 Structuring time

Time can be structured in a number of different ways. An important initial distinction is that between its *discrete* and *continuous* representation. In a practical sense, this distinction has little to do with the methodological problem of instrumentation. Indeed, the tools of information technology are discrete by their very nature; the representation of continuous time can only be approximated by recording facts at very fine intervals or through the use of functional representations. Even then, however, the results become discrete at the time of evaluation. It is only through the use of analogue methods that approximations to real continuity are possible. This is the notion of the granularity of time; the fineness with which time is partitioned governs the volume and complexity of information and limits the kinds of analyses that are possible.

The structure of order assumed in the representation of time is also important and has significant implications for handling the temporal dimension. The partial ordering of time typically encountered in linguistics (eg before, after, during, etc.) results in a lattice structure which creates difficulties in exploiting data bases using classical approaches. Tentative attempts have been made using temporal logic to represent simultaneously time and space (Vieu, 1991). The total ordering of time, the concept of dense ordering as used in physics, enables a number of important difficulties to be overcome.

The semantics of time are seldom made explicit but must also be considered. The concept of physical time is commonly used (eg that represented by the clock), but other interpretations are as important including the lifetime of a data base with its updates, archival mechanisms, and memory of its operation (its versions); the time of modelled reality; and the eventual transition between these especially when account is taken of known real events in the data base. For this study, only the second of these - time of modelled reality - is considered explicitely.

Two alternative but conflicting methods of representing physical time are classically used - the recording of points corresponding to events, and the definition of intervals - either explicitly (by recording the time of the birth and death of an event), or implicitly so that these can be deduced (by recording the time of birth and the duration of an event in a way that is analogous to run length encoding). Only the second of these makes it possible to reconstruct the collection of states at any given date, the logical sequences or transitions based on order in the timing of the local states of spatial objects, or to handle the process of discrete existence such as the duration of a storm or of a concert performance.

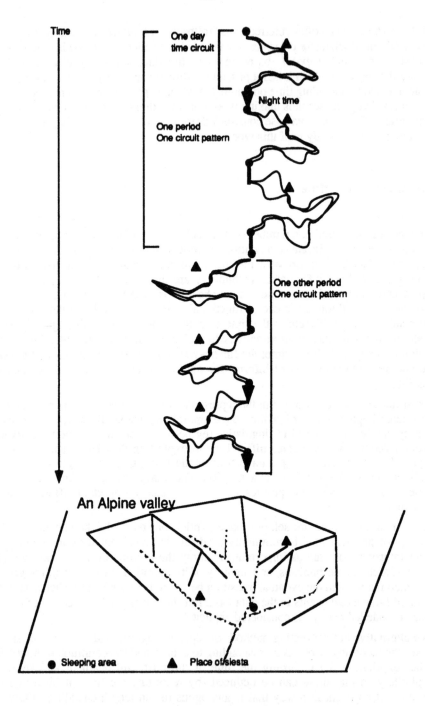

Figure 1: The time and space «band»

An important aspect of the problem addressed in this research is the cyclical nature of events and processes operating. These are more difficult to handle. They may either take the form of "physical" cycles - those inherent in calendars and timetables (eg the length of a day, the daily rotation of the sun, or the opening hours of an office), or of semantic cycles of periodicity the length of which are variable (eg circuits or circuit types used in the empirical part of this paper).The second of these can be deduced through the analysis of structural regularities - logical or statistical - observed in linear time and in cycles within timetables. This involves the translation from physical to logical time, the total order of which is not necessarily measured.(Figure 1 - The time and space band)

The representation and analysis of cyclical time assumes that time itself is amenable to measurement and that inferred time and weakly developed structures of order may be ultimately produced.

The problem of establishing the origin of time is another significant consideration. The classical way of doing this is to calibrate the timing of events using the concept of physical time. For example, in studying the process of the growth of cities an initial analysis may be couched in terms of calendar time. Is the process of growth then identical for each town even though their origins begin at different dates? If so, then it is possible to analyse the process by resetting the start of growthto a common origin. This is the case in analysing the daily patterns of grazing of the flock of sheep. The time each day begins varies seasonally and as a result of temporary pertubations. (Figure 2 - Alternative ways of representing a temporal sequence of activities).

The linearity of time, extending into the future and backwards into the past, is not an appropriate assumption in all situations but one that is used in this study. In the case of scenarios or simulation, recognising that time branches enables a better representation. Another common problem in temporal analysis is that of establishing the relationships between stocks and flows. Stocks can be considered to be the state at a particular instant in time representing the" balance sheet" of previous events - a temporal aggregation of flows based either on the rate time passes or on the composition of intervals. The representation of stocks and flows at any given moment in time is difficult because of the problems of measurement.

1.3 Dynamic situations

Information about dynamic phenomena results principally from the interaction of two relatively independant processes: that ocurring in reality - which is the object of analysis, and that actually recorded by an observer. The temporal relations between these two processes determine the sense and especially the temporal validity that can be ascribed to information.

Several ways of recording data relating to dynamic situations can be identified. One of the most common is to record observations on dates selected randomly as with a census of population or remotely sensed images. In this case, the observations are not in any way related to the underlying processes operating. The interval for which the information is valid is therefore null from a logical point of view. When a knowledge of the rate or periodicity of the process is available, it is possible to deduce the

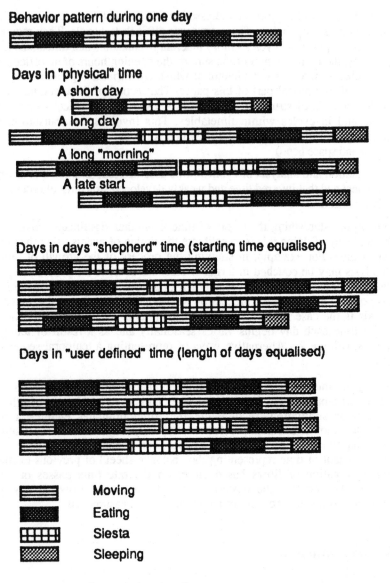

Behavior pattern during one day

Days in "physical" time
    A short day

    A long day

    A long "morning"

    A late start

Days in days "shepherd" time (starting time equalised)

Days in "user defined" time (length of days equalised)

≡≡≡ Moving

■■■ Eating

▦▦▦ Siesta

▨▨▨ Sleeping

Figure 2: Alternative ways of representing a
temporal sequence of activities

interval of validity although the result is purely semantic. The use of this procedure may lead to the conclusion that conditions during the defined interval remain unchanged until the date that subsequent observations are taken. Change is thus assumed to follow a stepped function.

An alternative is when observations are triggered or tied to explicit events. This involves a change in the method of observation - that from recording a version periodically to monitoring events in real time by linking observation to a specific event. In monitoring urban growth, for example, this is the difference between recording the total number of new buildings that have been completed at the end of a given period and successively recording each new building by the date its permit is issued.

The data on which this study is based was not collected by either of the above methods. Rather, the observations as recorded were linked in a continuous way to the process as it unfolded. This is the method known as *organically tied observation*. When each event that transforms a phenomenon is known, the reconstruction is complete. Thus in movies, for example, analogues are used to signify a change in state. Examples are observing individually sequenced building permits and the displacement of a flock of sheep as observed through the eyes of the shepherd. A modification of this way of taking observations is to record events periodically when there is a change of state. This is the technique used in cartoons; space and time are altered incrementally but only by small amounts. In the context of this study, the shepherd made observations of the sheep on the basis of his classification of the activities they were engaged in. The spatial envelope describing their disposition is marked by time intervals throughout which the sheep were engaged in the same activity.

In the case of observations of cyclical events, it is necessary to know the length of one (or half) cycle.This implies that several processes have already been observed to determine what the cycle is to begin with. Taking periodic observations without a knowledge of the underlying cycles or the long term trend is logically equivalent to taking observations on a particular date or at a particular time - from the point of view of logic, the interval of temporal validity of the observations is again null. Cyclical observation presupposes that the cycles can themselves be explained, regardless of whether or not they are tied to physical time (eg the shepherd's day - a cycle directly observed) or are reconstructed from the problem itself (eg the duration of a circuit).

2 The time-space behaviour of a shepherd and his flock

Against this conceptual background, the focus of this research is on the strategies used by a shepherd in managing a flock of sheep on an alpine pasture and the analysis of the timetables that are used to exploit the resources within a set of space and time constraints (Landais et al, 1988). The specific objective is to identify and understand the way in which the shepherd partitions space into compartments and the way these are used at different times (Landais and Deffontaines, 1989).

Factors influencing the way the shepherd does this include the characteristics and growth of the vegetation, the behaviour of the animals in satisfying their needs - particularly for food, the preservation of the alpine flora and fauna, and the maintenance of the grazing capacity of the area (Deffontaines, 1992). From a detailed understanding of the space-time behaviour involved, it should be possible to generate plans for the optimal exploitation of the resource and to identify the management practices that are required for their implementation (Osty and Landais, 1993).

Every summer the shepherd takes a flock of about 1,000 sheep, belonging to several owners in the valley, to a mountain pasture located in the Parc des Ecrins in the southern French alps (Cheylan et al, 1990). The mountain pastures are rented from the commune of Orcières from the beginning of July to the end of September before the onset of the winter snows.

The following objects need to be included in the model: (- the SHEPHERD: the actor who decides and intervenes; - the FLOCK: with needs to be satisfied during time, and which has some degree of manouvreability in space; - the MOUNTAIN: supplying the resources which vary in space and time during the season; - the SEASON: which governs the activities of the FLOCK on the mountain.

2.1 The seasonal pattern

At the outset, the shepherd divides up the alpine area available for grazing for the season into COMPARTMENTS, each one of which is used for a particular PERIOD. The July compartment, for example, is situated on the lower slopes which are covered by early pasture growth after the winter. In August, the flock is moved up to the higher pastures and in September the uppermost slopes are grazed before the flock is moved down again to the compartment previously used in July. Each compartment contains all the requirements needed by the shepherd and the sheep during the period it is in use: a cabin, watering points, folds, sleeping areas, etc.. The area of each compartment is typically delimited by natural features (eg ravines, ridges, scree slopes), and each comprises a number of SECTORS that are defined by the shepherd on the basis of the heterogeneous character of the vegetation in relation to the needs of the sheep (eg, vegetation type, amount of grass cover, abundance of species, rate of regeneration, etc.). Morphological features are also important with regard to the movement of the flock since the sheep need to graze in a sufficiently large area without too many slopes so that together they can be seen by the shepherd.A sector may belong to more than one compartment, for example because a cabin is located there or because by late summer the grass has regrown in certain sectors at lower elevation.

Data for 1990 is used for the particular analysis presented here which is based on the IGN 1:25,000 topographic sheet. Different kinds of information have been identified and mapped at this scale for the area: features supporting and constraining the movement and grazing of the sheep, the location of infrastructure (eg folds, sleeping and resting areas, salt rocks,watering points, etc.), as well as obstacles to the movement of the sheep, eg ravines, scree slopes, ridges, etc..., (Figure 3). Vegetation maps of pasture types enable the potential resource to be evaluated as well as the shepherd's perception of biomass from his dairy notes recording the preferences expressed naturally by the sheep, e.g "...that's good grass, the animals go to it spontaneously and graze peacefully" or "this is hard grass, the animals only accept it reluctantly". The boundary of the compartments delimited by the shepherd are also marked on the map.

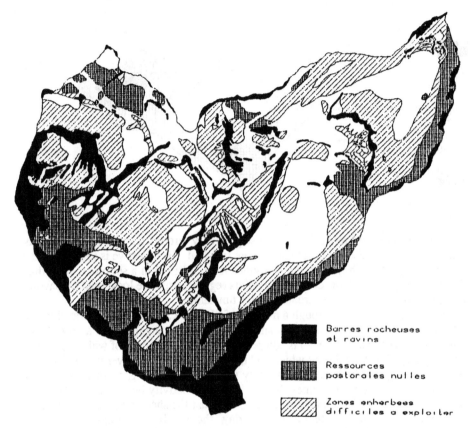

Figure 3 - The Saut du Laire alp, obstacles to the movement of the flock of sheep

The daily system. In determining the route for the day, the shepherd takes into account the needs of the animals. These can be expressed as a sequence of activities of the following type: move off, eat, rest during the middle of the day, move on, eat, sleep at night.Stops at watering point and when and where salt is distributred are also considered.

Figure 2   shows the sequence of activities for a typical day.Resting and sleeping places are given a priori or are determined during the day by the shepherd. Watering points, the location of salt licks, and obstacles such as narrow passages or bridges which make movement difficult are fixed whereas the areas actually grazed may vary depending on the state of the vegetation and the way the sheep spontaneously converge on a particular spot. The areas grazed by the sheep are therefore not necessarily determined in advance but emerge in the course of the day-to-day use of the mountain.

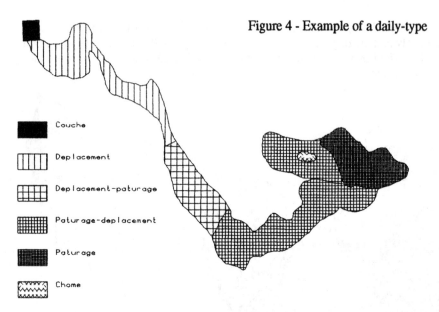

Figure 4 - Example of a daily-type

Couche

Deplacement

Deplacement-paturage

Paturage-deplacement

Paturage

Chome

Each day, the shepherd moves the sheep sequentially through a series of sectors thus ensuring that their needs to graze and rest are satisfied. His objective is to determine a route which enables these different activities to be fulfilled in a set order. It is thus neccessary for him to connect the fixed points (eg infrastructure locations) passing through particular areas in accordance with a timetable.

The pattern of movement through a number of SECTORS each DAY is termed here a DAILY CIRCUIT. The particular suite of activities engaged on a given day constitutes a daily-pattern which is composed of EPISODES, each of which is defined by the duration (TIME) during which and the area (POLYGON) over which the flock of sheep is engaged in a specified activity (eg grazing, moving, resting, etc.). Each episode is described by the times that the particular activity begins and ends: every change in activity defines a new episode. Given that the sheep may pass through the same area more than once during the day, the result cartographically is the superimposition of polygons. Alternatively, whenever the flock breaks up into smaller clusters the polygons become separated.

Figure 4 shows a typical daily circuit as recorded on a map by the shepherd as part of his dairy. The deployment of the sheep during each activity period is outlined on the map and the starting and ending times are shown. In addition, for each episode the shepherd makes an estimate of the portion of the daily intake of food ingested and records the behavoural and other events that occur during the circuit, for example intervention by the shepherd or his sheepdog and changes in the weather.

These descriptive facts are captured in the data base in a way that ensures the correspondance between the episodes describing the activities of the flock, the polygons delimiting a given activity space, and the day in question. The outlines of the polygons and the paths of the circuits are stored digitally. The data base thus contains the following entities: (EPISODES: - episode ID, - date, - polygon, - start time, - finish time, - activity type (eg grazing, moving, resting), - % ingestion by the sheep, - distribution of salt (yes, no), - intervention by the shepherd (yes, no), - weather events); POLYGONS (- ID, - sector, - area, - length of primary axis); DAYS (- ID, - weather conditions, - number of sheep).

Daily circuits are reconstructed for a particular day from the suite of episodes, each one of which is bounded by its start and end time: the end of one episode automatically becomes the start of the succeeding one. In this way, layers of information are built up for each day or a defined portion of it. Each coverage comprises the collection of polygons relating the the daily circuit together with attributes indicating what the sheep were engaged in at a particular time, the condition of the vegetation, etc.. Using this information it is possible to check that the reconstructed pattern matches a daily-pattern and to calculate its dimensions in terms of duration, area, grazing intensity (e.g number of animals by units of area and time), and total food intake.

2.2 The management of periods of grazing

The problem the shepherd faces is to work out the daily circuits for the period during which a particular compartment is to be used. He has to devise a strategy which will more or less enable the same daily circuit incorporating more or less the same sectors to be used for a number of consecutive days. To solve this, recourse is made to another level of spatio-temporal partitioning and sequencing of activities that is tied to the availability of resources.

A SUB-PERIOD is defined by the number of consecutive days during which a daily circuit involves the same area. Generalisation of all the daily circuits that utilise the same portion of space and time constitute a CIRCUIT TYPE. The concept of a circuit type enables each daily circuit to be fitted to a model regardless of the fact that a different set of areas may be involved. A number of reasons can give rise to this.(Figure 5 - A circuit type)

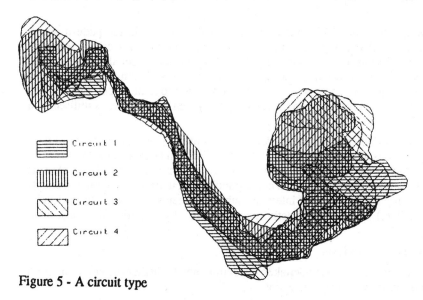

Figure 5 - A circuit type

Variations within a circuit type arise because of the rules the shepherd follows in deploying the sheep. Each day, the shepherd makes sure that the sheep are able to graze, in part, on a new area of pasture. Thus even within the same circuit the area grazed is slightly adjusted from day to day. The shepherd also avoids repeatedly trampling over a particular zone by varying the paths of convergence of the sheep. In this way, the spatial pattern of grazing is slightly altered even when the succession of activities retains the same temporal sequence. As a result, the boundaries of particular episodes vary although the suite of activities remains the same.

Variations in the strategy adopted by the shepherd are also based in part on the sequencing of episodes with respect to a pre-planned circuit and in part on occasional deviations from these, for example making a detour to access a slender resource thereby ensuring that only a limited number of paths are taken. In this way although the suite of activities remains identical, the area utilised varies and only corresponds in a global sense throughout the day rather than from hour to hour. The addition of another activity and hence a different area which supports it also results in variations.

Alternative strategies are also neccessary when the needs of the sheep cannot be completely satisfied on a particular circuit, for example if it lacks a watering point or suitable resting place. The shepherd is then obliged to couple together parts of two circuits. As a result, the circuits are not always identical every day even though they may generally correspond during a number of consecutive days. Typically, the suite of activities defining a circuit is repeated every 2-3 days within the same area.

Circuits and grazing patterns can also be disrupted by weather conditions - the interuption of grazing to seek shelter during a storm for example, and by other external factors. These can be either planned - returning the sheep to the fold for inspection by their owners from the village, or accidental - when the sheep are scattered and the circuit interupted by the intrusion of tourists.Circuits may therefore begin as planned, involving the usual sequence of activities and encompassing the same area, but are subsequently modified or even truncated as a result of these kinds of disruptions.

It is important to be able to differentiate between variations in the pattern which involve the same circuit type as opposed to the introduction of a completely new circuit pattern. Variants of the model are able to distinguish between these. Whenever the correspondance between variations in the spatial and temporal patterns is sufficiently strong it is hypothesised that they represent the same circuit type. In this way accounts can be kept based on the period that a given grazing area is utilised on a particular circuit type.

The collection of daily circuits undertaken throughout the season, together with their associated episodes, constitutes the CALENDAR. This can be represented in the form of a table showing the intensity of grazing activity by sector and day (Table 1). This can be broken down into components corresponding to circuit types. Partial accounts can also be compiled - eg by time intervals (sub-components of a circuit type), spatial divisions (sectors), or by type of activity engaged in - by summing the constituent episodes.

2.3 Methods of analysis

Two types of analysis can be undertaken depending upon whether the emphasis is on the spatial or on the temporal dimension.

Spatial analysis. The aim of this is to compute accounts by areas of the known availability of resources and to compare these with the potential resource as estimated by the shepherd. It is also possible to attempt to uncover the rules linking a particular activity with a particular area from an understanding of its characteristics for supporting a given type of activity. This can be done by overlaying the daily circuits and calculating the following for each episode: - the number of "passages"; - the total duration; - the level of frequentation (ie number of sheep x duration); - the grazing pressure (ie number of sheep x duration/surface area); - the level of intake (ie %ingestion x number of sheep x unit intake).

This involves a number of technical questions relating to the overlaying of coverages and data base management that are difficult given the limitations of currently available GIS software. Several solutions to the problem are possible depending on whether the representation of the data is in vector or raster form.

The vector solution is based on the construction of a generalised boundary (cartesian product) from the overlay of the coverages corresponding to the all of the episodes that make up a daily circuit. The problem in doing this is that slivers are generated which do not bear any relation to the original data. The management of the data is at the level of the attributes of the polygons. In addition, when partial accounts are to be computed, either by sub-period or by activity type, it is necessary to create new attributes. This is a complicated procedure and one that is at the limit of current technology.

A variation is to create a spatial level that is intermediate to the polygons and sectors called an episode-type and then assign to each episode the corresponding episode-type rather than the actual area within which is occurs.Although this involves the simplification of the original polygons, it has the advantage of not generating slivers.Part of the original information is lost and the procedure involves some preliminary work in delimiting the episode-types and a considerable amount of recoding of the data. However, the resulting visualisation is satisfactory (classic maps) and managing the data is at the same level (attributes for polygons).

A third solution is to use a raster approach by overlaying a fine grid over the study area. In this way a cube of data is generated, the x and y dimensions of which represent space and z the logical sequence of the episodes. Managing the data is fragmented - since there are as many grid cells as there are episodes their temporal sequence has to be preserved. The raster approach simpler and more straightforward.

An intermediate solution is to use a digital terrain model based on triangulation (TIN) to approximate the boundaries of the episodes. This approach makes a lot of sense because the digital terrain model reconstructs the morphological features of the area (eg crest lines, breaks of slope, micro-concavities, etc.) which are often the determinants of the way space is used.

These different approaches are more or less adapted to the characteristics of the data: an episode occupies a small area, successive polygons are planar, and the same polygon can be associated with several episodes. They operate on layers of information, each of which represents an episode or a daily circuit. Space is physically represented by overlaying layers of data. Time is reconstructed for the daily circuits from the contiguous activities. In addition, succession is not modelled in the data but by preserving order in the analysis. The ideal solution is to find an alternative way of representing the data which explicitly takes order into account - that is, a band which unfolds in space and time.

The accounts by sub-period, either by activity type or by sector, are created in the data from intermediate calculations. The results can be mapped and compared with other cartographic representations of the data. In this way it is possible, for example, to study the pasture zones at the beginning of the season or the intensity of movement along the paths of convergence in the compartment used in August. The global picture of accounts of ingestion constitute the intake of vegetation by the sheep. This can be compared with the potential vegetation, that estimated by the shepherd, or with the maps of obstacles or those used for planning.

Temporal analysis. In this, the object is to reproduce the sequences of daily-patterns and circuit- types from a knowledge of the episodes and daily circuits. It is the combination of these logical sequences and the analysis of the topologiocal paths to achieve this which constitutes the basic aim of the research. Extracting the rules by which the shepherd moves the sheep from place to place would make it possible to simulate other solutions and to create alternative grazing strategies for use by other shepherds.

In doing this, however, it is necessary to identify in the suite of episodes during a day the existence of a daily-pattern even when certain events are suppressed, inverted, or added. Similarly, for the daily circuits it is necessary to identify that they belong to the same circuit type by daily cycle even if their starting and finishing times are offset or they are of different duration.

Representing the timing of episodes does not present any problems. Time bounds each change of state. An episode is defined by the homogenous comportment of the sheep; it has a known starting and finishing time, and each is rythmed by the sequence of days. This makes it a totally ordered structure - episodes follow each other hour by hour and the daily circuits day by day.

The structure of the data discussed above means that it is possible to represent the sequence of episodes during the day by the contiguity of the polygons in a coverage. To represent the sequence of daily circuits it is necessary to represent them by the sequence of overlayed coverages.

The analyses consist firstly of describing the observed circuits followed by a second stage in which other possible circuits are simulated. For this, it is intended to use grammatical analysis and graph theoretic constructs. Conceptually, a daily circuit can be described in the form of a "word" each letter of which stands for a particular activity. The analysis of the words generated in this way based on the original observations makes it possible to formulate grammatical rules which can then be used to generate new "words" corresponding to other possible situations. The daily circuits corresponding to a circuit type are the repetitions permitting the analysis. The reconstruction of the calendar can be conceived as a chain which is based on the concept of bands of time and space discussed above.

It is necessary, however, to take account of the topological relationships of the paths traversed by the sheep - the polygons are contiguous. Therefore a neighbourhood constraint must be introduced to take account of contiguities in the graphs. In this way it is possible to formulate the problem in the following way. Construct a double chain, the components of which are firstly, a sequence of activities in which the order is preserved compatible with the daily-pattern which can be realised in parts of the area making that activity possible. Second, construct a sequence of polygons in which the graph of contiguity is preserved and which permit the realisation of the corresponding activity. Since there is a finite number of activities and areas, even after taking account of the evolution of charactersitics through time and in particular of the vegetation.

3 Conclusion

The practices of a shepherd managing the grazing of a flock of sheep have been observed and an attempt has been made to formulate these at the different scales of space and time that are involved.It has been possible to illustrate some of the methodological difficulties encountered in doing this and to suggest ways of specifying the necessary properties to represent and analyse the differrent concepts involved. The interest in such application is thematic: an attempt has been made using new tools and conceptual thinking to grapple with the problem of articulating at different levels of time and spatial scale phenomena that have inbuilt logic and which mutates from one level of organisation to another. Other studies have already attempted to tackle these problems either through the use of methods of analysing spatio-temporal facts but without taking into account the topological properties of space (Hubert et al, 1989), or by representing phenomena in space using daily accounts based on predefined time and functional sequences (Meuret et al, 1992). Most studies of processes have been based on changes of states in spatial structure (Deffontaines and Lardon, 1989) or have been based on the sequence of functions in a given space (Guerin and Bellon, 1989) without combining at the same time both structure and function and space and time.

The application domain and generalisation of the analysis discussed here is very large. It involves the class of problems of calendars in the utilisation of natural resources and the driving of production processes from the level of the farm to that of rural community (Osty and Lardon, 1992). The analysis of the way individuals behave in space, whether it be animals in the wild hunting for food or consumers in towns shopping for goods and services, can be just as easily studied using the above approach.

One of the primary goals of this reseacrh has been the emphasis on the handling of time in spatial enquiry. On the basis of the discussion of the conceptualisation of time and the explicit and implicit application of temporal concepts in the empirical part of the paper, a classification based on a tentative exploration of the temporal and semantic relationships between processes as they occur in reality and the way these are observed may be proposed.

3.1 The timing of spatial units

Random observation:

Few observations: The cartesian product of time layers enables basic operations to be performed (union, intersection, etc.) to establish differences between states as well as operations that restore changes and trajectories in the collection of states - at least within the framework of the assumptions about validity that should have been carefully established.

Many observations: Technically equivalent to the following category because of the abundance of polygons resulting from generalised intersection.

Referenced observation

Static units : Changes can only be explained in terms of attributes and their values. Changes in the values of attributes are expressed in the form of multiple attribute values at points in time, for intervals of time, and in a table of versions of states. Changes in attributes per se are expressed by the appearance and/or disappearance of a system of description which involves the conceptual problem of missing values and the translation between systems of description.

Dynamic units: two situations can be identified.

(a) Reconfiguration (partitioning and regrouping) - the example of the temporal cadastre, a process of a collection of discrete states with semantic temporal relationships. The disappearance of a feature generates new ones (subject to the constraint of completeness). What are the decision criteria for handling new separations using transformations based on constant identifiers and mutations producing new identifiers? Maintaining the cartesian products of all versions in real time is sufficient to permit analysis of and transition.

(b) Displacement - the object changes its spatial position and form. Two situations may be envisaged..

(1) The way that observations are taken produces space-time positioning of disjoint objects (or contiguous objects in the case of continuous displacement). Time is thus discretised in organic relation to the phenomenon into disjoint intervals. This is the case of a flock of sheep. Each period of uniform activity engaged in corresponds to an enveloping polygon bounded in time and space that is spatially linked to the events that take place before and after it. This creates a time-space object that has particular properties - the spatial and temporal contiguities are interdependent and, by implication, reciprocal. Thus it is possible to envisage topological relations linking them in which planar space is replaced by planar time-space.

(2) Methods of observation which do not reconstitute this planar time-space. An example is the monitoring of a cloud of atmospheric pollution or a swarm of locusts by taking observations at regular intervals. Neither time nor space are divided into disjoint intervals. Consequently the problem is again that of solving the cartesian product of relative positions.

3.2 A classification of query types

It should be possible to address these in absolute as well as relative time.

States referenced at instant t. A precondition for this is that time is structured at least by total order in which the intervals of temporal validity are associated with objects. This is the case of reconstructing the cadastre for a particular date from records of all the changes in ownership through time. Hence in this case the state is a planar topological coverage overlaying the spatial reference which becomes the locus of the process and its collection of states. In the case of the flock of sheep, the result is of a different kind because the unit of observation is the flock itself behaving in a static space. Only the polygon corresponding to the position of the flock at time t describes its state.

The lineage of an object. This is described by the table of transition states provided that the object itself exists. It is assumed that the collection of the states of a process reveal the qualities of the object. If the collection of states is extended to other semantic objects descending from the first, for example new land parcels which have a static spatial intersection with the first, then one returns to the situation of the lineage of space.

The preconditions for this are that time is totally ordered, the intervals are associated with objects, and that there are relations between descendants. Space only exists in terms of its identifiers - the history of ownership in the case of land parcel or the history of activities engaged in by a flock of sheep. The analysis of a subset of mutations, analogous to transitions, without spatial conditions is based on the characteristic linkages in transition matrices. From an operational viewpoint, this involves Markov methods, transformational grammars or space-time critical path analysis. These fall within the same formal universe as approaches to decoding DNA. It is the timetables, more or less clearly represented spatially, that need to be represented.

The lineage of space. The preconditions required for this are the relations between descendants together with spatial operators and spatio-temporal operators (spatial operators within a time interval creates problems of non-planarity). Examples are the history of a cadastral section or, in the case of the flock of sheep, the history of the utilisation of pasture resources. This is analogous in filmaking to zooming on a static portion of space.

Multiple processes and interactions. The flock of sheep eat the grass which has its own dynamism tied to the seasons and meterological events. The state of the grass is therefore the sum of two processes at any given moment in time. In this case, it is the coordination of processes that need to be addressed.

Cyclical processes. Treatment of these assumes two interrelated times - that of cycles considered as parallel intervals having the same time origin and then physical time in which the succesive effects of the cycles are accumulated.

3.3 Prospects

The lack of a spatio-temporal index necessitates experimentation with various partial solutions. These include the *a priori* systematic division of formal space thus simplifying the evolution of its structure using, for example, a grid system, TIN, or other user defined units and which returns to the case of eternal spatial units in which it is only the values of the attributes that change.

An alternative approach might be to derive a conceptual data model based on the cartesian product of time and space. Although formally coherent, this approach quickly becomes limited by its complexity and the performance of existing data base models employed in GIS. There are also problems of managing updated permanent cartesian products reconstructed for each delayed modification. A practical solution appears to be possible only in the case of the analysis of trajectories of converging processes. When there is effective updating (as a result of which the information is modified) propogation of the updates becomes difficult. More specific approaches based on data models designed specifically for time-space analysis need to be explored including temporal object-oriented data bases and those incorporating extended relations.

As Frank (1991.) has shown, the object-oriented approach is particularly promising. It has been demonstrated that this is capable of handling the integration of diverse spatial representations and its application conceivably can be extended to handle the representation of change provided that a method can be found to ensure the coherence of spatio-temporal integrity. Newer, more experimental object-oriented approaches - in exploratory situations and without having *a priori* knowledge of them, could possibly be developed based on other criteria than objects. One remaining problem is the dependence of data models on applications (local concepts) and the lack of a method for harmonising exchange and the transfer of information. Only a complete formalisation of spatio-temporal processes will ensure in time the emergence of generic appraoches to the problem.

The combination of observations and their formalisation in models of knowledge representation and simulation provide a solid base on which to develop decision support and reasoning systems for more effective environmental management.

Acknowledgements

This research is supported by a grant from the CNRS as part of the French Environment Programme (96 10 30).

Thanks are extended to J.P.Deffontaines for his helpful comments and constructive advice on the research; to André L., the shepherd, for sharing with us his knowledge and experiences, to P. Thinon and I. Savini for theit help; and to Barry J.Garner for undertaking the task of translating the manuscript from French into English and his perceptive and relevant comments.

REFERENCES

1. Allen J.F. 1983 «Maintaining knowledge about temporal intervals» CACM, vol 26(11) Nov.
2. Barrera R., Frank A., Al-Taha K., 1991 «Temporal relations in GIS, a workshop at the University of Maine», ACM, SIGMOD, Vol 20, N° 3.
3. Cheylan J.-P. Desbordes-Cheylan F. 1980 «From data sources to questioning in temporal data bases with cognitive aims», in Data bases in the Humanities and social sciences, North-Holland
4. Cheylan J.-P. Desbordes-Cheylan F. 1982 «Banques de données et informations administratives pour une observation continue des espaces péri-urbains» in Cahiers de l'OCS, Vol 10, ed: CNRS.
5. CHEYLAN J. P., DEFFONTAINES J. P., LARDON S., SAVINI I., 1990. «Les pratiques pastorales d'un berger sur l'alpage de la Vieille selle : un modèle reproductible ?» Mappemonde 90/4, pp 24-27.
6. DEFFONTAINES J. P., LARDON S., 1989. «Grasslands and agrarian systems. methodological considerations on space in the management of grasslands.» in 'Grassland systems approaches. Some french research proposals.' Etudes et recherches sur les systèmes agraires et le développment. INRA Paris, pp 209-218.
7. Deffontaines J.-P., 1992, «Berger et troupeaux dans le Parc des Ecrins», Cahiers Agriculture, 1, 278- 282.
8. Frank U.A., 1991, «Properties of geographic data: requirementsfor spatial access methods», in Advances in Spatial Databases, Günther O. and Schek eds, LNCS, Springer.
9. GUERIN G., BELLON S., 1989. «Analysis of the functions of pastoral areas in forage systems in the mediterranean region.» in 'Grassland systems approaches. Some french research proposals.' Etudes et recherches sur les systèmes agraires et le développment. INRA Paris, pp 147-156.
10. HUBERT B., MEOT A., HAVET A., LASSEUR J., COPPEL B., 1989.«Grazing systems and flmock management : patterns of land use practices revealed by a holistic approach». in 'Grassland systems approaches. Some french research proposals.' Etudes et recherches sur les systèmes agraires et le développment. INRA Paris, pp 157-169.
11. LANDAIS E., DEFFONTAINES J. P., LEROY A., 1988. «Un berger parle de ses pratiques.» Série Doc de travail INRA, Versailles, 111p.
12. LANDAIS E., DEFFONTAINES J. P., 1989. «Analysing the management of a pastoral territory. The study of the practices of a shepherd in Southern french Alps.» in 'Grassland systems approaches. Some french research proposals.' Etudes et recherches sur les systèmes agraires et le développment. INRA Paris, pp 199-207.
13. Langran G., 1992, «Time in GIS», Taylor and Francis.
14. MEURET M., MIELLET P., MAITRE P., MAZUREK H., 1992. «Diagnostic sur une pratique de gardiennage de troupeau caprin en milieu boisé». in Buche P., King D., Lardon S. Editeurs, 'Gestion de l'espace rural et système d'information géographique'. INRA, Paris, pp 109-119.
15. OSTY P. L., LANDAIS E., 1993.«Fonctionnement des systèmes d'exploitation

pastorale». Rapport au 4e Congrès International des terres de Parcours, Montpellier, France, Avril 1991.

OSTY P. L., LARDON S., 1992. « Systèmes techniques et gestion de l'espace. Les élevages ovins du Causse méjan (Lozère).» in Buche P., King D., Lardon S. Editeurs, 'Gestion de l'espace rural et système d'information géographique'. INRA, Paris, pp 39-51.

16. Snodgrass, R. 1987: «the temporal query language TQuel» ACM trans. on Dadabase Systems, vol 12, pp 247-298.

17. Vieu L., 1991, «Sémantique des relations spatiales et inférence spatio-temporelles. Thèse, Univ P. Sabatier, Toulouse.

18. Vrana R., 1989 «Historical data as an explicit component of land information systems» I.J.G.I.S., Vol 23, N° 1.

19. Yeh T.S., Viémont Y.H. 1991 «Temporal aspects og geographical databases. EGIS 91

The Cognitive Structure of Space: An Analysis of Temporal Sequences

Stephen C. Hirtle
Thea Ghiselli-Crippa
Michael B. Spring

Department of Information Science, University of Pittsburgh
Pittsburgh, PA 15260

Abstract. This paper discusses some methods for representing the structure of cognitive space as revealed through temporal sequences. The temporal sequences most commonly arise either through recalls of landmarks in memory experiments or through choice sequences in variants of the traveling salesman problem. Methods discussed include ordered trees, which capture hierarchical relationships, and path graphs, which include the explicit spatial information. The paper concludes with comments on the visualization and presentation of spatial information.

1 Introduction

The problem of representing time and space simultaneously has continued to be problematic as indicated by several recent conferences (e.g.,[4]), bibliographies (e.g., [1]), and workshops (e.g., [13]) devoted to this issue. Time and space depiction is inherently difficult due the multidimensional nature of simultaneously representing two (or three) dimensions of space with the additional dimension of time [10].

Furthermore, spatio-temporal reasoning can take on several distinct tenors, depending on whether the reasoning processes concern spatial or temporal processes, and whether the reasoning processes occur over time or over space. These two distinctions (process and occurrence) define four distinct research paradigms, as noted in Table 1 below. In the upper right cell (cell 1), one is reasoning about spatial processes over temporally distinct periods. This would include research on changes in land formation over time or the development of a city (e.g., [3]). The upper left corner (cell 2) is one in which reasoning occurs about space over space. This would be a typical wayfinding task, where one asks where should I go, now that I am here (e.g., [5,11,19]).

	Reasoning over space	Reasoning over time
Reasoning about space	2	1
Reasoning about time	3	4

Table 1. Classification of spatio-temporal reasoning

In the bottom row, the emphasis is on reasoning about temporal processes. In cell 3, one is reasoning is about time over space. This cell would include scheduling problems of the form: now that I am at location X, what temporal event should I initiate next? The final cell is one in which reasoning occurs about time over time. Here the problem is one of scheduling without any spatial component.

Some tasks transcend several cells in the Table 1 matrix. Finding the fastest route, rather than just the shortest route, would be reasoning about space (cell 2) and time (cell 3). In addition, all the reasoning tasks may occur in a static form or a dynamic form. In a static form, reasoning operates on a fixed representation of the environment, whereas in a dynamic form, reasoning operates on a continually changing environment. In the latter case, presumably the cognitive load is reduced by the addition of perceptual cues and reduction in memory load. However, it also difficult, if not impossible, to backtrack in dynamic reasoning tasks.

2 Background

In this paper, we are interested in how temporal sequencing through space reveals the cognitive structure that a subject has of that space. We begin by reviewing strategies for representing temporal sequences through space. In the next section, we propose several new solutions to the problem of representing temporal sequences, which consider both the problem of data reduction and the related problem of data visualization.

The problem of representing time and space simultaneously can be seen in Figure 1 from Parkes and Thrift [14]. Their daily "choreography" shows the complete information about one family's location for one day. While this record is complete, it is also difficult to read given the three-dimensional nature of the plot. Furthermore, the extraction of meaningful conclusions is even harder. In a very different domain, Yarbus [23] chose to represent eye movements while studying pictures as a two-dimensional trace. As seen in Figure 2, this information can be viewed as a projection of the choreography onto two-dimensions. Information is lost in this representation of the actual sequence of the eye movements, but the overall pattern of study and several repeated sequences, such as the shifting of attention from one area to another is clearly seen in the figure.

The conclusion from these two disparate examples is that it is often critical to reduce the data in a meaningful way, so that visualization (or representational) techniques can highlight important differences and subtle relationships.

3 Structure of Cognitive Spaces

The representation of time and space is an important, yet understudied, area within the discipline of cognitive science. The interaction of these two variables can provide insight into the structure of cognitive space. Whereas a computer scientist might be interested in using time/space representations to find an optimal wayfinding path, the cognitive scientist would turn the problem around and examine the non-optimal human-generated solutions. Such solutions can provide an indication of the internal representation of space.

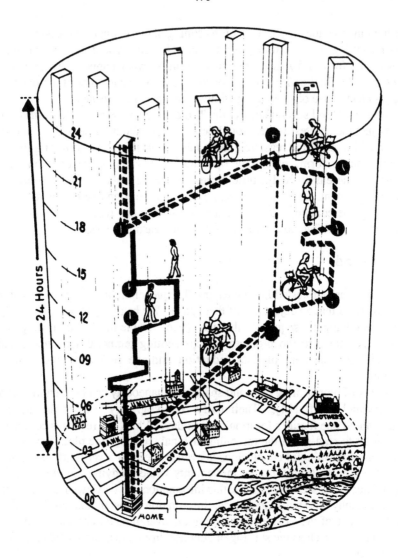

Fig. 1. Daily "choreography" by Parkes and Thrift (1980).

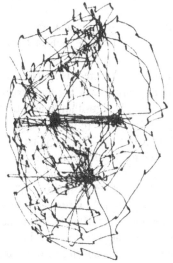

Fig. 2. Eye movement record for looking at a photograph by Yarbus (1967).

Below, we review two collections of previously reported data, both comprised of temporal sequences in a space. In the first experiment, subjects were asked to recall, repeatedly, one of three sets of cities in the United States [6]. In the second experiment, subjects were asked to solve a variant of the traveling salesman problem, in which the start and end locations were fixed [8]. Each of these datasets will be discussed, in turn.

3.1 Recall of Cities

The first set of data comes from a memory experiment, in which 48 subjects memorized, then recalled, one of three sets of U.S. cities [6]. Each of the sets contained 28 city names and were recalled 20 times by each subject. The three sets of cities varied in terms of the distributions of prominence and spatial locations. That is, the first set had cities of equal prominence and was evenly distributed across the United States. The second set had cities of equal prominence, but was heterogeneous in spatial distribution. The third set was heterogeneous in both spatial location and prominence of the city. One purpose of the experiment was to examine the role of clustering in cognitive maps, with the confirmed hypothesis that the third city set would result in a greater degree of clustering and that the clustering obtained would be more consistent across subjects.

The resulting recall orders were examined using two related data analysis tools: ordered trees and path graphs. Each of these tools is described in turn.

Ordered tree algorithm. The ordered tree algorithm was developed by Reitman and Rueter [15] to represent regularities in free-recall, particularly in verbal data. Hirtle and his colleagues have applied this algorithm to account for regularities in the recall of spatial landmarks in several studies (e.g., [6, 9, 12]). The algorithm, which has been described in detail elsewhere (e.g., [7, 15]), parses a set of linear orders, typically recall-orders, into chunks of items recalled contiguously. The chunks are then written as an ordered tree, in which the nodes at any level can be one of three types: unidirectional, bidirectional, or nondirectional. The ordered tree algorithm is implemented in a program called TIGER, written in RATFOR by Henry Rueter, and is available in an executable version for an IBM PC from the first author.

The ordered tree algorithm was applied to the sets of recall orders to produce one ordered tree for each subject. The resulting ordered trees could be classified into different recall strategies, including alphabetical and spatial strategies. For the present discussion, we focus only the spatial strategies. One such ordered tree is shown in Figure 3. Note that there is an overall strategy of recalling from the east coast to the west coast. It also appears that the subject, who comes from Pittsburgh, shows a richer internal structure in the eastern clusters (e.g., Boston, Buffalo, New York, Pittsburgh, Columbus), than in the western clusters (e.g., Denver, Albuquerque, Tulsa, Salt Lake City, Minneapolis).

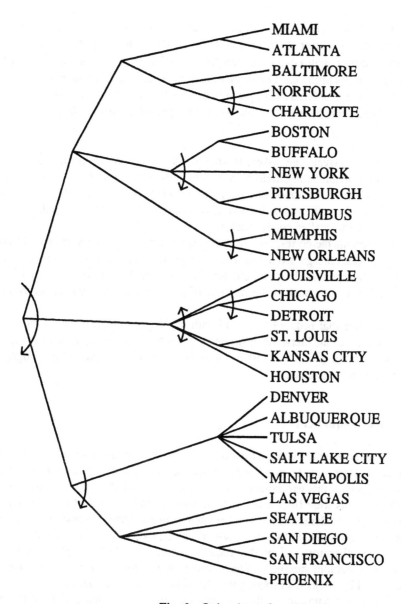

Fig. 3. Ordered tree for subject 7.

The ordered tree reflects the hierarchical and ordered structure of each subject's cognitive map. Additional research has shown that subjects judge within-cluster distances as closer than equally distant between-cluster distances [9, 12], and that within-cluster pairs prime each other more than between-cluster pairs in a recognition task [12]. Thus, the ordered trees appear to capture the cognitive structure of the space for each subject, which, in some sense, is independent of the spatial distribution of the points.

Path Graphs. Given the inherent spatial nature of the stimuli, can another alternative representation highlight the structure of the free-recalls? As one alternative, Hirtle [6] introduced the notion of a path graph, where the cities are taken to be nodes and the width of a path connecting two nodes reflects the number of trials in which two locations were recalled contiguously[1].

Figure 4 presents the corresponding path graph for the ordered tree shown in Figure 3. The clusters emerge from the visual inspection of the path graph by small connected subgraphs, such as the cluster {Seattle, San Francisco, and San Diego}. Other spatial strategies, such as the chaining of locations, are clearly seen, as shown in Figure 5.

It is our belief that ordered trees and path graphs complement each other. The ordered trees reflect the hierarchical structure, whereas the path graphs show the spatial distribution of the points. Neither alone is sufficient to extract the cognitive representation of the subject.

3.2 Choice Sequences

The second experiment was not a memory experiment, but rather subjects were asked to plan the shortest route from a fixed start location to a fixed end location [8]. There were 24 subjects in this experiment, each of whom solved 48 problems. Half of the set of 48 problems were filler items. The remaining 24 consisted of four replications of six key maps. Each replication was a rotated and/or reflected version of the original. The six key maps included two maps with 6 points, two with 10 points and two with 18 points. Furthermore, all of the maps were impoverished, simply indicating a location by a random number. The purpose of the experiment was to examine heuristics used by subjects in solving the traveling salesman problem.

In analyzing the data, unlike the previous experiment, we first collapsed the data across all subjects. In addition, we modified the path graphs to include order information so that any link which is traversed consistently in one order will be indicated with an arrow. One such graph is shown in Figure 6 for one of the 10 point maps. Note that the direct links from the start node and to the stop node must be directed, but that all other links may be either directed or undirected.

---

1. A program is available from the first author, written in Turbo Pascal, to generate Postscript code directly from the free recall data to draw the path graphs.

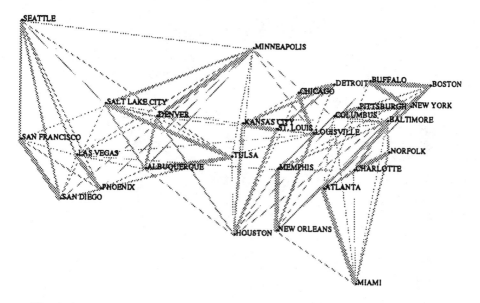

Fig. 4. Map path graph for subject 7 showing evidence of spatial clustering.

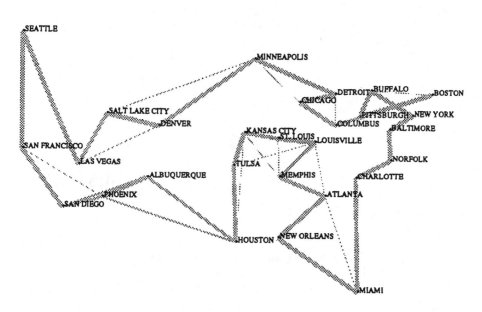

Fig. 5. Map path graph for subject 25 showing evidence of a unidirectional recall pattern based on spatial proximity.

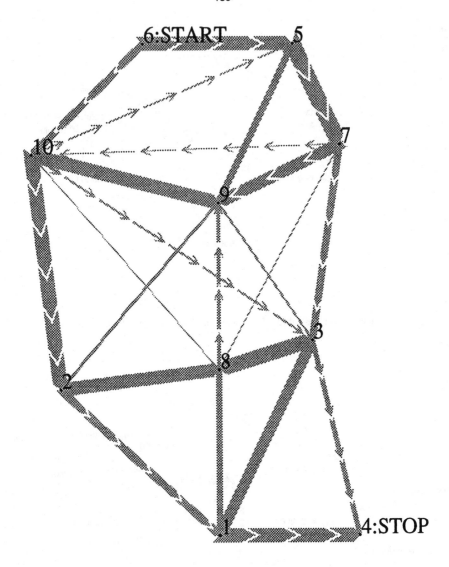

Fig. 6. Ordered path graph for Map 10a.

Overall, the graph indicates no one single strategy in solving this problem. However, we were able to parse the solution set into distinct substrategies using an extension of the ordered tree algorithm. The lattice-theoretic method, which is described in detail by Hirtle and Gärling [8], results in parsing the complete set of linear orders into like subsets. By looking at the path graphs for these reduced subsets, we can identify distinct heuristics. For example, one subset, shown in Figure 7, indicates that subjects treated the top five points as a cluster in addition to using a general "horizontal" sweeping across the points. Figure 8, drawn from another subset, indicates a general "vertical" sweeping[2].

4 Discussion

In this paper, we have discussed two methods of representing temporal sequences through space. The ordered tree algorithm is a deterministic method, which reflects the regularities of the repeated free-recalls without regard to the spatial distribution of the points.

The path graphs are based on the principle of scientific visualization by overlaying the temporal sequences on the true spatial locations. Unlike previous approaches (e.g., [14, 23]), we retained the order information within a two-dimensional solution to ease the process of assimilating the data.

The potential for analysis based on visualizing numerical data has been discussed by [2, 16, 18, 21, 22], among others. In general, the visualization of data endeavors to make greater use of the parallel processing mechanisms of the human visual system to present data and patterns. As Spring and Jennings [18] have pointed out, the research on human perception provides some support for decision making about the mapping of abstract data to various sensory dimensions. In the case of the figures presented above, all of the data has been mapped to a limited set of visual dimensions, i.e. location, shape, saturation. In conducting this mapping, important data dimensions should be mapped to visual dimensions which are processed quickly [20]. Equally important, attention should be paid to the absolute and relative discriminability of the sensory dimension in terms its mapping to the variability of the dataset. Finally, attention

---

2   The graphical methods used reflect the authors' interest in developing methods that are easily replicable in a wide variety of environments and across multiple media. To this end, we have relied on the capabilities of the Postscript Page Description Language which has the dual capabilities of being a widely implemented interpreter in laser printers, phototypesetters, filmsetters, and workstations. Equally important, Postscript is a capable full function programming language. The path graphs were all generated by simple programs that use two basic procedures. One procedure was written to place names at specified locations. A second procedure was written to draw lines of various types between two specified locations. These procedures may then be automatically applied to a variety of data sets using trivial driver programs to produce the needed Postscript output [17]. The use of Postscript has been extended by the authors to eliminate the need for driver programs. In the current form, simple Postscript prologs may be attached to ASCII data files which are absorbed at the Postscript output device and imaged.

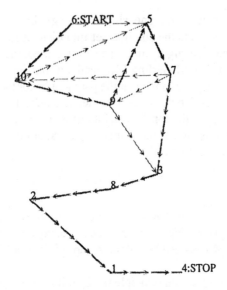

Fig. 7. Ordered path graph showing the cluster heuristic on Map 10a.

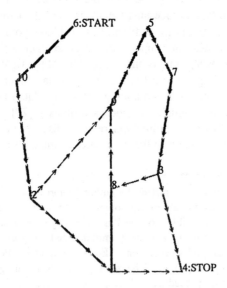

Fig. 8. Ordered path graph showing the zig-zag heuristic on Map 10a.

needs to be paid to the type of the data and the sensory dimensions. For example, while it might be argued that hue is an ordinal dimension, there is not an automatic ordering of colors for all individuals. (We recognize the relationship between the colors in a spectrum, but do not find red to be less than blue or green for most individuals.) Thus, hue is a natural dimension for data that are nominal in nature. On the other hand, saturation is a dimension that is at least ordinal, and would be suited to a data dimension that is of the same type. Thickness is a subdimension of shape that would appear to have interval and possible ratio characteristics and would be a good choice for comparable data. Within this framework, we are beginning to look at additional data visualization techniques that expand either the number of locations and paths, or the number variables to be plotted.

References

1. Al-Taha, K. K., Snodgrass, R. T., Soo, M. D. Bibliography on spatiotemporal databases. (Available via anonymous FTP from cs.arizona.edu). (1993)

2. Defanti, T.A., Brown, M.D., McCormick, B.H. Visualization: Expanding scientific and engineering research opportunities. Computer, 22(6), 12-26 (1989)

3. Egenhofer, M. J., Al-Taha, K. K. Reasoning about gradual chages of topological relationships. In: A. U. Frank, I. Campari, U. Formentini (eds.): Theories and methods of spatio-temporal reasoning in geographic space. Lecture Notes in Computer Science 639. Berlin: Springer-Verlag 1992, pp. 196-219

4. Frank, A. U., Campari, I., Formentini, U. (eds.). Theories and methods of spatio-temporal reasoning in geographic space. Lecture Notes in Computer Science 639. Berlin: Springer-Verlag 1992

5. Gopal, S., Klatzky, R.L. Smith, T.R. Navigator - A psychologically based model of environmental learning through navigation. Journal of Environmental Psychology, 9, 309-331 (1989)

6. Hirtle, S. C. Knowledge representations of spatial relations. In: Doignon, J.-P., Falmagne, J.-C. (eds.): Mathematical psychology: Current developments. New York: Springer-Verlag 1991, pp. 233-249

7. Hirtle, S. C. Ordered trees: A structure for the mental representation of information. In: Advances in Classification Research, ASIS monograph series. Medford, NJ: Learned Information 1991, pp. 77-84.

8. Hirtle, S. C., Gärling, T. Heuristics rules for sequential spatial decisions. Geoforum, 23, 227-238 (1992)

9.   Hirtle, S. C., Jonides, J. Evidence of hierarchies in cognitive maps. Memory and Cognition, 3, 208-217 (1985)

10.   Mark, D. M. (September, 1992). Spatio-temporal reasoning and G.I.S.: A geographer's view. Keynote address of the international conference on GIS - From space to territory: Theories and methods of spatio-temporal reasoning, Pisa, Italy.

11.   McDermott, D. Davis, E. Planning routes through uncertain territory. AI, 22(2), 107-156 (1984)

12.   McNamara, T. P., Hardy, J. K., Hirtle, S. C. Subjective hierarchies in spatial memory. Journal of Experimental Psychology: Learning, Memory, and Cognition, 15, 211-227 (1989)

13.   National Center for Geographic Information and Analysis (NCGIA). (May, 1993). Workshop on Initiative 10: Spatio-temporal reasoning. Lake Arrowhead, California.

14.   Parkes, D., Thrift, N. J. Time, spaces, and places: A chronogeogrphic perspective. New York: Wiley 1980

15.   Reitman, J. S., Rueter, H. R. Organization revealed by recall orders and confirmed by pauses. Cognitive Psychology, 12, 554-581 (1980)

16.   Spring, M.B. Informating with virtual reality. Multimedia Review, 1(2), 5-13 (1990)

17.   Spring, M. B. Dubin, D. S. Hands on Postscript. Carmel, IN: Hayden 1992

18.   Spring, M. B., Jennings, M. C. Virtual reality and abstract data: Virtualizing information. Virtual Reality World, 1(1), (1993)

19.   Timpf, S., Volta, G. S., Pollock, D. W., Egenhofer, M. J. A conceptual model of wayfinding using multiple levels of abstraction. In: A. U. Frank, I. Campari, U. Formentini (eds.): Theories and methods of spatio-temporal reasoning in geographic space. Lecture Notes in Computer Science 639. Berlin: Springer-Verlag 1992, pp. 348-367

20.   Treisman, A. Attention, and object perception. Scientific American, 255(5). 114B-125 (1986)

21.   Tufte, E.R. The visual display of quantitative information. Chelshire, CT: Graphics Press 1983

22.   Tufte, E.R. Envisioning information. Chelshire, CT: Graphics Press 1990

23.   Yarbus, A. L. Eye movements and vision. New York: Plenum Press 1967

Hierarchies of Space and Time

P.A.Whigham

CSIRO Division of Water Resources
Canberra, A.C.T. AUSTRALIA

Abstract. The ability to have a clear representation of what has happened and where it has happened are basic to decision making and prediction. We investigate a structure to represent a class of problems that are slowly changing with respect to the viewer. A hierarchical structure for both time and the events associated with time is proposed. The concept of persistence for an event is introduced, and shown to be a useful tool for management and decision making. A simple linkage system is described which allows causal associations to be made between events. We conclude that this type of system is a useful addition to the current tools available for storage and representation of spatio-temporal concepts.

1 Introduction

Space and time are fundamental to our existence. The ability to have clear representations of *what* has happened and *where* it has happened are basic to decision making and prediction [3]. Most research into spatio-temporal representations has focussed on extending relational database models [7] and the issue of time as a point or interval [4]. The aim appears to be to create structures that can allow a *complete* set of queries. Inevitably, time is referenced as a numerical value to achieve these requirements. This paper approaches the problem from a different perspective, namely that a symbolic description of time, and events associated in time, will create an intuitively simple and clear system for a large class of problems. Although the system cannot handle many of the traditional queries associated with spatio-temporal structures, it supports the storage of information relevant to problems such as environmental management and land use monitoring.

2 Temporal Resolution

People perceive. This seems fundamental to any understanding of what we perceive and why we perceive it. If we consider what it is to perceive time, comments range from the theoretical physicist to the poet. Time, and our perception of it, is *relative* [9]. This point seems even clearer when we consider how a person structures and views a set of events, such as a typical work day. The fact that the temporal metaphor *work day* was used to describe the collection of events for discussion shows our acceptance of time as a *perceivable object*.

Time appears as a structured object. We may view a series of events as a group of differing seasons., or as the week and week-end, or on an hourly basis, building into days, weeks and so on. However we structure events in time one point is *always clear*, there is an imposed order; before, simultaneous and after.

The imposed ordering of events leads to the description of a resolution for time. Some events (E_i) appear to be structured from events that are completely *subsumed* by some event E_j. For example, a working day may be described in terms of the morning, lunch and afternoon. The morning may be described as pre-morning-tea, and after-morning-tea. These events could be further described in hourly terms, or as the typical set of events that occurs during each of these extended moments. This duration we refer to as the *temporal resolution* of an event.

Our use of metaphors for time and temporal association leads use to conclude that a system which embodies concepts of time must allow a mixed representation. Time, and its description, cannot be fixed.

3 Buffered Systems

Time and action seem inseparable. It is difficult to imagine an event without associating a value (or set of values) for its temporal relation to the viewer, and other events which we perceive to be *related* to this event. Many events that we find of interest have different resolutions, and in particular, many natural events appeared buffered in terms of our view of their dynamics [8] . The changing condition of an erosion gully, viewed from the managers perspective, takes time. On a daily basis there does not appear to be a great deal of change, however over the period of several years a hillslope and creek system may be turned into an unpassable area of gullies and ruts. The question of representing this type of change becomes important, both in a management and historical context.

We will consider a representation for those events that appear to change slowly in relation to the view of the world. In other words, where the change in the system is slow in comparison to the events of the system. The system may be described as buffered - changes of significance are infrequent but important. Hence the system developed to represent these ideas would not require large numbers of updates over a small period of (viewers reality) time.

4 The Reference Background

An event appears to have significance due to its relation with other events and the overall background of *current conditions*. These current conditions may be viewed as the static background over which change occurs [7]. The implications of this description are exploited in most fields where a large amount of changing *information* is being described and/or transmitted. The use of data compression algorithms is fundamental to the techniques used by devices such as fax machines and graphical displays. Measuring change, rather then an entire image, allows many types of data to be represented in an efficient way [1]. In a similar fashion, we are interested in describing a slowly changing system - dynamic in space over time. The spatio-temporal background may be as simple as a spatial database (ie. a map showing significant positions for reference) that captures the current condition for some past moment.

The temporal background is associated with the events that occur at some spatial location - the background is defined by the way the user has perceived the structure of time and the objects that are referenced to these structures. There may be the need for several reference backgrounds as the number of recorded events become large - the overall description may be condensed into a new *current condition*. The storage of temporal events appears to be totally defined *if and only if* the current condition (ie. background reference) and events are shown simultaneously.

5 Ordered Hierarchies

We have described temporal events without describing how these events are represented. To satisfy the description of time that we have introduced, a hierarchy of events must be recorded. A hierarchy (such as a family tree, or the note duration system used with musical scores), has the useful properties of *order* and *inheritance* [12] . A (child) member further down a branch of the tree is *deemed later* in temporal position to its *parent* nodes. The child, however, is intimately *related* to the parent nodes. Its existence and meaning are only fully described by enumerating the tree which defines it. These type of events may,however, be treated in isolation, and typically the viewer perceives the events in an unattached way. The structure of the hierarchy becomes part of the perceived structure of the event. A system that is used to store events must allow the user to create their own description of time, and to base queries on this structure. We therefore embody the viewer as an integral part of the description of time and space. An example of some temporal and event definitions will make this point clearer.

6 Temporal Objects

We consider the simple problem of describing a farmers view of time. Typically they would be interested in the current season, and these would likely be grouped into years. Figure 1 shows the years 1993 - 1994 as described by this farmer. Note that as we trace down the hierarchy from TIME to Spring 1994, we trace a path through time. Figure 2 shows a different view of 1993-1994. The description could be used by an academic or student working in a university. Note that the way the years have been separated appear almost like objects rather then a temporal description. The distinction between time and action has become vague. These examples do not yet capture *what* has happened during these temporal moments, only that these *divisions of time are significant*.

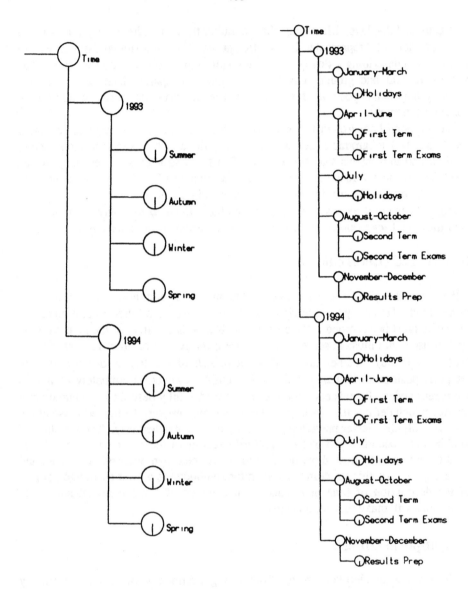

Fig. 1. Seasonal View of Time

Fig. 2. The Academic View of Time

7 Event Objects

Describing an event allows the connection of our temporal divisions to actions or conditions that are to be recorded. An event appears just like a temporal description - it is part of a hierarchy [6] . Figure 3 shows the structured event *Erosion Control*. This high level task is further broken into the classes Stream Control, Drainage, Road Construction, Dam Constructing, Gully Stabilisation and Tunnel Erosion. The actual events that may be recorded are represented by the tips of this tree. Any tip may be sketched onto the background reference indicating existence for the selected time and location. Although the tip of the tree defines the event, it is only fully explained by tracing the path down the complete tree hierarchy. For example, the event *Sediment Retention* is described by the path:

Erosion Control -> Dam Construction -> Gully Construction -> Sediment Retention

The context of each event is therefore shown by its location in the hierarchy, and its location on the reference background. This description is intuitively appealing, as time and events are represented by the same *language*.

8 Associating Events with Time

We appear to need two hierarchies when we describe an event in time - the temporal structure and the event hierarchy. Associating these two structures creates a tree whose tips are events, and subtrees may be either event or temporal structures. For example, if we link the *Erosion Control* tree to the temporal node *Summer* we create a structure that defines *what* may happen during a certain temporal division (ie. what may happen during *Summer*). The ability to sketch a location (ie. position and extent) onto a background reference map allows the user to describe *where* and *when* a significant event has occurred. This is different to the current trend of representing events using a GIS [2] , in that the description of the event *need not* be accurately referenced spatially to be a useful carrier of information. The emphasis on accurate, clean digitised data seems inappropriate when we consider the scale at which the events are being registered. One need only think of the farmer who deals with events such as *ploughing the bottom paddock* to realise the lack of necessity in creating a system which emphasises accuracy over general location. Figure 4 shows one possible system that allows these hierarchies to be defined. The temporal and event structures are shown in separate windows - the map shows the current events defined for the month of January.

If we displayed each tip of the event structure connected to each temporal node, a large number of events that no longer exist, or are out of *date*, may be displayed. Creating a new map showing *current condition* may not always be appropriate. The lifetime of an event (ie. its duration of effect) can be incorporated with each event to allow change in a dynamic fashion. This property of an event we refer to as *persistence*.

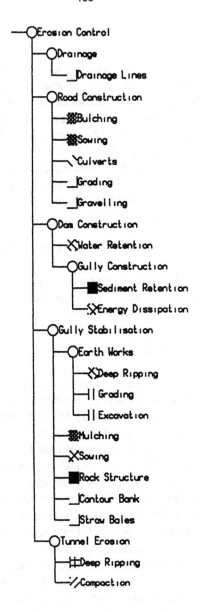

Fig. 3. A Structured Event - Erosion Control

196

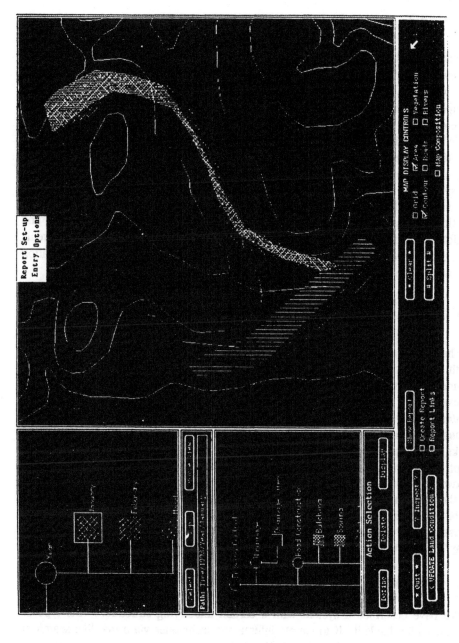

Fig. 4. A prototype system for spatio-temporal hierarchies

9 Persistence

Persistence is a property fundamental to temporal descriptions [4]. The *lifetime* of an event has meaning only in association with a temporal structure. This structure *is* the way time is referenced. Hence when an event hierarchy is connected to a temporal structure, the persistence of an object relates to a node on the temporal tree where this object no longer has significance. Note that although many nodes could satisfy this criteria, we would expect the most useful definition to be the highest position in the tree where this is true. A system which connects the various tips of the event tree to locations in a temporal structure may be used to represent the persistence of an event. When an event is defined, this predefined duration may be shown as further documentation towards the purpose and effect of the event In particular, event persistence may be used to show what conditions are *current*, and which conditions have *expired*. This is useful in a management context, as events that expire tend to indicate a need to update the information for the specific location (ie. what is the condition now ?).

10 Causal Association

The system that has so far been described allows events over time to be defined with certain spatial extents. There is no way in which two or more events may be connected or associated as having a linked meaning. What types of events are we considering ? A simple linked example may be that a location is described as badly guttered, and subsequently the area is ploughed. The events *badly guttered* and *ploughed* have an association - one has subsequently led (or *caused*) the other. This linkage may be represented by an arc that connects one event from a hierarchy to another event in the same, or a different, hierarchy. Note that the linkage is always between instances stored on a background reference. This linkage may now be used to show why a certain event has occurred, and forms another type of documentation to be associated with an event or collection of events. The recursive display of these causal links from an event instance shows a trace of spatio-temporal events which may be exploited for future decisions, justification of action, and training.

11 Inference

We have described a system with a certain underlying structure. This type of representation lends itself to simple inferencing techniques via a tree-like search of the temporal and event instances [12]. Although the system is not complete in a query database context, several useful operations may be performed which are meaningful to the user as manipulations of their personal description of events over time.

11.1 Simple Tree Search

Selecting a particular node in the tree and displaying all instances attached to the children nodes performs the operation

display everything subsumed by this temporal/event node

In other words, show what we know about this time and class of events. This operation is a basic tree search. Note that the inverse operation,

display everything which subsumes this temporal/event node

collects the events that have happened before a certain temporal division.

12 Combining Persistence

The use of persistence enables the tree search to include an indication of past and future effect from some current viewpoint. The operation

display everything which subsumes this temporal/event node AND
exists at this node

with inverses does not exist and nodes which are subsumed traces what changes have occurred, and which effects have not been updated.

13 Causal Linkage and Common Tree Structures

There are two main operations that build on the tree operations. They are the use of the *causal linkage* and *common tree* structures for display. Causal linkage may be used to show the links from an instance or set of instances. These linkages may be recursively applied to show all previous (or subsequent) events that have occurred based on some initial event.

Common tree structures imply when the same class of events (or division of time) occur in a parent or child of the current node. These common constructions indicate a repetition of conditions or work performed, and would be of interest to users wishing to display all events *of a particular type* from (or previous to) a current event.

14 Multi-valued Events

Our schema for representing spatio-temporal events has not yet considered the problem of representing multi-valued instances. The tree hierarchy structure is suitable for all class representation where the possible enumerations of an event class is finite (and low in cardinality). An instance that takes an *integer* or *real* value cannot be represented by a tip of the tree, unless this particular value is stored explicitly on the tree, which in many situations may not be possible or desirable. Hence, the tips must be capable of carrying more information.

One solution is to allow the instances (when applied to a background reference) to be given additional attribute values, similar to a vector-based polygon system which stores spatial data as a set of records [10] . This would imply that the instances are typed, so we could view them as a collection of objects with properties similar to parameters of an expert system [5]. This addition could lead to using the inferencing capabilities of the expert system to traverse the tree hierarchy. A similar approach has been used for creating inferencing systems such as *SPECS* [11]. This is beyond the scope of this paper, but is noted for future consideration.

15 Conclusion

We have shown that a large number of interesting buffered systems may be captured by a hierarchical definition for time and events, placed on a spatial background reference. The use of persistence and causal links adds to the overall information content of the description, and leads to tools for decision making and prediction. Although the system is not complete in its possible queries the symbolic nature of the representation is intuitively appealing and comprehensible.

Current systems for spatial information tend to emphasis the accuracy of the spatial data that they describe. Often, however, this is not required, or possible, for many problems in the real world. The hierarchical structure for time and events carries information both through its own components, and the instances which are attached to the background reference. The instances need not be accurately defined (in the way that GIS data is reference) to carry useful meaning. This allows the system to be relatively fast at data input (such as using the mouse to define the spatial extent of an event instance) which promotes use and, subsequently, useful data collection. The users view of time and events is formed from their own definitions. This makes the system a flexible tool for storage of diverse buffered events such as land management and general data collection.

The use of expert system concepts to enhance the storage of instances implies certain inferencing techniques which may be applied to the tree structure. These operations should be investigated.

The hierarchical system, as described, is a useful addition to current spatio-temporal modelling concepts. The user defined view of time and events allows the system to be flexible and meaningful to the user. The removal of the need for costly spatial referencing methods appears suitable for many types of problems.

Acknowledgements

The author would like to thank Andy Mckinnon and Ross Batten for invaluable discussions which led to this design. This paper has been partially funded by the Australian Army.

References

1.Abel, D.J. (1985) Some Elemental Operations on Linear Quadtrees for Geographic Information systems. The Computer Journal, Vol 28, No 1 pg73-77.

2.Albert,T.M. (1988) Knowledge-Based Geographic Infromation Systems (KBGIS): New Analytic and Data Management Tools. *Mathematical Geology*, Vol. 20, No. 8, 1988. pp 1021-1035.

3. Alexander E.R. (1982) Design in the decision making process.
Policy Sciences 14,3:279-292.

4.Allen,J.F.(1983)Maintaining Knowledge about Temporal Intervals. Communications of the ACM, Vol 26, No. 11, November, 1983.

5.Forsyth, R (ed) (1984). EXPERT SYSTEMS: Principles and case studies. Chapman and Hall, London , New York. 232p.

6. Hadzilacos T. and Tryfona, N. (1992) A Model for Expressing Topological Integrity Constraints in Geographic Databases. Lecture Notes in Computer Science 639. Springer Verlag. Frank,A.U. Campari I. and Formentinin,U. Eds .pp 252-268

7. Langran,G. and Chrisman,N.R. (1988) A Framework for Temporal Geographic Information. Cartographica Vol. 25 No. 3 pp1-14, 1988.

8. Papagno, G. (1992) Seeing Time contribution to the discussion on space time dynamic. GIS From Space to Territory: Theories and Methods of Spatio-Temporal Reasoning - additional material. Frank,A.U.,Campari,I. and Formentinin,U. Eds.

9. Salmon,W.C. (1975). Space, Time and Motion: A philosophical introduction. Dickenson.

10. Smith,T.,Peuquet, D. Menon,S., and Agarwal, P., 1986, KBGIS-II: A Knowledge-Based Geographic Information Syste: Dept. Of Computer Science, University of California, Santa Barbara, 36 p.

11. Stanton,R.B. and H.G.Mackenzie.Expert System techniques in spatial planning problems. Applications of Expert Systems. Proceedings, 4th Conference on Applications of Expert Systems, R.Quinlan, ed. University of Technology, Sydney, Australia. 1988. pp 57-66.

12. Wirth, N. (1976] Algorithms + Data Structures = Programs. Prentice-Hall, Inc. Englewood Cliffs, New Jersey.Chapter 4.

The Voronoi model and cultural space: applications to the social sciences and humanities

Geoffrey Edwards

Chaire industrielle en géomatique appliquée à la foresterie
Centre de recherche en géomatique
Université Laval, Sainte-Foy, Québec, G1K 7P4
edwardsg@vm1.ulaval.ca

Abstract. The Voronoi model of space is becoming more and more important as a tool in the mathematical modelling of space for many application domains. In the Voronoi model, space is neither an empty void within which can be found occasional objects (the vector model), nor a lattice of arbitrary cells (the raster model). Rather, space is a continuous medium filled with proximity fields generated by objects. This representation of space has important implications for domains where geographic space is endowed with cultural characteristics or values. The Voronoi model of space concords fairly closely with the perceptual and linguistic spaces of humans and hence Voronoi zones around objects are meaningful. The Voronoi model of space is also a closer fit to qualitative data representation and analysis than other models. Finally, the Voronoi model permits nested hierarchical relations between entities which increases the richness of the querying capabilities. It is shown that the Voronoi model favors a rich qualitative database of the kind which will be found in culturally intensive application domains.

1 Introduction

Although GIS has found widespread use in numerical applications of geography, ressource management and decision support, its use for more culturally oriented domains has remained relatively limited. What GIS presently has to offer is an object-based division of space (either discrete objects in empty space or tiled objects which collectively fill space) with an associated database and some fairly primitive spatial operators (overlay, corridor, point in polygon and assorted metric operations). These operations are of some use to the social sciences and humanities as an archiving tool (such as for characterising and mapping archeological dig sites, etc.), but its usefulness in the social sciences has been limited for more demanding analysis, because of several factors — a poor understanding of how these domains use space as well as technical difficulties for how to represent and exploit space in "non-linear" ways. Within the discipline of geography, some attention has been focussed on these issues (see [17] for a survey of some of this work). However, a great deal more work needs to be done to explore these capabilites.

For many culturally-oriented domains, quantitative modelling has little interest. Some features of interest might be quantifiable and indeed some researchers have been attempting to use fuzzy set theory to characterise such features, but these attempts are not unlike the proverbial round peg in the square hole. What is really needed is a tool which allows one to express qualitative values and relationships and to incorporate space into such an expressive structure. In fact, GIS may have a real potential as an "exploration" medium for cultural information, in ways not unlike "visualisation" software which is used in the physical sciences to help understand complex datasets.

Existing GIS, whether raster or vector, require a "freezing in" of the spatial nature of the data. Modifying the spatial data is relatively difficult, unlike the modification of the non-spatial database information [13]. These data structures affect the ways in which users "think about" space, mostly in limiting ways. Hence many GIS users are unable to imagine what another space model might permit them to do.

Experience to date with the Voronoi spatial model has supported this idea. In forestry, the Voronoi model of space has led Nantel [15] to reformulate the digitizing problem completely, with very successful results. The Voronoi model of space has also led Edwards [4] to rethink image analysis and to come up with new possibilities for the manipulation of images. The Voronoi model appears to allow for the integration of both raster and vector information within a common data structure [5]. Furthermore, the possibility exists to create a hierarchical map structure in such a way that parts of a map may be "folded away" until needed [5]. Most of these capabilities can be realised within existing (raster and vector) data structures (not necessarily easily, however), but many of these approaches to data handling and representation either would not have been considered reasonable or would not even have been imagined within the framework of these existing spatial models. The study of the Voronoi model of space has proven to be fertile ground for extending our ideas about handling and representing spatial information.

In section 2 we examine several features of the Voronoi model of space and their consequences for "qualitative" as opposed to "quantitative" modelling. The notion of proximal query first introduced by Gold [11] is explored in some detail. In section 3, we develop some ideas about how to use these features within the framework of the social sciences and humanities and give some examples. In section 4, we address other possible spatial operators based on the similarity between the Voronoi model of space and linguistic structuring of space. These ideas were first addressed by Gold [10]. In section 5, we discuss implementations of these ideas within the Voronoi data structure in comparison with vector and raster data structures and present overall conclusions.

2 The Voronoi Model of Space

A "model of space" is a set of characteristics associated both with space and with the representation of objects and attributes within that space. Each major data structure can be associated with a distinct model of space, however, the concept of model of space is more general than the concept of data structure. Indeed, most features of a particular model of space could be implemented in any of the standard spatial data structures. However it is not unreasonable to assert that a given data structure has a "natural" model of space. Many more models of space exist than do data structures which are used today, but we shall confine our attention to those which characterise the spaces of the most well-known data structure families, the raster and vector models. Other variants on these models exist (such as the quadtree for the raster, and the TIN for the vector) and they may affect the model of space fairly strongly, but this effect usually consists of increasing the restrictiveness of the model while adding somewhat increased functionality. Hence, for example, the quadtree retrofits a hierarchical structure onto the raster model of space, but may further limit the ability to define neighbourhoods than the pure raster model. Note also that this idea of model of space is not the same as "data model" as discussed by Peuquet [16], but is closer to "concept of space", discussed by [2].

The Voronoi region of an object is quite simply the region of space closest to the given object than to any other object in the space. Voronoi regions can be determined for any arbitrary shape. Most commonly they are generated around points. Recent work has seen the extension to line segments, curves and faces [1]. The set of Voronoi regions for a set of objects in space is called the Voronoi diagram for the space and objects (it is also called a Dirichlet tesselation, and Voronoi regions are alternatively called Thiessen polygons). If the existence of a Voronoi boundary between two given objects can be established, then these objects can be said to be neighbours, and we can say they are adjacent. If the adjacency relations are represented by a line segment connecting the objects, then the set of all such line segments will form a Delauney triangulation (for a set of points), called the dual of the Voronoi diagramme. What we are calling the Voronoi model of space here consists of a set of objects (half-lines and points) in two-dimensional space (for the time being), their associated Voronoi regions and the dual which contains the information on adjacencies. The Voronoi regions around the objects are here also called proximity fields. Objects are defined as consisting of collections of half-lines and points. We use the term half-line to denote a side of a line or, alternatively, a line and an orientation. All lines are composed of pairs of half-lines, although it is conceivable that a single half-line might be modelled in some circumstances. Finally, it should be noted that although different kinds of Voronoi diagram can be generated by assuming different distance metrics, the set of neighbourhood relations expressed by the dual of the Voronoi is not unique to a given Euclidean metric, but to a set of conformally equivalent spaces. The Voronoi model of space (as distinct from a given kind of Voronoi diagram) does not require a fully defined metric to function (see [13] for more details).

In contrast, other models of space include the raster model and the vector model (here we assume the vector model knows about line-intersection topology). In the raster model, space is broken into regular, discrete units called cells. Cells may be grouped into agglomerates to which a unique object identification is given, called raster objects, they may be kept as single objects, which we might call pixel objects, or they may be made to represent empty space (although this latter is rarely done). Neighbouring raster objects can be defined as those raster objects which share boundaries between outer cells. Topology is represented implicitly in this model via cell adjacencies, but no formal structure for expressing these neighbours is usually created in implementations of the raster model. In the vector model, space is defined by a Euclidean metric and an appropriate coordinate system. Objects are defined as collections of points, line segments and polygons. Adjacencies can be defined only when two objects share a common point, line segment or polygon. Neighbourhood relations can be defined additionally via a metric relation.

The Voronoi model differs from the other two models in many fundamental ways. Both raster and vector models are coordinate based models or fully metric models of space (although they use different metrics). The Voronoi model is not a fully metric model of space in the sense that coordinates are not needed to determine adjacency relationships (although they are used in the computer implementation of the model). Instead, it is possible to construct and use the Voronoi using only compass and straight edge (and a notion of minimum distance). Furthermore, the vast majority of the Voronoi operations can be expressed as purely logical relations and not as "spatial" relations at all. Hence the Voronoi model of space is closer to a "qualitative" view of space than existing spatial models used by GIS. For example, the vector model of space must define a neighbourhood in terms of a metric distance (whether fuzzy or not). In a sense, the raster model also defines neighbourhood in metric units (where the atomic metrical unit is the cell size). The Voronoi model of space allows definitions of neighbours which are based purely on topological information (i.e. logical information). A related example concerns the notions of "near", "far", "in-between", etc. Again, in both the raster and vector models of space, these notions can be defined only in terms of metric quantities, with varying degrees of success.

With the Voronoi model of space, on the other hand, these notions are unambiguously and directly determined (see Section 4 of this paper). Furthermore, the Voronoi answers to questions about such relations appear to be very close to those which a human interpreter is likely to pick. Finally, the Voronoi appears to correctly respond to most of the "closed element" spatial relations in linguistic structures discussed by [18].

Secondly, the Voronoi model of space is fundamentally dynamic, again unlike the other models. In the raster model of space, change is realised by changing the attribute values of the cells or modifying their links to object IDs on a discrete (i.e. cell by cell) basis. In the vector model, dynamic movement of the objects (with corresponding changes in the topology) can occur (although implementation of such movement on a computer is difficult). However, only objects which share common atoms (points, line segments or polygons) know anything about the movement of the other object. Movement largely consists therefore of a change in coordinates. In the Voronoi model, on the other hand, no object may move without several other objects knowing it has done so and, indeed, object motion could be defined largely in terms of the topological changes (changes in adjacency relations between objects) and does not need to be expressed in terms of a coordinate change. Once again, this model supports unambiguously the possibility of changing relations between objects without needing to resort to a metric or coordinate representation of the information.

Furthermore, it is possible to determine a "weighting" relation expressing the "influence" of any one neighbour on a given object. This is achieved with the use of the area-stealing interpolation method developed by Gold [11,13]. In the usual quantitative application of this method, a weight is obtained for each neighbour and used to weight a common attribute value in order to obtain a new (interpolated) value at an intermediate location. However, it has been recognized [11] that these weights could be used independently of each other to specify the relative "influence" of purely nominative information associated with each neighbour. An alternative "influence" function is the amount of common boundary. This function is less satisfactory than the area-stealing function, but still likely to be useful under some circumstances. We shall see how such an "influence" function could be used in a few moments.

A final feature of the Voronoi is that it is profoundly hierarchical. Space, in the Voronoi model, is nested into increasingly levels of detail, each level of which is embedded in the previous level. This can be true of both the raster and vector models, also, but there are real limitations in these cases. In the raster model of space, there is a fine limit to level of detail which can be represented. In the vector model of space, objects can be grouped into higher levels, but space itself cannot be so grouped. In the Voronoi model of space, objects and space are intimately connected. Hence grouping finer and finer object features (individual points and line segments, for example) implies necessarily grouping their associated Voronoi regions.

In the next section, we shall see how these advantages of the Voronoi model can be exploited in the context of culturally-rich GIS databases.

3 Applications of the Voronoi Model of Space to Qualitative Analysis

We have observed that the Voronoi model of space appears to lend itself more readily to the study of qualitative relations between objects distributed in space than traditional coordinate based models of space. Is this simply a curiosity or can this characteristic actually be used to advantage? We believe that the Voronoi model of space is sufficiently different from the raster and vector models of space that just understanding its characteristics can lead to new ways of perceiving space and new ideas about how to use GIS under different conditions.

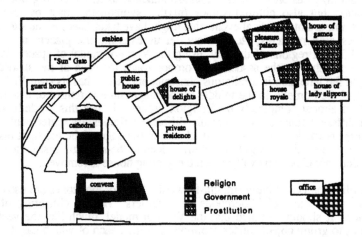

Fig. 1. A fictive city which will be used to illustrate some of the Voronoi analysis capabilities

In this section, we shall examine several examples of the kinds of applications which might be realised using the Voronoi data structure in a GIS and exploiting its associated model of space. It is hoped that these examples will provide a better "feel" for the Voronoi model of space as well as possibly provoking new ideas on the use of GIS in non-quantitative domains.

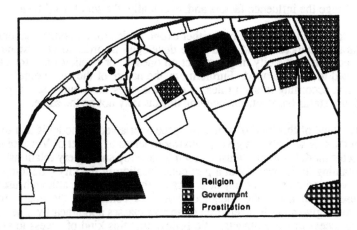

Fig. 2. An example of how the Voronoi model of space could be used for "proximal queries" in historical or anthropological contexts. Each "important" building has its own zone of influence. A query point collects a set of neighbours and their respective influences (proportion of area "stolen" from each neighbour's zone of influence, shown in dotted line).

In Figure 1 is shown a fictional map of a historical city. Shown in patterns are three kinds of key building in the city: religious buildings, governmental buildings and buildings devoted to prostitution and games of chance. Let us assume that each of these buildings has an extensive database attached to it, containing not just factual information but also cultural values (myths, symbolisms, past owners and their histories, important historical events that might have taken place at the site, anecdotes, interviews with residents, etc.). Now, let us examine the placement of buildings within different regions of the city, using the Voronoi model of space. In Figure 2, we have sketched in the associated Voronoi zones for each of the shaded objects. Note that the gate also is a religious construct and hence has its own Voronoi zone. Clearly, if the bordello in the middle of the diagramme is removed, it's zone will be "filled" in almost completely by "religion" zones. Hence this bordello is likely to have a different function and possibly different attitudes associated with it than the other three bordellos on the right, which are on the edge of the red light district. The gaming house on the far right, however, is deep in the red light district and hence the region does not change its value when the gaming house is removed.

This application is relatively straight-forward. The Voronoi zones are used to explore the notions of city neighbourhoods or "quartiers" via the clustering of zones of influence. It should be noted that the Voronoi model, much like the vector model, allows us to group objects under common "themes", each of which has its own set of proximity fields.

Another application within the same scenario is the following: Let us query the database by pointing to any arbitrary location anywhere in the city (Figure 2). The answer we get is a list of "influence weights" and objects. For each of these neighbouring objects, we can access, in an "exploratory mode", for example, its cultural values or attributes, for any number of themes we desire (see Table 1). For example, the query point indicated in Figure 3 will get a strong influence from the gate itself, a relatively weak influence from the church and the bordello respectively and almost no influence from the baths. Changing the position of the query point would change the influence factors and, eventually, the set of neighbours.

In order to exploit these relations, either explicit rules for combining several symbolic or cultural influences would need to be developed in order to understand how one's spatial location in the city might reflect on oneself, or explicit links for each context would need to be developed. Furthermore, the nature of the query object may change the context according to such rules. Hence the rules for a building's status versus an individual's status based on surrounding objects may not be the same.

The advantage of this kind of model of space over a vector model is that querying any point in the city will always give an answer in terms of the set of Voronoi neighbours. In a vector model, a metric must be used to develop such query relationships. Rather than obtaining an answer about a particular location in the city ("building A", "street B", or even "no significant object is located there"), which, although useful, doesn't tell a user much about the "neighbourhood" or context of the query, the Voronoi query would give a multi-layered set of answer classes about nearby objects and allow database access to these objects. We believe that this kind of access to neighbouring objects provides a new kind of "spatial operator" of widespread interest which considerably enriches the capabilities of GIS.

For example, the idea of such proximity queries could be extremely useful for software devoted to the task of mobile tourist guiding, or to an individual on such a tour. His or her location could be used to generate the proximity lists and influence factors, indicating what the surrounding entities and their history are. Causal links might also be expressed in the database between different objects [6], and these could also be brought out into the open by the querying process.

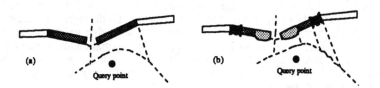

Fig. 3. Different levels of detail can be represented in the Voronoi model of space. Shown in (a) is a magnification of the Sun Gate in Figure 1 at the level of detail represented in that figure, and with a query point. In (b) is shown the embedded detail of three panels in each part of the gate, with their associated Voronoi zones. Case (a) is a generalization of case (b).

Furthermore, the hierarchical grouping of space can be used to further enrich the querying power for these applications. Hence, for example, the gates shown in Figure 1 might be composed of panels, each of which has a different religious meaning. The Voronoi model allows us to represent the gate at a much higher level of detail, if we choose to do so (Figure 3a). Indeed, the nested and tiled nature of Voronoi space allows one to effectively remove any piece of the map, "fold it away" and replace this piece with a simplified (i.e. generalised) version of the object at a lower level of detail (Figure 3b). Hence the proximal query space can be navigated "vertically" through different levels of generalisation, as well as "horizontally" from one location to the next, allowing a user to query additional levels of detail (provided the database contains this information). Note also that it is not simply the object itself which is nested at different levels of detail, but also the accompanying Voronoi zone or space which neighbours the object. Hence, in the Voronoi model, the "gate" consists not only of the line segments which compose the gate, but also of the regions on both sides of the gate. Viewed a different way, the gate itself transfers meaning to the space associated with the gate. Another way of thinking about this would be to imagine a mural painted on the city wall. The space in front of the mural clearly acquires a different meaning than the space in front of the wall on either side of the mural. Hence the Voronoi model of space is useful for providing a framework for representing and manipulating cultural space, as one would like to do in domains such as anthropology [3].

Finally, the extensions of the Voronoi model into the temporal dimension also hold the promise of many useful operations. The Voronoi model as it is being implemented by Gold [9,12] permits one to insert a floating query point (or any number of such points) into the database. This point can be arbitrarily moved around inside the space of the database in a fully dynamic sense (immediate updating of all topology). Furthermore, the effects of the movement can be kept track of in a separate log or archiving file. This has, among other consequences for temporal GIS, the consequence that events can be managed by a Voronoi-based GIS, as opposed to simply keeping track of static "snapshots" of the spatial database. This means that it would be possible to "thread" a query point by laying down a continuous "path" or "rail" through the spatial database, along which specific moments in time could be identified and labelled. We can now accumulate a "contextual history" of the movement of the point through the database or, alternatively, we could select interactively parts of the contextual database as the floating point is moved along its track. Such threading could be of use for tourist databases It would also be possible to use it within a new kind of interactive fiction, especially if one were allowed to implement several such threads (one for each character) and to permit interactions between threads. Indeed, the same kind of database could be used for literary analysis of existing texts. Imagine, for example, incorporating an old map of Dublin with an extensive database information (possibly in a multi-media sense) containing background information on buildings, sites, etc. Now the path of Bloom's wanderings (in James Joyce's Ulysses [14]) could be "threaded" through the database and information incorporated both via the threaded path and other existing objects in the database with reference to the book. Exploring such a spatial database and interacting or following Bloom's path, the path of Stephen Daedalus and other important characters would increase immensurably one's understanding and appreciation for different elements of the book.

This spatial-temporal browsing capability of the implentation of the Voronoi model of space proposed by Gold [9, 12, 13] is therefore likely to be of interest in almost any area where qualitative information about objects in space are stored in a database. This will apply throughout the social sciences and humanities, but also to the management of more qualitative information within organisations which presently use GIS to manage their quantitative information only.

4 The Voronoi Model of Space Versus Linguistic Structuring of Space

The "proximity query" as discussed above is certainly one interesting way to approach more qualitative spatial analysis. Other proximal queries which have been discussed earlier in the literature are based on linguistic concepts of space. Hence, for example, Frank [7] and Frank and Mark [8] have written about operators such as "near", "between", "among", etc. One of the advantages of the Voronoi model of space over other models is that it can be used directly to develop map operators to implement a large number of linguistic concepts of space, with the additional feature that the same invariance properties found in linguistics hold for the Voronoi model. This discussion is fashioned after the linguistic analysis of Talmy [18].

Talmy indicates that spatial structuring in language is done principally through closed-class grammatical elements such as prepositions, demonstratives and inflections. He further states that the majority of spatial content is contributed by open-class lexical elements, such as nouns, verbs and adjectives. Furthermore, Talmy claims that four invariances hold for linguistic structuring of space: magnitude neutrality, shape neutrality, closure neutrality and continuity neutrality. According to traditional models of space used in GIS, magnitude neutrality and shape neutrality are preserved, whereas closure neutrality and continuity neutrality are not. Hence, for example, the mechanics of representating in a GIS an ant crossing a palm or of a bus crossing a country are essentially equivalent (for all three models of space discussed here). Another way of stating this is that scale is not obvious from a given representation; it must be specified as additional information. Secondly, the mechanics of representing the path of an object across another object of arbitrary shape is equivalent for all shapes and paths, hence constituting shape neutrality, again valid for all three models of space (vector, raster, Voronoi). Talmy points out, however, that deciding whether an object is "inside" another depends, within standard GIS models of space, on whether the container is closed or not, whereas linguistically, we may say a bean counter is both "in" a bowl and "in" a closed box. Furthermore, in GIS models of space, it is relatively difficult to say that a bird is "in" a birdcage, unless the birdcage is represented as a closed and continuous polygon.

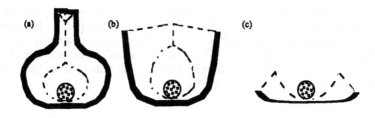

Fig. 4. Is the ball in the "container"? We would probably say the ball is both in container (a) and (b), but we might be more hesitant about container (c). The dotted lines show the boundaries of the Voronoi zones, where the "objects" are considered to be the "interior" surfaces of each "side" of the container. The Voronoi model would claim the ball is "in" both containers (a) and (b). The Voronoi model provides only partial support for the idea of the ball being "in" container (c).

Unlike the traditional GIS models of space, the Voronoi model handles both these latter neutralities in ways similar to linguistic methods. Hence, in Figure 4 is shown three different closure cases, whereas in Figure 5 is shown three different continuity cases. It is seen that the Voronoi "answer" is very close to the human perceptual/linguistic answer. In fact, humans seem to judge "inside" and "outside" by "guessing" according to the same proximity relations that the Voronoi uses.

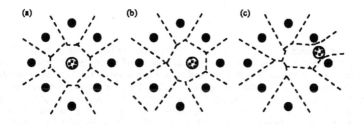

Fig. 5. Is the ball in the "cage"? We would probably say the ball is in the cage in situation (a) and situation (b), but we probably not say the ball in in the cage in case (c). The dotted lines again show the boundaries of the Voronoi zones (approximately), where the "objects" are considered to be the "bars" of the cage. The Voronoi model would claim the ball is "in" the cage in both situationss (a) and (b), but not "in" the cage in case (c).

Now let us examine spatially related prepositions used in English, again taking our examples from Talmy's work. In Figure 6, several cases of "nearness" are presented. The degree of nearness is determined by comparing the Voronoi boundary between the two objects with the boundary that would be found for simple objects (i.e. points). If the boundary is very similar to such a simple boundary, then the objects are said to be "near" each other, otherwise they are "far". This is close to the linguistic sense of proximity, where the concept of near is not the same for a man standing a metre away from a wall than for an ant located a metre away from a crumb. In Figure 7 the case is presented for the preposition "between", where the Voronoi answer is expressed in terms of length of common boundary between the different objects. The notions of "among" and "amidst" would be similar to the case of the birdcage, shown in Figure 5.

Relative magnitude in the target object compared with other objects might be used to indicate the difference between "among" (i.e. target object smaller or of similar size as surrounding objects) and "amidst" (i.e. target object considerably larger than the surrounding objects). Other possibilities exist for differentiation between these concepts. This is not meant to be an exact analysis, but only to show the potential of the Voronoi for exploiting these concepts.

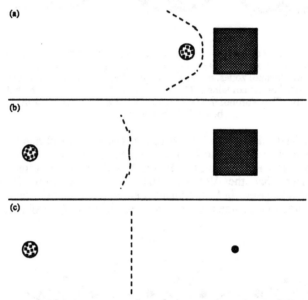

Fig. 6. The Voronoi idea of the concept "near" is expressed as the similarity of the Voronoi zones of the actual objects compared to the Voronoi objects of simplified objects. In situation (a), the Voronoi boundary between the ball and square is not easily replaced by a simplified object. In situation (b), on the other hand, the boundary between the two Voronoi zones is much closer than that we would expect to see for a simple point object seen at a distance. Hence the idea of "nearness" is expressed implicitly with respect to the mutual sizes of the objects compared to their relative distance. This is also close to the linguistic sense of "near".

In this paper, lack of space prevents a more complete analysis of such correspondances. Other issues include motion prepositions, nouns with associated spatial qualities such as boundedness, plexity and continuity, and demonstratives. In all such cases, mappings between the Voronoi model of space and these linguistic properties appear to be not only possible, but directly derivable (i.e. without requiring complicated reasoning or arbitrary structure). In fact, several mappings appear to be possible in most cases. The concept of "between", for example, might be examined by comparing length of common boundary between two objects as another object is introduced, as shown in Figure 7, or it might be mapped using area-stealing interpolation [13]. The concept of "in" might be handled be either of these methods, or by examining the angular range of enclosure of the Voronoi boundary separating the object from the bowl. More work needs to be done to decide which mapping is most appropriate for different linguistic concepts.

Two other specific features of the Voronoi model are worth mentionning. The Voronoi contains "sidedness" information, and hence can handle requests such as "above" versus "below", "in front" versus "behind", etc. Secondly, the Voronoi handles demonstratives "this object over here", "that object over there" since "pointing" becomes an important and straightforward selection mechanism in the

Voronoi model. Pointing does not require "hitting" the object, but only the proximity field of the object and hence is much easier to achieve in the Voronoi model.

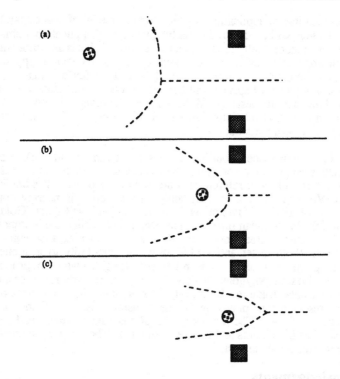

Fig. 7. Is the ball between the two rectangles? The notion of "between" can be determined to be the amount of common boundary which exists between objects. In case (a), the presence of the ball does not greatly influence the shared boundary between the rectangles. Furthermore, the rectangles contribute different boundary lengths to the ball's Voronoi zone. In case (b) each rectangle contributes roughly equally to the boundary of the ball, as is also true in (c). Hence the Voronoi answer would be more or less "no" for case (a), "yes" for case (b) and a definite "yes" for case (c). Note that the Voronoi quite naturally gives a handle on "certainty" or "likelihood" of the response.

From this discussion we can conclude that the Voronoi model of space has sufficient power to express a very wide range of spatial relationships and hence that natural language-like queries, with much of their inherent ambiguity, could be implemented within a Voronoi model.

5 Discussion and Conclusions

The above discussion, I believe, demonstrates the expressive power of the Voronoi model of space for spatial relations. Not a single linguistic concept of space has yet been identified which does not seem to have a role in the Voronoi structuring of space (although other language groups may have unusual features which will need further elaboration). Furthermore, the Voronoi model of space permits a special kind of "proximal" query whereby the full set of neighbours of a floating query point embedded in the database can be collected and accessed dynamically. It is posited that the full class of such proximity queries constitutes a new kind of "spatial operator" for GIS which are particularly appropriate for application to databases which contain more qualitative information, such as are found throughout the social sciences and humanities.

Indeed, non-quantitative applications of the Voronoi model of space may be of very wide interest even within the traditionally quantitative application domains, since much of the information really used in practice in these fields is also more qualitative. The Voronoi approach may be of very high utility for navigation applications, for example. Any application requiring natural language interface may benefit from using the Voronoi model of space if not the full Voronoi implementation. We believe that natural language requests could be mapped relatively directly to a Voronoi implementation of spatial relations. The full applicability of a GIS built around these concepts has yet to be understood.

The Voronoi model of space as outlined here is structured around work implementing a Voronoi data structure on a computer, work essentially carried out over the past ten years by Gold (see full set of references at the end of this paper). The implementation of the basic Voronoi concept in a dynamic data structure is nearing completion. Before the end of 1993, a functioning Voronoi "engine" will exist (Gold, personal communication). In the mean time, and has been stated earlier, the Voronoi model of space can be realised, albeit only partially, within other data structures. Many commercial GIS (both raster and vector) offer the possibility to generate Thiessen polygons for point data sets. It is relatively straightforward to programme the generation of Thiessen polygons (Voronoi zones) around arbitrarily shaped objects within the raster data structure. Hence other researchers who are interested in further pursuing this research can "play with the ideas" using existing GIS and, indeed, they are encouraged to do so. The Voronoi concept has gained widespread support in a very large number of domains and is here to stay [1]. There is far too much potential for any one researcher to exhaust.

6 Acknowledgements

The author would like to thank Dr. Christopher Gold, along with Dr. Kim Lowell, for extended discussions which formed the background for this paper, as all the concepts discussed in this paper are rooted in common discussions between the three of us. I would also like to thank the Association des industries forestières du Québec and the Natural Science and Engineering Research Council of Canada for their financial support for some of this work via the creation of the Industrial Research Chair in Geomatics applied to forestry which brought Chris, Kim and myself together as a single team.

References

1. Aurenhammer, F., 1991, "Voronoi Diagrams — A Survey of a Fundamental Geometric Data Structure", *ACM Computing Surveys*, Volume 23, 345-405.
2. Bruegger, B.P., R. Barrera, A.U. Frank, K. Beard and M. Ehlers, 1989, "Research Topics on Multiple Representation", *Proceedings of the NCGIA Specialist Meeting on Multiple Representation*, Technical Paper 89-3, NCGIA, Buffalo, New York, 53-67.
3. Catedra, M., 1991, " 'Through the Door': A View of Space from an Anthropological Perspective", in Mark, D., and A.U. Frank (eds), *Cognitive and Linguistic Aspects of Geographic Space: Proceedings of a NATO ASI*, Kluwer Academic Publishers, Boston, 53-63.
4. Edwards, G., 1992, "Error Minimization in Integrated GIS/IAS System Design", *Proceedings of the 5th International Symposium on Spatial Data Handling*, Charleston, 20-29.

5. Edwards, G., 1991, "Remote Sensing Image Analysis and Geographic Information Systems: Laying the Groundwork for Total Integration", *Proceedings of the NCGIA Initiative 12 Special Session on Remote Sensing and GIS integration*, Baltimore, March.
6. Edwards, G., P. Gagnon and Y. Bédard, 1993, "Spatial-temporal topology and causal mechanisms in time-integrated GIS: from conceptual model to implementation strategies", *Proceedings of the Canadian Conference on GIS*, Ottawa, 842-857.
7. Frank, A.U., 1992, "Qualitative Spatial Reasoning about Distances and Directions in Geographic Space", *Journal of Visual Languages and Computing*, Volume 3, 343-371.
8. Frank, A.U., and D.M. Mark, 1990, "Language issues for Geographical Information Systems", in Maguire, D., D.W. Rhind and M.F. Goodchild (eds.), *Geographic Information Systems: Principles and Applications*, London: Longman.
9. Gold, C.M., 1992a, "An Object-based Dynamic Spatial Model and its Application in the Development of a User-friendly Digitizing System", *Proceedings of the 5th International Symposium on Spatial Data Handling*, Charleston, 495-504.
10. Gold, C.M., 1992b, "The Meaning of 'Neighbour'", in "Theories and Methods of Spatio-Temporal Reasoning in Geographic Space", *Lecture Notes in Computing Science* No. 639, Springer-Verlag, Berlin, 220-235.
11. Gold, C.M., 1992c, "Surface Interpolation as a Voronoi Spatial Adjacency Problem", *Proceedings of the Canadian Conference on GIS*, Ottawa, 419-431.
12. Gold, C.M., 1992d, "Dynamic Spatial Data Structures — the Voronoi Approach", *Proceedings of the Canadian Conference on GIS*, Ottawa, 245-255.
13. Gold, C.M., 1991, "Problems with Handling Spatial Data — the Voronoi Approach", *CISM Journal*, Volume 45, 65-80.
14. Joyce, J., 1960, *Ulysses*, Penguin Books, Harmondsworth, Middlesex, England, 720 pp.
15. Nantel, J., 1993, *The Forestry Chronicle*, in press.
16. Peuquet, D.J., 1984, "A Conceptual framework and comparison of spatial data models", *Cartographica*, Volume 21, 66-113.
17. Nunes, J., 1991, "Geographic Space as a Set of Concrete Geographical Entities", in Mark, D., and A.U. Frank (eds), *Cognitive and Linguistic Aspects of Geographic Space: Proceedings of a NATO ASI*, Kluwer Academic Publishers, Boston, 9-33.
18. Talmy, L., 1983, "How Language Structures Space", in *Spatial Orientation: Theory, Research and Application*, edited by H. Pick and L. Acredolo, New York, Plenum Press, 225-282.

TABLE 1: AN EXAMPLE OF THE PROXIMITY LISTS ONE MIGHT ACQUIRE
FOR QUERY POINT SHOWN IN FIGURE 2

Theme Attribute	Neighbour #1	Neighbour #2	Neighbour #3	Neighbour #4...
Major symbols				
Title of object	Sun Gate	Cathedral	Bath House	House of delights
Type of symbol	Purification	Purification	Purification	Gratification
Influence factor	60%	20%	15%	5%
Clientele	All	All	Upper class	Upper class
Owner	City	Church	Church	Church
Physical blds				
Title of object	Guard house	Gates	Stables	Public house
Type of activity	Control/def.	Control/def.	Sup./transp.	Sup./hotellerie

Interaction with GIS Attribute Data Based on Categorical Coverages*

Gary S. Volta and Max J. Egenhofer

National Center for Geographic Information and Analysis
and
Department of Surveying Engineering
Boardman Hall, University of Maine, Orono, ME 04469-5711
max@mecan1.maine.edu

Abstract. The human-computer interface is a crucial element in the design of the next generation of Geographic Information Systems (GISs). We discuss the user interface design process by separating the *formalization* of the problem domain (identifying the objects a user manipulates, and their pertinent operations) from its *visualization* (describing human-computer interaction techniques such as windows and dialog boxes). This framework is used to examine the process of manipulating *attribute data* in a GIS on the basis of the common cartographic concept of a *categorical coverage*. The characteristics of categorical coverage data and the user requirements for interacting with this data are formalized in the form of a set of fundamental objects and operations. A visualization for a windows-icons-menus-pointing devices (WIMP) interface is presented.

1   Introduction

Throughout the history of Geographic Information Systems (GISs), most of the research attention has concentrated on methods to store, retrieve, process, and analyze geographic data efficiently (Maguire *et al.* 1991). Only recently, attention has moved toward the user interaction components of a GIS (Egenhofer and Frank 1988; Kuhn and Egenhofer 1991; Mark and Gould 1991; Mark and Frank 1992). For example, investigations of metaphors appropriate for dealing with geographic data (Gould and McGranaghan 1990; Jackson 1990; Kuhn and Frank 1991; Mark 1992; Egenhofer and Richards 1993) have influenced the design of some GIS user interfaces. In order to implement a successful system, GIS designers must have a clear understanding of the user requirements, including the concepts users will apply when they deal with geographic data. This is a particularly important concern for the design of GIS user interfaces, as it should not just focus on the way people interact with computers, but

---

* This work was partially supported by NSF grant No. IRI-9309230 and a grant from Intergraph Corporation. Additional support from NSF for the NCGIA under grant SES-8810917 is gratefully acknowledged.

should be based on how people go about problem solving when interacting with geographic information itself (Mark and Gould 1991).

GISs are typically thought of as containing two very different, but interrelated kinds of data and information, spatial and attribute. *Data* are measurements of the "real world" that are used to create useful *information* about the world for use in analysis. For this paper, the "real world" is *geographic space*. The term *spatial data* describes position or location within geographic space. *Attribute data* (also known as aspatial or non-spatial data) consist of the descriptive or statistical characteristics of spatial data. Research into modeling the geometry of geographic data has been abundant. For overview articles see (Peuquet 1988; Egenhofer and Herring 1991; Frank 1992). On the other hand, modeling of the attribute-data component of a GIS has been secondary to the spatial component. Attribute-data modeling has been examined primarily within the usual frameworks developed in computer science for general-purpose database management systems such as the hierarchical, network, or relational data model, and most recently also object-oriented methods have been favored (Worboys *et al.* 1990; Egenhofer and Frank 1992). A second perspective of attribute data that has received attention is the classification of attribute data into nominal, ordinal, interval, and ratio (Stevens 1946), for instance to model spatial data quality (Clapham 1992). While these views of attribute data modeling are useful for a discussion of some aspects of geographic data, they consider attribute data as separate from the spatial domain. This intimate link between attribute and spatial data, however, is the essence of many geographic applications and its correct modeling is critical for the design of the interaction with attribute data in a GIS.

This paper uses the cartographic concept of a *categorical coverage* (Chrisman 1982) as a basis for interacting with attribute data. Categorical coverages are particularly well suited as a geographic framework for attribute interaction as they organize the spatial domain according to properties of the attribute domain and manipulations of the attribute domain trigger changes in the spatial domain. This is different from the behavior of choropleth maps, which have a similar visual appearance to categorical coverages. They arrange attribute information for enumeration areas on the ground (Robinson *et al.* 1984); however, unlike in a categorical coverage, the zones on a choropleth map are predetermined mapping units for which a representative value (e.g., average or maximum) is measured and depicted.

The user interface design is based on the formalization of the objects and operations present in a *family of categorical coverages* (Frank *et al.* 1992), which provides both a consistent, compact description of the problem and a means of communication of functionality between the user and the developer. This is followed by the visualization of the objects presented to the user at the screen surface and the implementation of the pertinent operations.

The following section introduces a user-interface design paradigm that is tailored to complex applications such as GISs in which an interplay exists between domain-expert knowledge and human-computer interaction methods. It emphasizes the formalization of the problem at hand prior to the discussion about its visualization and interaction concerns. The remainder of this paper documents the design of a GIS user interface for attribute data within this framework. Section 3 describes informally the cartographic concept of a categorical coverage. Section 4 formalizes the relevant parts and section 5 presents one possible visualization/interaction with a WIMP (windows-icons-menus-pointing devices) user interface. The paper concludes with a summary of the results.

2   Formalization and Visualization

"Current commercial GISs widely disregard fundamental aspects of human-computer interaction. GIS users need extensive and expensive training prior to using a particular system, due to the researchers' and designers' concentration on functionality and implementation rather than usability. Systems tend to evolve from a small set of commands to hundreds of features without the necessary considerations of how users learn them and interact with them" (Frank *et al.* 1991). Exceptions to this trend are the design of Arc/View, parts of which had undergone extensive theoretical investigations prior to its implementation (Jackson 1990; Kuhn *et al.* 1991), and user interfaces that build on Tomlin's (1990) map algebra (Kirby and Pazner 1990; Egenhofer and Richards 1993).

To overcome some of this *ad hoc* mentality in GIS user interface design, we propose a three-tier framework for the design of GIS user interfaces (Fig. 1). The three steps are interleaved, but different enough to warrant such a separation. Each step requires from the user-interface designers distinct qualifications and command in a number of areas.

- First, formal systems with well-defined behavior and properties are investigated with respect to the problem. This process uncovers the pertinent objects, operations, and behavior of the problem that must be dealt with in the user interface. During this phase, the designers lay the ground for what the system will be able to do and what it will not.

- Second, different interaction procedures are investigated to allow prospective users to perform the intended manipulations. The formalism is used as a consistent foundation from which various user interfaces can be developed according to different interaction techniques such as command-line, windows, and icons. The formalism of objects and operations identifies the core functionality necessary for all the various user interfaces. Therefore, the evaluation of the user interface can focus on the different interface techniques.

- Third, each visualization- and interaction-design is implemented on a particular platform with its operating system and specific user-interface management system, toolbox, graphics library, etc.

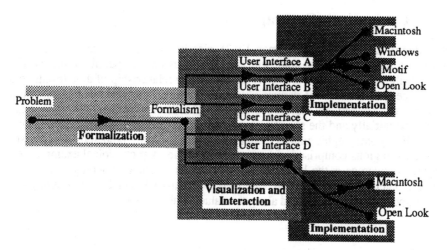

Fig. 1. User interface design process.

The background knowledge necessary in each of the three areas is fairly distinct and the fallacy of many GIS user interfaces is that too often a single user interface designer is expected to master all of them.

- In order to formalize the objects and operations with which a future user will deal, a designer has to know about the problem domain and needs sufficient knowledge of and practice with the formalization tools (e.g., specification languages, algebras); therefore, we will call them the *domain experts*. Without their active involvement in the design process, any user interface will be something made by a designer society for a user society.

- During the second phase, the designer must be proficient in cognitive analysis, psychology, human-computer interaction techniques, and graphics art. This may be what is commonly understood by the *user interface designer* (Mulligan *et al.* 1992). The focus during this phase is on the translation of the domain experts' specifications into a particular interface environment and it involves giving feedback to the domain experts so that they can improve their design in close cooperation with the user interface designers.

- The third phase of the user interface design requires extensive expertise in user-interface programming, graphics packages, and user-interface management systems. We call these experts the *user-interface software engineers*.

We will employ this framework here to design a user interface for the interaction with geographic attribute information.

3   Categorical Coverages

The *categorical coverage* (Chrisman 1982), also known as area-class map (Bunge 1966), is a type of thematic map, which concentrates on the spatial variations of a single attribute (Robinson *et al.* 1984) and shows the relationship of an attribute to a specific geographic area. It depicts a specific bounded area by a set of mutually exclusive and collectively exhaustive regions. The theme and its subcategories are determined initially and the specific area is then evaluated according to the thematic criteria. Every geographic unit is evaluated according to the predetermined classes resulting in regions comprised of homogeneous value. A prototypical example of a categorical coverage is the land use map (Fig. 2), in which a taxonomy of land use classes is determined (Residential, Commercial, Industrial, and Transportation) and then the specific area is evaluated along the values of this theme.

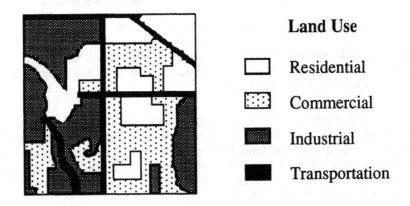

Fig. 2. Simple categorical coverage for land use.

Categorical coverages describe a subset of space thought of as a 2-dimensional plane, possibly embedded in 3-dimensional geographic space. This space is conceptualized as regular Euclidean space and typically mapped onto $\mathbb{R}^2$. For the categorical coverage, the spatial data model—raster or vector—is of limited importance. Either model can represent the necessities of a categorical coverage. We will proceed with the assumption of a vector model, thereby focusing on the identity of the zones belonging to attribute classes and thus ignoring individual pixel values.

The zones of a categorical coverage (Fig. 3) are an exhaustive subset of space and that not overlap (Beard 1988).

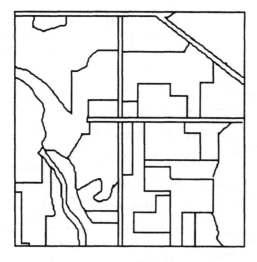

Fig. 3. The zones of the categorical coverage.

Categorical coverages have a theme that is based on the purpose of the examination. The theme of the categorical coverage is a domain of attribute values (categories) that describe the salient properties of a given space. The domain consists of a finite or infinite number of attribute values, which are at least discernible on a nominal scale, but often have more internal structure. The set of values that are differentiated are often called the *categories of the coverage* (Beard 1988). The categories form a partition of the domain of attribute values. Attribute values are often ordered hierarchically, typical for taxonomies.

Conceivably, more than one categorical hierarchy can be built from the same initial coverage in order to address a different map purpose (Fig. 4). The collection of these categorical hierarchies has been called a *family of categorical coverages* (Frank *et al.* 1992). A family of categorical coverages is defined as all possible categorical coverages that can be deduced from a single, most detailed coverage. Each member of the family will preserve various levels of the original categorical detail. The cardinality of the domain of attribute values (finite or infinite) of the theme determines the cardinality of the complete family of categorical coverages.

The operations discussed in this paper for the construction of a categorical coverage apply only to the categories of the attribute values. These operations induce changes in the complexity (the number of zones) of the spatial domain. There are no operations that are directly applicable to the geographic domain except the cartographic viewing operations, such as the selection of symbology, which are omitted from this discussion. However, if we allow that the display of the categorical coverage is only an approximation to its. conception then additional operations become necessary to show the coverage on any given display device. These generalizations are better considered as part of the cartographic rendering process than the formation of the categorical coverage.

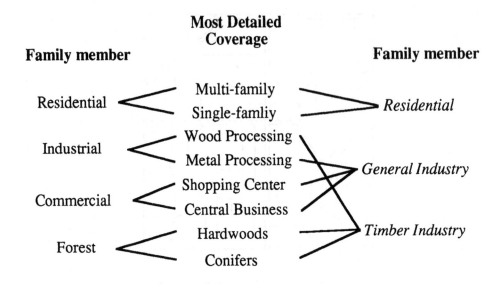

Fig. 4. Partial family of categorical coverages.

4 Categorical Coverages and Partition Formalism

The basic concept of a categorical coverage is such that the attribute categories are defined first and the spatial zones are then constructed corresponding to these categories. There are distinct connections between the attribute and geographic domains of a categorical coverage and partitions. The common objects of this relationship are described in Section 4.1 and followed by a discussion of common operations.

4.1 Objects

An attribute domain or theme, A, corresponds to a "universal set" of attribute categories or values, $\{a_0, a_1,..., a_n\}$. The set of attribute categories for a categorical coverage defined by the data collection and input procedures is finite and by default, defines the universal set of attribute categories. While in reality additional categories for a particular theme can always be created, they may be deemed unnecessary for the purposes of a particular categorical coverage mapping. The union of all a_i equals A, and the set is said to be collectively exhaustive of A. This must be the case for a categorical coverage because geographic areas of undefined attribute values may occur. Therefore, the attribute domain of a categorical coverage is characterized as follows:

$$A = \{a_0, \text{K} , a_n\} \tag{1}$$

$$\bigcup_{i=0}^{n} a_i = A \tag{2}$$

For a categorical coverage, the mapping of each unit of the geographic domain is restricted to exactly one attribute category. Therefore, all elements of the set of attribute categories must be mutually exclusive.

$$\forall_{i,j\,|\,i\neq j}\quad a_i \cap a_j = \varnothing \tag{3}$$

The intersection of all the attribute categories must be empty to insure that there are no geographic overlaps. For instance, reality may show a geographic unit to be both wooded and wet, but it can not be mapped both Forest and Wetland in a categorical coverage. The most distinguishing characteristic must prevail (Forest or Wetland) or a new category, for example Wooded Wetland, must be created.

If these conditions are met, the set of attribute categories form the thematic component, a partition of A of a categorical coverage, denoted by P_A.

The definition of the attribute domain directly affects the representation of the geographic domain, G. The geographic domain corresponds to a "universal set" of geographic positions. Every position (x, y) in G is mapped onto exactly one attribute value that exists in the attribute partition, P_A (Fig. 5).

attribute partition (P_A) **mapping** **geographic domain (G)**

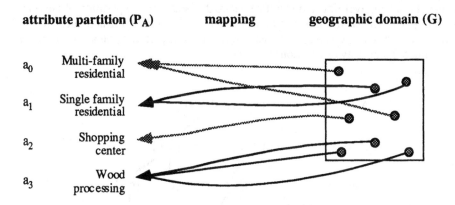

a_0 — Multi-family residential

a_1 — Single family residential

a_2 — Shopping center

a_3 — Wood processing

Fig. 5. Creating a categorical coverage.

These points are grouped by equivalent attribute value to form homogeneous zones, $\{g_0,...,g_m\}$ (Fig. 6). These zones become the elements of the geographic domain. The mutually exclusive and collectively exhaustive properties of the attribute category set also hold for these zones. Therefore, the set of zones form a partition of the geographic domain, P_G.

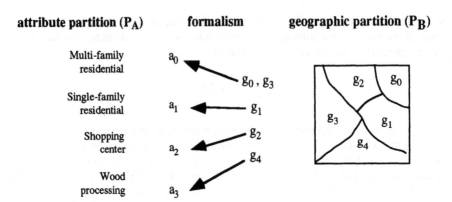

| attribute partition (P_A) | formalism | geographic partition (P_B) |

Fig. 6. Formal components of a categorical coverage.

Although, both the geographic and attribute components are essential to a categorical coverage, the geographic domain is only a spatial representation, which is dependent on the definition of the attribute domain.

4.2 Operations

The attribute domain and the geographic domain of a categorical coverage are both forms of a set and partition and therefore, share the behavior of a partition. As defined by refinement of partition, a partition of a set can be further refined such that an additional partition is created. Successive refinement of partitions on a set will result in an ordered hierarchy of partitions (Fig. 7).

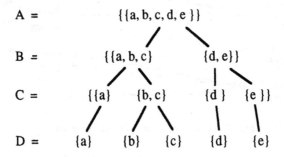

Fig. 7. Ordered hierarchy of partitions.

Spatially, this refinement can be visualized by the decomposition of a space into a set of mutually exclusive and collectively exhaustive polygons (Fig. 8).

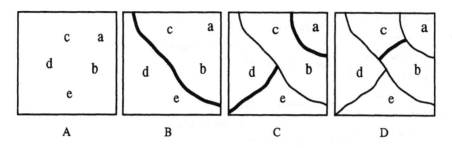

Fig. 8. Corresponding spatial partitions.

Partition *D*, in both the attribute example (Fig. 7) and the geographic example (Fig. 8), which was constructed from *A* through *B* and *C* respectively, is the most refined partition of this set since all the elements of the initial set are themselves blocks of the partition. This partition, *D* in Fig. 7, is analogous to the categories of a categorical coverage. In a GIS, a most detailed partition (a set of most detailed categories) is predetermined by the data stored in the system. This partition is the link to the formal notion of refinement of partitions. This categorical coverage contains the maximum number of categories, which determines the maximum number of zones possible for this coverage.

Additional, less detailed coverages can be created from this initial coverage by classifying the categories into more general categories. This may induce an aggregation of zones in the geographic domain. This classification operation, defined as coarsening, must abide by the same rules as for refinement of partitions. Chrisman (1982) and Beard (1988) have used the term *coarsening* with respect to categorical coverages to describe a spatial operation that is intended to remove sliver polygons generated from an overlay. Their definition of coarsening is meant as a spatial, attribute, and topological generalization tool. Coarsening, as it is used in this paper, is strictly an attribute classification mechanism. For the coverage to remain a categorical coverage, the categories can only be coarsened into one higher category (maintains mutually exclusivity) and all categories must be included in at least one higher category or it becomes one itself (maintains collectively exhaustive property). The categorical coverage that results from a coarsening will have fewer categories and potentially fewer zones. The relationship between set partitioning and attribute coarsening is such that in the case of the categorical coverage, the blocks are built from merging the individual attribute categories (elements), whereas the blocks in set partitioning are built from decomposing an initial set into individual elements.

As refinement partially orders partitions, coarsening partially orders the categories of the attribute domain. Therefore, once a categorical coverage hierarchy is built through coarsening (Fig. 9), it is possible to state that a categorical coverage is a refinement of another.

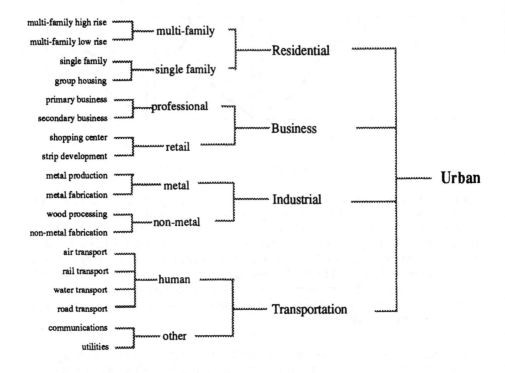

Fig. 9. Multiple attribute coarsenings

By the nature of the coarsening operation, the initial set of attribute values becomes a refinement of the partition resulting from the first coarsening (Residential, Commercial, Industrial, and Transportation) and this partition is then a refinement of the partition resulting from the second coarsening (Urban).

The refinement of a partition relation is such that a partition can be a non-proper refinement of itself if $A \leq B$ and $B \leq A$, implies that $A=B$. Proper refinement, on the other hand, states that at least one subset of the refined partition must be a proper subset of the initial partition. The inverse, or *proper coarsening*, as applied to the attribute domain of a categorical coverage, could then be defined to be any coarsening where at least two elements of the previous attribute partition are classified together to create a "new" attribute category. In this case, it can be said that the geographic representation will contain less than or equal to the number of zones of the previous partition.

The geographic property that determines the number of zones is *contiguity*. If the geographic representation after a proper coarsening does not change (same number of zones) then the zones involved in the attribute coarsening are not contiguous. Contiguous zones are defined as any two zones sharing a common edge such that they exhibit the "1-meet" relation between regions (Egenhofer and Herring 1990). Any two or more contiguous zones whose corresponding attribute values are coarsened to the same category are also geometrically merged to form one zone. This merging operation increases zone areas, which graphically results in a less detailed geographic representation.

Coarsening will usually result in partitions consisting of fewer, but larger zones and fewer and less descriptive attribute categories. This is sufficient for organization of the attribute domain, but is insufficient for proper exploratory analysis of the domains. A selection operator that is be constrained by the rules of refinement, is needed to enhance this analysis ability. Such an operator should provide for varying limbs of an attribute hierarchy to be selected for display, thus allowing varying levels of attribute detail, without explicitly requiring another attribute coarsening.

Once any attribute hierarchy is completely built, an additional operator is needed to directly control the display and facilitate exploration of the hierarchy.

4.3 Family of categorical coverages

In a family of categorical coverages, all family members $(M_1, ..., M_f)$ are attribute hierarchies and are built from the same initial set of attribute values and corresponding geographic zones by coarsening the attribute domain. The complete set of hierarchies constitutes a family of categorical coverages.

Within a member of a family, it is possible to distinguish a refinement without detailing whether it is a refinement of the attribute domain or a refinement of the space partition (it is automatically a refinement of both). However, this is only true for refinement, not proper refinement: a partition of the attribute domain, which is a proper refinement of another one, only induces a refined partition of space (not a proper refinement of the partition of space).

5   A Visualization

The previous section detailed the objects and operations that linked the categorical coverage to the mathematical concepts of the partition. This section discusses the visualization of a user interface based on the principles set forth by the formalism. The user interface presents the objects and operations of the categorical coverage concept to the user through the framework established by the formalism.

5.1   Objects

The separation and dependence of the attribute domain and the geographic domain fit nicely to the user interface concept of a dual representation. An example of this concept using two windows of the respective components of a categorical coverage is shown in Fig. 10. This example of a dual representation using windows is continued throughout the remainder of this section to illustrate additional elements of the categorical coverage interface. The attribute domain is shown textually by attribute category. The geographic domain is shown geometrically. Since the categorical coverage is based solely on attribute manipulation, the attribute window will be where all the user interaction takes place. The geographic window is simply a passive window used for viewing the attribute interaction.

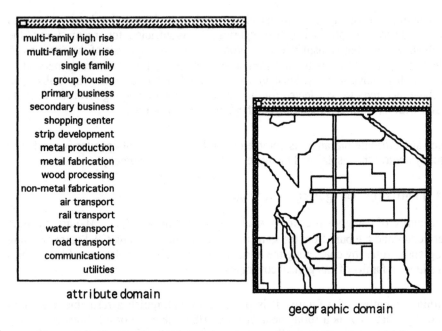

attribute domain

geographic domain

Fig. 10. Window interface for categorical coverage.

5.2 Coarsening

The primary attribute manipulation operation is the coarsening operation. Fig. 11 visualizes this concept of attribute coarsening in the context of the window interface shown in Fig. 10. In this example, "new" attribute categories are created to lessen the thematic or geographic complexity of the categorical coverage.

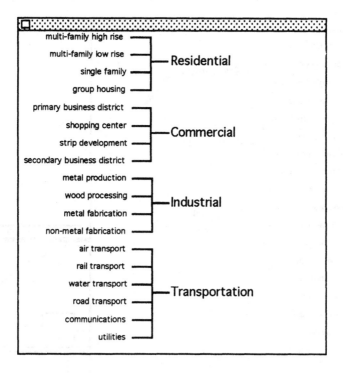

Fig. 11. Attribute coarsening.

In order to enhance exploratory analysis of an existing categorical coverage partition hierarchy, a selection tool can be added to the interface. Such a tool could select/unselect a portion of the hierarchy for viewing. In Fig. 12, the higher level category of Transportation has been selected and hidden in favor of the more detailed categories of air transport, rail transport and so on. This operation carried out by a pointing device is shown progress on the category Residential. This operation allows a user to quickly obtain the desired level of complexity of an existing hierarchy without the nuisance of coarsening the coverage from the initial set of attributes for each session.

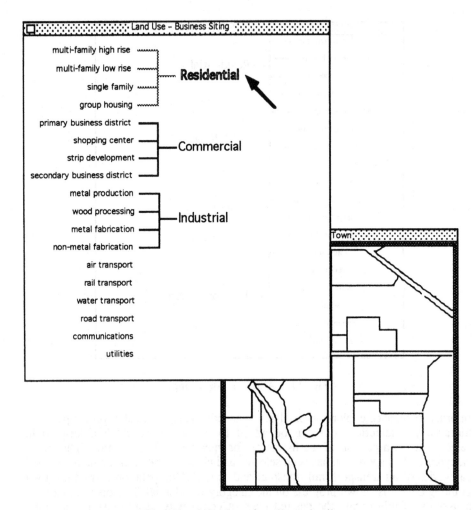

Fig. 12. Selection of attribute detail.

5.3 Family of categorical coverages

The user interface for the family of categorical coverages incorporates all of the above functionality and must also provide the means to choose a family of coverages and a particular coverage. In Fig. 13, another window is added to handle this functionality. The family of coverages is chosen by a click and select of a "pop-up" menu. Other families of coverages for the geographic location of Old Town could be Soils, Parcels and so on. The number of initial attribute values of any particular family is given as ancillary information. Within any single family, many coverages may exist. A similar "pop-up" menu is provided for this operation as well. The number of existing partitions and zones is given in this case to aid the user in the selection.

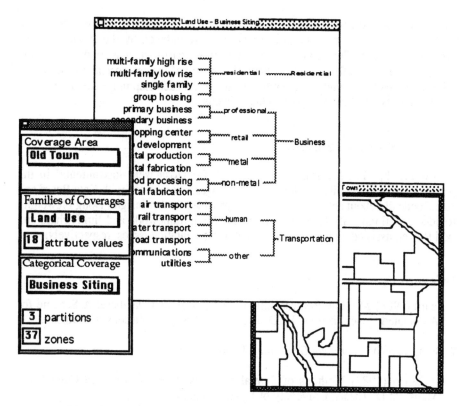

Fig. 13. User interface to select family of coverage and member.

6 Conclusions

This paper discussed GIS user interface design. The novel approach was the conceptual separation of the entire process into *formalization, visualization/interaction,* and *implementation.* During the formalization, designers identify the objects and pertinent operations a user wants to perform. Of particular importance is here the intense knowledge of the application domain and what prospective users will do with the planned system. During a second phase, designers identify interaction techniques that are best suited to implement the intended users interface. Here, particular considerations of cognitive and ergonomic issues are important for the successful design of a user interface. Finally, user interface designers who master the particular software implementation aspects will implement one or more user interfaces on a variety of hardware and software platforms. There is some overlap among the three steps and certainly members of the different subsets of the design team will interact and provide feedback to each other.

Within this framework, we discussed the design of a user interface for interacting with attribute data in a GIS. First, the well-known cartographic concept of a categorical coverage was formalized for the purpose of designing a human-computer interface. The formalism relies heavily on partition theory found in mathematics and serves as the conceptual model for functionality of the intended user interface. It was shown how the geographic zones in a categorical coverage are induced by the creation of the attribute categories. The refinement operation in set partitions was found to provide the fundamental rules for expanding the functionality of the categorical coverage in a GIS environment to include a hierarchy of categories, each step in the hierarchy itself a partition in the attribute and geographic domains.

7   Acknowledgments

Andrew Frank, Matt McGranaghan, and Bob Franzosa were instrumental in the formalization of the family of categorical coverages. Discussions with Kate Beard and Mike Goodchild about categorical coverages provided further insight. Doug Flewelling provided useful input about the user-interface design framework, and Jim Richard and Todd Rowell assisted with visualization and implementation issues, respectively.

References

K. Beard (1988) Multiple Representations from a Detailed Database: A Scheme for Automated Generalization. Ph.D. Thesis, University of Wisconsin, Madison.

W. Bunge (1966) *Theoretical Geography.* Lund, Sweden.

N. Chrisman (1982) Methods of Spatial Analysis Based on Error in Categorical Maps. Ph.D. Thesis, University of Bristol, England.

S. Clapham (1992) A Formal Approach to the Visualization of Spatial Data Quality. *GIS/LIS '92*, San Jose, CA, pp. 138-139.

M. Egenhofer and A. Frank (1988) Designing Object-Oriented Query Languages for GIS: Human Interface Aspects. *Third International Symposium on Spatial Data Handling*, Sydney, Australia, pp. 79-96.

M. Egenhofer and A. Frank (1992) Object-Oriented Modeling for GIS. *Journal of the Urban and Regional Information Systems Association* 4(2): 3-19.

M. Egenhofer and J. Herring (1990) A Mathematical Framework for the Definition of Topological Relationships. *Fourth International Symposium on Spatial Data Handling*, Zurich, Switzerland, pp. 803-813.

M. Egenhofer and J. Herring (1991) High-Level Spatial Data Structures for GIS. in: D. Maguire, M. Goodchild, and D. Rhind (Eds.), *Geographical Information Systems, Volume 1: Principles.* pp. 227-237, Longman, London.

M. Egenhofer and J. Richards (1993) Exploratory Access to Geographic Data Based on the Map-Overlay Metaphor. *Journal of Visual Languages and Computing* 4(2) (in press).

A. Frank (1992) Spatial Concepts, Geometric Data Models and Data Structures. *Computers and Geosciences* 18(4): 409-417.

A. Frank, M. Egenhofer, and W. Kuhn (1991) A Perspective on GIS Technology in the Ninties. *Photogrammetric Engineering & Remote Sensing* 57(11): 1431-1436.

A. Frank, G. Volta, and M. McGranaghan (1992) Formalization of Families of Categorical Coverages. Technical Report , Department of Surveying Engineering, University of Maine.

M. Gould and M. McGranaghan (1990) Metaphor in Geographic Information Systems. *Fourth International Symposium on Spatial Data Handling*, Zurich, Switzerland, pp. 433-442.

J. Jackson (1990) Developing an Effective Human Interface for Geographic Information Systems Using Metaphors. *ACSM-ASPRS Annual Convention*, Denver, CO, pp. 117-125.

K. C. Kirby and M. Pazner (1990) Graphic Map Algebra. *Fourth International Symposium on Spatial Data Handling*, Zurich, Switzerland, pp. 413-422.

W. Kuhn and M. Egenhofer (1991) CHI '90 Workshop on Visual Interfaces to Geometry. *SIGCHI Bulletin* 23(2): 46-55.

W. Kuhn and A. Frank (1991) A Formalization of Metaphors and Image-Schemas in User Interfaces. in: D. Mark and A. Frank (Eds.), *Cognitive and Linguistic Aspects of Geographic Space*. pp. 419-434, Kluwer Academic Publishers, Dordrecht.

W. Kuhn, J. Jackson, and A. Frank (1991) Specifying Metaphors Algebraically. *SIGCHI Bulletin* 23(1): 58-60.

D. Maguire, M. Goodchild, and D. Rhind, Eds. (1991) *Geographical Information Systems*. Longman, London.

D. Mark (1992) Spatial Metaphors for Human-Computer Interaction. *Fifth International Symposium on Spatial Data Handling*, Charleston, SC, pp. 104-112.

D. Mark and A. Frank, Eds. (1992) Research Initiative 13 Report on the Specialist Meeting: User Interfaces for Geographic Information Systems. National Center for Geographic Information and Analysis, University of California, Santa Barbara, Technical Report 92-3.

D. Mark and M. Gould (1991) Interaction with Geographic Information: A Commentary. *Photogrammetric Engineering & Remote Sensing* 57(11): 1427-1430.

R. Mulligan, M. Altom, and D. Simkin (1992) User Interface Design: When You *Do* Know How. *Communications of the ACM* 35(9): 19-22.

D. J. Peuquet (1988) Representations of Geographic Space: Toward a Conceptual Synthesis. *Annals of the Association of American Geographers* 78(3): 375-394.

A. Robinson, R. Sale, J. Morrison, and P. Muercke (1984) *Elements of Cartography* John Wiley & Sons, New York, NY.

S. Stevens (1946) On the Theory of Scales of Measurement. *Science Magazine* 103(2684): 677-680.

C. D. Tomlin (1990) *Geographic Information Systems and Cartographic Modeling.* Prentice-Hall, Englewood Cliffs, NJ.

M. Worboys, H. Hearnshow, and D. Maguire (1990) Object-Oriented Data Modelling for Spatial Databases. *International Journal of Geographical Information Systems* 4(4): 369-383.

The Semantics Of Relations In 2D Space Using Representative Points: Spatial Indexes

Dimitris Papadias[1] and Timos Sellis[1,2]
Computer Science Division
Department of Electrical and Computer Engineering
National Technical University of Athens
157 73 Zographou, Athens, GREECE
{dp,timos}@theseas.ntua.gr

Abstract. The paper describes *spatial indexes*, a 2D array structure which can be used for the representation of spatial information. A spatial index preserves only a set of spatial relations of interest, called the *modelling space*, and discards visual information (such as shape, size etc.) and information about irrelevant spatial relations. Every relation in the modelling space can be defined using a set of special points, called the *representative points*. By filling the index array cells with representative points we can gain adequate expressive power to answer queries regarding the spatial relations of the modelling space without the need to access the initial image or an object database.

1 Introduction

Several representational structures have been proposed for the representation of spatial knowledge in different areas. Depending on the particular viewpoint the goals have been explanatory and predictive power, in the case of psychological models of Vision and Imagery, expressive power and inferential adequacy, in the case of Artificial Intelligence knowledge representation schemes, or the efficient manipulation of large amounts of geographic and geometric data, in the case of Spatial Databases. Not all of the representation systems however distinguish spatial relations from visual information and some of them treat them as one inseparable entity. For example, computational structures, like the quadtrees [9] and the R+-trees [10], involved in geographic and geometric data management store both spatial and visual information in a unified form. The algorithms in this case usually retrieve spatial relationships by examining object co-ordinates, since relationships are not explicitly stored. In this sense we can say that the previous representational structures are *object-based* because they are concerned with the representation and storage of object characteristics. We will use the term *relation-based* for representational structures concerned with the explicit representation of spatial relationships, and not of individual objects.

[1] Research supported by the ESPRIT basic research project 6881 (*Amusing* Project).

[2] Research supported by the National Science Foundation under Grant IRI-9057573 (PYI Award).

Consider, for example the image of Figure 1. The image conveys information about the objects' appearance (e.g., shape, size, colour) which we will call *visual information*, as well as, *spatial information* about the relations in 2D space among the objects.

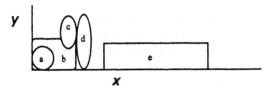

Fig. 1. Original image

We could classify the queries which we can answer using this image in:
- queries involving visual information regarding the objects (e.g., what is the shape of object b? is object c larger than object d ?)
- queries involving spatial relations (e.g., does object b overlap with object c? is e to the left of a?)

Relation-based representations of the image discard visual information and irrelevant spatial relations and preserve only a predefined set of spatial relations which is called the modelling space M; they can be used to answer queries of the second category. Buisson [1] argues that the spaces of interest in spatial reasoning are *topological spaces* which include only concepts of connectedness and continuity, *vector spaces* which deal with vectorial dimensions and directions, *metric spaces* which deal with the concept of distance and *Euclidean spaces* which admit notions of scalar products, orthogonality, angle and norm. In this paper we will only deal with the relation-based representation of modelling spaces consisting of topological and direction relations.

We will use the notation $s \models^M r(p, q)$ to denote that image s implies the spatial relation r ($r \in M$) between objects p and q. Two images s and t are said to be equivalent with respect to a modelling space M ($s \equiv^M t$) iff they imply the same subset of M for every pair of objects, that is : $s \models^M r(p, q) \Leftrightarrow t \models^M r(p, q)$. For instance, the image in Figure 2a is equivalent to the image of Figure 1 with respect to M_{D1}={north, east, ...} (i.e., direction relations in 2D space). Figure 2b is equivalent to Figure 1 with respect to M_T={meet, disjoint, overlap, contain, cover, equal} (i.e., topological relations). Notice that depending on their definitions, direction and topological relations might be related. Topological properties, for instance, might be inferred from direction relations as in [2]. Furthermore if two images are equivalent with respect to a modelling space M, they are also equivalent with respect to any subset of M.

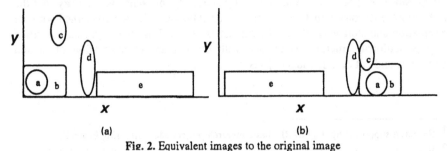

(a)   (b)

Fig. 2. Equivalent images to the original image

Let S^M the relation based representation which preserves the spatial relations of M that exist in an image s. Each spatial relation $r \in M$ between two object representations p and q in s is mapped onto a relation R between two symbolic object representations P and Q in S^M. For instance, using logic-based representations we could map the spatial relationships among 2D pixel sets representing objects in Figure 1 onto relation predicates whose arguments are symbols denoting distinct objects. According to [7], the relation-based representation of spatial knowledge (Hernandez uses the term *relative representation based on comparative relations*) avoids the falsifying effects of exact geometric representations by not committing to all aspects of the situation being presented; in this sense a relation based representation is underdetermined since it might correspond to many situations. Using our terminology we can say that each representation does not correspond to a single image, but to a class of equivalent images with respect to a modelling space i.e., $s \equiv^M t \Leftrightarrow S^M = T^M$.

Spatial representational formalisms which can be characterised as relation based include graph-based representations [8], logic-based representations [5], the 2D strings [2] and the symbolic arrays [6]. These representations incorporate spatial information about symbolically represented objects, and not visual information about the object characteristics. What we gain by using separate spatial representational structures is storage size, since irrelevant information is discarded, and computational efficiency in spatial knowledge retrieval, since the representation structures make the spatial relations of interest explicit. Furthermore, the extraction of spatial relations involves symbolic, and not numerical, computation and avoids the usual problems of geometric representations such as finite resolution and geometric consistency. Notice that we do not claim that relation based representations are adequate for all the applications involving spatial knowledge. For instance, they can not be used in problems that involve visualisation, quantitative reasoning etc. Nevertheless there exist several potential application domains where relation-based representations can be used independently (e.g., qualitative reasoning) or in conjunction with other representation structures (e.g., spatial databases).

This paper describes another representational structure, called a spatial index, which can be characterised as relation-based. Section 2 presents the spatial indexes and describes the notion of the construction process. Section 3 describes how spatial indexes can be used to capture direction relations using one and four points per object, while Section 4 enhances the expressive power of spatial indexes by incorporating topological relations. Section 5 concludes with some comments and a discussion about future work.

2 Description of Spatial Indexes

A spatial index is an array structure which can be used for the representation of spatial information. The spatial index is equivalent to the initial image from which it is generated with respect to the spatial relations of the presumed modelling space but discards visual information and irrelevant spatial relations. In order to construct a spatial index array from an input image, we assume a construction process that detects a set of special points in the image, called representative points. Every spatial relation in the modelling space can be defined using only the representative points. By filling the array cells with representative points we can gain adequate expressive power to answer queries regarding the spatial relations of the modelling space without needing to access the initial image or the object database.

The complexity of spatial indexes can be increased or decreased to match the representation and processing goals of a given application domain. A complex modelling space though (i.e., one involving a large set of spatial relations), implies a complicated construction process and expanded storage requirements. Figure 3 illustrates several relations whose formal definitions using representative points, will be given later in the paper. The first line denotes the direction relation between objects p and q while the second line denotes the topological relation between the two objects.

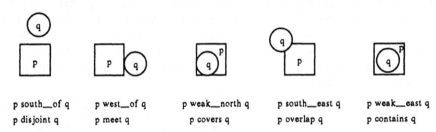

p south__of q   p west__of q   p weak__north q   p south__east q   p weak__east q

p disjoint q   p meet q   p covers q   p overlap q   p contains q

Fig. 3. Some direction and topological relations

A spatial index is a 2D array in which each object occupies a set of cells; each cell of the index can be empty, or it can be occupied by one (or more) point(s) belonging to one (or more) object(s). We use subscripts (e.g. S_i) to denote the individual cells of a spatial index S. The predicate $S_i(\partial P)$ means that the cell S_i is occupied by P's boundary. Similarly $S_i(P^O)$ denotes that the cell S_i is occupied by P's interior. We will use the predicate $S_i(P)$ to denote the fact that the cell S_i is occupied by object P. That is:
$S_i(P) = S_i(\partial P) \vee S_i(P^O)$.
The exterior of an object is denoted by the object's absence from a cell. That is:
$S_i(P^{-1}) = \neg\, S_i(P)$
A point that belongs to one of the previous sets (i.e. ∂P, P^O, P^{-1}) is denoted with a subscript, that is, P_1 is a point that belongs to P (to the boundary or to the interior) and ∂P_u is a point of P's boundary (subscripts can be characters, in the case of direction representative points, or numbers). The predicate S_i can also be used with points instead of sets; for instance, $S_i(\partial P_k)$ denotes that the cell S_i of a spatial index is occupied by the point ∂P_k. More than one objects might occupy the same cell; for instance, it is allowed to have $S_i(\partial P) \wedge S_i(\partial Q)$.

We denote with $S^{\partial P}$ the set of cells occupied by object P's boundary and with S^{PO} the set of cells occupied by object P's interior; that is:
$S^{\partial P} = \{\ S_i\ /\ S_i(\partial P) = \text{true}\ \}$,
$S^{PO} = \{\ S_i\ /\ S_i(P^O) = \text{true}\ \}$.
With S^P we denote the set of cells occupied by object P: $S^P = S^{\partial P} \cup S^{PO}$.
Notice that when two objects P and B are not disjoint we have: $S^P \cap S^Q \neq \varnothing$.
We follow the same notation to denote the cell occupied by an individual point; for instance $S^{\partial P_k}$ denotes the cell occupied by point ∂P_k.
Each spatial index corresponds to a class of equivalent images. This property allows for efficient retrieval of equivalent images. If we do not use spatial indexes, in order to determine equivalence between two images we would have to search for a transformation chain that would transform one image into the other. The construction

process helps avoid the combinatorial search for transformation chains connecting the two expressions by computing the corresponding spatial indexes and comparing them for identity. This feature is of great importance in cases, like visual scene matching, where equivalent images should have the same internal representations. Notice that when there exist more than one spatial indexes corresponding to the same equivalence class the construction process is responsible for unambiguously creating the "correct" one. Multiple options arise in cases where two or more points satisfy the conditions to be representative points. In such cases one point should be chosen in a deterministic way, in order to avoid ambiguities. For instance, the construction process should pick the one with the smaller y co-ordinate and in the case of equality of y co-ordinates, the one with the smaller x co-ordinate. We will follow this approach in the paper, that is, the construction process can be thought of, as a line that sweeps the plane from bottom to up and from left to right detecting the representative points. The first representative point marked for each object is kept in the resulting spatial index. That is:

```
Construction Process(s, M)
for y=1 to ymax
        for x=1 to xmax
                If detect (representative_point, M) then
                        If first (representative_point) then
                                mark (representative_point)
                        else
                                discard (representative_point)
        next x
next y
generate_array(S)
put(S, representative_points)
return(S)
```

A similar process, called the *cutting function*, was used in [2] in order to create symbolic object projections on x and y axes. Our construction process, instead of detecting and recording differences of object projections, marks representative points needed to represent the spatial relations of interest. After the representative points are marked, they are positioned in the proper cells in a spatial index. By keeping the relative arrangements of representative points in spatial indexes, and discarding all the other points, we capture the spatial relations of the assumed modelling space. The representative point detection mechanism depends on the input image representation. Although we do not commit to a particular representation we assume that the format of the input renders the positions, the boundaries and the interiors of the distinct objects explicit. With this assumption the detection part requires time $O(N)$ (N is the number of pixels in the initial image) when the detection does not involve pre-processing (e.g., in case that edge points, like the northern etc. are used as representative points). Otherwise the complexity of the detection mechanism depends on the process that pre-computes the desired representative points. In the next two sections we will demonstrate how the construction process detects representative points and generates spatial indexes that preserve the spatial relations of the modelling space.

3 Representation of Direction Relations in Spatial Indexes

For the representation system of this section we assume modelling spaces of direction relations, that is, the relations of interest include, *north_of, east_of, south_of, west_of, northeast, same_level* etc. Notice, that the meaning of these relations is not obvious. For instance, most people will agree that England is north of Portugal, but what about the relation between Portugal and Spain? There are parts of Spain that are directly north of any part of Portugal, but is it enough for stating that Spain is north of Portugal? In the rest of the section we will specify the meaning of the direction relations by using representative points and we will show how we can map direction relations onto relations among representative points stored in spatial indexes. We will use two sets of definitions for these relations: the first one defines the direction relation between two objects by using one representative point per object while the second set of definitions uses four representative points per object.

3.1 Representation of Direction Relations Using One Point per Object

It is common for some cases in everyday life to determine the spatial relation between two objects by examining the objects' centers. We will denote the center of an object P by P_*^O. The term center though, is not well defined; it might imply the center of mass, the center of symmetry etc. For instance, if we use mathematical morphology for the implementation of the detection mechanism, we can define the center as the last pixel that remains if we continuously erode P with a structuring element. Using one representative point (the center) per object, we can define a mapping relation that maps spatial relations among objects in the initial image, to relations among the object centers in a spatial index.

We will use the term *primary object* for the object to be located and the term *reference object* for the object in relation to which the primary object is located. Order on axis x determines west-east direction and order on axis y determines south-north direction. When one point is used for the representation of the reference object the plane is divided into nine partitions. The symbol * at the beginning of the co-ordinate axes in Figure 4a denotes the reference object, while the numbers correspond to the possible positions of the primary object with respect to the reference object (Figure 4b). Let X a function that returns the x co-ordinate of a cell S_i and Y a function that returns the y co-ordinate of a cell S_i. Then the direction relations of the modelling space M_{D1} can be defined as follows:

$p \ north\text{-}west \ q \Leftrightarrow X(S^{P*}) < X(S^{Q*}) \ \wedge \ Y(S^{P*}) > Y(S^{Q*})$

$p \ north\_of \ q \Leftrightarrow Y(S^{P*}) > Y(S^{Q*}) \ \wedge \ X(S^{P*}) = X(S^{Q*})$

$p \ north\text{-}east \ q \Leftrightarrow X(S^{P*}) > X(S^{Q*}) \ \wedge \ Y(S^{P*}) > Y(S^{Q*})$

$p \ west\_of \ q \Leftrightarrow X(S^{P*}) < X(S^{Q*}) \ \wedge \ Y(S^{P*}) = Y(S^{Q*})$

$p \ same\_pos \ q \Leftrightarrow X(S^{P*}) = X(S^{Q*}) \ \wedge \ Y(S^{P*}) = Y(S^{Q*})$

$p \ east\_of \ q \Leftrightarrow X(S^{P*}) > X(S^{Q*}) \ \wedge \ Y(S^{P*}) = Y(S^{Q*})$

$p \ south\text{-}west \ q \Leftrightarrow X(S^{P*}) < X(S^{Q*}) \ \wedge \ Y(S^{P*}) < Y(S^{Q*})$

$p \ south\_of \ q \Leftrightarrow Y(S^{P*}) < Y(S^{Q*}) \ \wedge \ X(S^{P*}) = X(S^{Q*})$

$p \ south\text{-}east \ q \Leftrightarrow X(S^{P*}) > X(S^{Q*}) \ \wedge \ Y(S^{P*}) < Y(S^{Q*})$

The previous relations correspond to the highest resolution we can achieve using one point per object. Additional direction relations can be defined using the basic relations of M_{D1}. For instance, we can define a relation: $p \ same \ level \ q \Leftrightarrow p \ east\_of \ q \vee p \ west\_of \ q \vee p \ same\_pos \ q$. Figure 4c illustrates how direction relations among objects on the plane are mapped onto relations among object symbols in a spatial index. The symbol * in 4c denotes the reference object symbol and the numbers refer to the direction relation depending on the position of the primary object symbol in the index array. The relation between two objects does not change if we add or remove from the index array lines and columns that do not contain the objects.

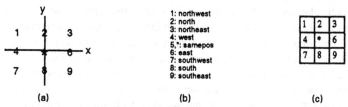

(a)　　　　　　　　　(b)　　　　　　　　　(c)

Fig. 4. Direction relations using 1 point per object

Figure 5b illustrates the spatial index $S^{M_{DI}}$ that preserves the direction information of Figure 1 using the centers of the objects as representative points. The symbol * in Figure 5a denotes the centers of the of the objects in the initial image and the grey lines correspond to the instances that the construction process has detected a representative point. The number of horizontal lines equals the number of rows, and the number of the vertical lines equals the number of columns of $S^{M_{DI}}$.

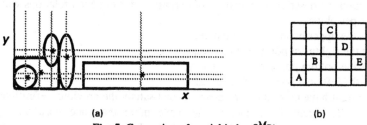

(a)　　　　　　　　　　　　　　　(b)

Fig. 5. Generation of spatial index $S^{M_{DI}}$

In an actual system there might be a name index of the objects pointing to the spatial index in order to facilitate object retrieval within the index array. Using this name index we can find an object in the array in time O(log m) (m is the number of objects), instead of O(n) when we have to search for the object in the whole array (n is the number of cells in the array; the maximum value of n is m^2 when there is exactly one object in each row and each column). Furthermore there should be pointers from the spatial index to the object records in the database in order to enable the system to answer queries involving additional information. Such an implementation of spatial indexes is illustrated in Figure 6.

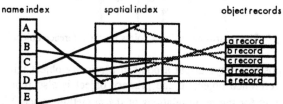

Fig. 6. A possible implementation of spatial indexes

Let, for example the query "is there a red telephone east of C". Using the name index we locate object C in the spatial index and then we select the objects east of C (the objects that exist in the last two columns). Using the pointers from these objects (i.e., D and E) to the object database we retrieve the additional information concerning the query (i.e., colour and use of the objects).

If we use one point per object the generated index $S^{M_{DI}}$ is similar to the grid file organisation (but lacks the notion of distance) and resembles the symbolic array structure (but lacks the nesting capability). The assumed modelling space M_{D1} is adequate if we deal with objects of regular shape (e.g., convex objects) and in cases where we do not want to store much information. For instance we might want to preserve the information that object c is north of b, without the fact that some parts of b are north of some parts of c. But consider that we want the ability to answer queries of the form "are there parts of c, which are south of some parts of b", or that we deal with spatial arrangements of irregularly shaped objects (e.g., objects with holes), lines etc. In this case the positions of the centers do not convey adequate spatial information and we need representations that preserve more than one representative points for each object.

3.2 Representation of Direction Relations Using Four Points per Object

Depending on the application domain, there are several options in case that we want to use more than one points per object for the definition of direction relations. For instance, we could use the two points within the object, whose Euclidean distance is maximum. In this subsection we describe a second set of definitions for direction relations which uses four points per object.

P_u denotes the upper point of P's boundary,
P_r denotes the rightmost point of P's boundary,
P_b denotes the bottom point of P's boundary,
P_l denotes the leftmost point of P's boundary,

An object might have more than one elements that satisfy the previous definitions of the edge points (for example a country can have more than one northmost points). Furthermore the same element might denote more than one of the special points (for example the eastmost point of a country might also be the southmost).

When we use four points for the representation of the reference object the plane is divided into 81 partitions (Figure 7). The symbol * in Figure 7 denotes the four representative points, and the numbers the partitions of the plane.

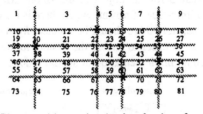

Fig. 7. Plane partitions using 4 points for the reference object

Thus there are 81 possible ways to put a point of a primary object with respect to the four points of the reference object. If the primary object is also represented by four points the upper bound for the basic relations of the modelling space M_{D4} is 81^4 (i.e., the number of possible ways of putting four points into 81 partitions when more than one points are permitted in each partition; not all of the configurations, however, are valid). In general, if k is the number of points used to represent the reference object, then the plane is divided into $(2k+1)^2$ partitions; k^2 of the partitions are points, $2k(k+1)$ are open line segments and $(k+1)^2$ are regions. If l is the maximum number of points used to represent the primary object, then the upper bound of the basic direction relations that we can represent is $(2k+1)^{2l}$.

Of the possible relations that M_{D4} contains, only a subset might be needed for an application domain. This subset might not require the highest possible resolution, i.e., the relations might not correspond to a single partition, but to a set of partitions. For instance we can give the following definitions:

p north_of $q \Leftrightarrow Y(S^{\partial P}u) > Y(S^{\partial Q}u)$

p east_of $q \Leftrightarrow X(S^{\partial P}r) > X(S^{\partial Q}r)$

p south_of $q \Leftrightarrow Y(S^{\partial P}b) < Y(S^{\partial Q}b)$

p west_of $q \Leftrightarrow X(S^{\partial P}l) < X(S^{\partial Q}l)$

When we use four points instead of one we can represent more detailed spatial knowledge. For instance, we can define several refinements of the north relationship such as:

p strong_north $q \Leftrightarrow Y(S^{\partial P}b) > Y(S^{\partial Q}u)$ (Figure 8a)

p weak_north $q \Leftrightarrow Y(S^{\partial P}u) > Y(S^{\partial Q}u) \land Y(S^{\partial P}b) \geq Y(S^{\partial Q}b) \land Y(S^{\partial P}b) < Y(S^{\partial Q}u)$ (Figure 8b)

p north_south $q \Leftrightarrow Y(S^{\partial P}u) > Y(S^{\partial Q}u) \land Y(S^{\partial P}b) < Y(S^{\partial Q}b)$ (Figure 8c)

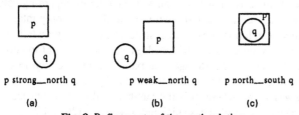

p strong—north q   p weak—north q   p north—south q

(a)   (b)   (c)

Fig. 8. Refinements of the north relation

According to the previous definitions Spain is *weak_north* of Portugal while England is *strong_north* of Portugal. Similar refinements might be necessary for many applications and cannot be achieved if we use one point per object. Although the basic relations of M_{D4} are not given specific names they are preserved in $S^{M_{D4}}$. Figure 9a illustrates the representative edge points for the objects of the image of Figure 1 and Figure 9b contains the corresponding spatial index $S^{M_{D4}}$. In case that, two or more, points satisfy the conditions to be representative points (e.g., all the points of the lower line boundary of objects b and e can be used as bottom points) the construction process picks the one that is detected first, that is the one with the smaller y co-ordinate, and in the case of equality of y co-ordinates, the one with the smaller x co-ordinate. Notice that for objects b and e there exist only three representative points for each, since the construction process considers the first point (bottom left) as being both bottom and leftmost point, the second representative point (bottom right) as the rightmost and the third (upper left side) as the upper point and as a consequence a fourth point is not needed. Since all the points belong to object boundaries we can omit the symbol ∂ from $S^{M_{D4}}$.

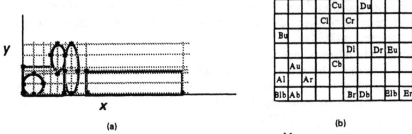

(a)   (b)

Fig. 9. Generation of spatial index $S^{M_{D4}}$

Notice that although the image of Figure 2a will generate the spatial index of image 5b using the centers as representative points, it will not generate the index of 9b using four representative points per object. That is, even when we deal with direction relations only, image equivalence depends on the choice of representative points. By increasing the number of representative points per object we extend the number of spatial relations that we can represent. In any case the spatial knowledge preserved in a spatial index is a subset of the information stored in the initial image. By using spatial indexes though, we gain storage size, since irrelevant information is discarded, and efficiency in retrieval, since the indexes make direction relations explicit. In order to demonstrate the usefulness of spatial indexes consider the following example. Assume that we are given a map that contains the countries of Europe (e.g., 10^3 x 10^3 pixels) in the form of a segmented image. With this image we can answer queries regarding relative sizes of countries, relative distances and spatial relationships. If we are only interested in queries involving 2D direction relationships, then instead of keeping the whole image, we can construct and store a spatial index, which does not preserve information about size, shape or distances but makes the storage and extraction of direction relationships more efficient. The maximum size of the spatial index corresponding to the input image of the map of Europe is 4m x 4m, where m is the number of distinct countries; that is, for 30 countries the maximum size of the spatial index will be 120 x 120 and there will be exactly one non-empty element in each row and each column. This happens because the definitions of direction relationships, involve only four points and the spatial index should contain only these four points for each object in order to preserve these relationships.

Consider now that we want to answer the question whether Greece is north of Germany. In the initial image we would have to find and compare the points used in the definition of the *north_of* relation for each country (time complexity O(N) where N is the number of pixels in the initial image). In a spatial index we just find for each country, using a name index, the points of interest involved in the definition of the *north_of* relation, and check whether they satisfy the definition (time complexity O(log m) to find the representative points of an object using a name index).

4 Representation of Topological Relations in Spatial Indexes

Usually spatial queries, in addition to direction relations, involve neighbourhood problems, inclusion problems etc. Thus a system which can only answer queries involving direction relations might be inadequate for many practical applications. In this section we will show how we can incorporate a space M_T of topological relations in the modelling space M_{D4} of the previous system (i.e., the system that represents direction relations using four points per object). In particular we will show, how spatial indexes can be used to capture the topological relations *disjoint, meet, equal, overlap, contains* and *covers* defined in [3].

The topological relationship between any two objects is described by the intersections of one's boundary and interior, with the boundary and the interior of the other. The following definitions are equivalent to the 4-intersection definitions (the relations *inside* and *covered_by* are not included since they are the converse of *contains* and *covers* respectively).

p disjoint q $\Leftrightarrow (S^{\partial P} \cap S^{\partial Q} = \varnothing) \wedge (S^{P^O} \cap S^{\partial Q} = \varnothing) \wedge (S^{\partial P} \cap S^{Q^O} = \varnothing) \wedge (S^{P^O} \cap S^{Q^O} = \varnothing)$

p meet $q \Leftrightarrow (S^{\partial P} \cap S^{\partial Q} \neq \varnothing) \wedge (S^{P^O} \cap S^{\partial Q} = \varnothing) \wedge (S^{\partial P} \cap S^{Q^O} = \varnothing) \wedge (S^{P^O} \cap S^{Q^O} = \varnothing)$

p equal $q \Leftrightarrow (S^{\partial P} \cap S^{\partial Q} \neq \varnothing) \wedge (S^{P^O} \cap S^{\partial Q} = \varnothing) \wedge (S^{\partial P} \cap S^{Q^O} = \varnothing) \wedge (S^{P^O} \cap S^{Q^O} \neq \varnothing)$

p contains $q \Leftrightarrow (S^{\partial P} \cap S^{\partial Q} = \varnothing) \wedge (S^{P^O} \cap S^{\partial Q} \neq \varnothing) \wedge (S^{\partial P} \cap S^{Q^O} = \varnothing) \wedge (S^{P^O} \cap S^{Q^O} \neq \varnothing)$

p covers $q \Leftrightarrow (S^{\partial P} \cap S^{\partial Q} \neq \varnothing) \wedge (S^{P^O} \cap S^{\partial Q} \neq \varnothing) \wedge (S^{\partial P} \cap S^{Q^O} = \varnothing) \wedge (S^{P^O} \cap S^{Q^O} \neq \varnothing)$

p overlap $q \Leftrightarrow (S^{\partial P} \cap S^{\partial Q} \neq \varnothing) \wedge (S^{P^O} \cap S^{\partial Q} \neq \varnothing) \wedge (S^{\partial P} \cap S^{Q^O} \neq \varnothing) \wedge (S^{P^O} \cap S^{Q^O} \neq \varnothing)$

The previous definitions did not use representative points but intersections among sets of elements. In order to transform them to definitions using points we have to replace each equation of the form: $S^P \cap S^Q \neq \varnothing$ with a formula of the form: $\exists S_i (S_i(P) \wedge S_i(Q))$ and each equation of the form $S^P \cap S^Q = \varnothing$ with a formula of the form $\neg \exists S_i (S_i(P) \wedge S_i(Q))$.

Thus, the definitions of the topological relations of M_T are transformed to:

p disjoint $q \Leftrightarrow \neg \exists S_i (S_i(\partial P) \wedge S_i(\partial Q)) \wedge \neg \exists S_i (S_i(P^O) \wedge S_i(\partial Q)) \wedge$
$\qquad \neg \exists S_i (S_i(\partial P) \wedge S_i(Q^O)) \wedge \neg \exists S_i (S_i(P^O) \wedge S_i(Q^O))$

p meet $q \Leftrightarrow \exists S_i (S_i(\partial P) \wedge S_i(\partial Q)) \wedge \neg \exists S_i (S_i(P^O) \wedge S_i(\partial Q)) \wedge$
$\qquad \neg \exists S_i (S_i(\partial P) \wedge S_i(Q^O)) \wedge \neg \exists S_i (S_i(P^O) \wedge S_i(Q^O))$

p equal $q \Leftrightarrow \exists S_i (S_i(\partial P) \wedge S_i(\partial Q)) \wedge \neg \exists S_i (S_i(P^O) \wedge S_i(\partial Q)) \wedge$
$\qquad \neg \exists S_i (S_i(\partial P) \wedge S_i(Q^O)) \wedge \exists S_i (S_i(P^O) \wedge S_i(Q^O))$

p contains $q \Leftrightarrow \neg \exists S_i (S_i(\partial P) \wedge S_i(\partial Q)) \wedge \exists S_i (S_i(P^O) \wedge S_i(\partial Q)) \wedge$
$\qquad \neg \exists S_i (S_i(\partial P) \wedge S_i(Q^O)) \wedge \exists S_i (S_i(P^O) \wedge S_i(Q^O))$

p covers $q \Leftrightarrow \exists S_i (S_i(\partial P) \wedge S_i(\partial Q)) \wedge \exists S_i (S_i(P^O) \wedge S_i(\partial Q)) \wedge$
$\qquad \neg \exists S_i (S_i(\partial P) \wedge S_i(Q^O)) \wedge \exists S_i (S_i(P^O) \wedge S_i(Q^O))$

p overlaps $q \Leftrightarrow \exists S_i (S_i(\partial P) \wedge S_i(\partial Q)) \wedge \exists S_i (S_i(P^O) \wedge S_i(\partial Q)) \wedge$
$\qquad \exists S_i (S_i(\partial P) \wedge S_i(Q^O)) \wedge \exists S_i (S_i(P^O) \wedge S_i(Q^O))$

For instance, we can say that two objects are *disjoint* if there does not exist a cell which stores points that belong to both objects. Two objects *meet* if there is one or more cells which store representative points that belong to both object boundaries and there do not exist any other common cells. Similarly we can proceed with the rest of the definitions for topological relations. The previous definitions using representative points allow for the straightforward development of procedures which scan the index array cells and retrieve common cells (i.e., cells which store points that belong to two or more objects). The intersections of the horizontal and vertical lines in Figure 10 illustrate the instances that the construction process has detected a representative point for direction or topological relations. Some of the representative points for direction relations are also used as representative points for topological relations; the points ∂A_b and ∂A_l, for instance, intersect with ∂B and we use them for the intersection $\partial A \cap \partial B$. Since the direction representative points do not suffice for all the intersections (e.g., the intersection $A^o \cap B^o$) the construction process has to detect additional points used for the intersections that do not involve edge points. The construction process will keep the first point for each of the rest intersections and will skip the subsequent points involved in intersections. Since the direction representative points are illustrated in Figure 9a we omit them in Figure 10 and we denote with * the representative points involved only in topological relations.

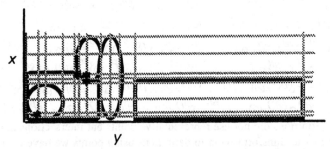

Fig. 10. Detection of representative points involved in intersections

Notice that while the number of points per object needed for the direction relations is independent of the relations of the object with the other objects (i.e. one or four points for the first and the second case of the previous section), the number of points per object for the topological relations depends on the configuration. Figure 11 illustrates the spatial index S^{MTD4} which incorporates both direction and topological information about the image of Figure 1.

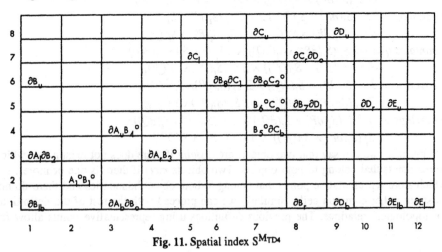

	1	2	3	4	5	6	7	8	9	10	11	12
8							∂C_u		∂D_u			
7				∂C_l				$\partial C_r \partial D_o$				
6	∂B_u					$\partial B_8 \partial C_1$	$\partial B_9 C_2{}^o$					
5							$B_6{}^o C_o{}^o$	$\partial B_7 \partial D_1$		∂D_r	∂E_u	
4			$\partial A_u B_4{}^o$				$B_5{}^o \partial C_b$					
3	$\partial A_l \partial B_2$			$\partial A_r B_3{}^o$								
2		$A_1{}^o B_1{}^o$										
1	∂B_{lb}		$\partial A_b \partial B_o$					∂B_r	∂D_b		∂E_{lb}	∂E_l

Fig. 11. Spatial index S^{MTD4}

We can use the spatial index to generate the matrix I(A,B) corresponding to the topological relation between objects a and b in Figure 1 (in the parentheses there are the co-ordinates of the cells that store the representative points for the corresponding intersections):

$S_i(\partial A) \cap S_i(\partial B) \rightarrow (1,3),(3,1)$ $S_i(\partial A) \cap S_i(B^0) \rightarrow (3,4), (4,3)$

$S_i(A^0) \cap S_i(\partial B) \rightarrow \varnothing$ $S_i(A^0) \cap S_i(B^0) \rightarrow (2,2)$

Thus the representation formalism is equivalent to Egenhofer's 4-intersection matrices with respect to topological relations. Notice that the array $S^{M_{TD4}}$ preserves more information than the set of direction relations of M_{D4} plus the topological relations of M_T. That is, the array cannot be represented using, for instance, the array of $S^{M_{D4}}$ and a set of intersection matrices, since it also contains the relative positions of intersections with respect to the direction representative points. This property allows for the definition of relations that belong to M_{TD4}, but not to M_T or M_{D4}. For example, we can define a relation west_touch as:

$$p \ west\_touch \ q \Leftrightarrow \exists \ S_i \ ((S_i(\partial P) \wedge S_i(\partial Q)) \wedge (X(S_i) = X(S^{\partial P}l)) \wedge (X(S_i) = X(S^{\partial Q}l)))$$

Objects a and b in Figure 1 satisfy the relation west_touch since there exists a point in the leftmost part of their boundaries where the two objects touch. In the array of Figure 11 this point corresponds to the cell (1,3). When the positions of the intersections in the array are not important, further compression of the array is possible. For instance we can move the contents of column 2 to another column and remove the column from the index array. Eventually we can generate an array with the dimensions of the array used for the direction relations (i.e., array of Figure 9b).

5 Discussion

The paper describes spatial indexes, a 2D array structure which can be used for the representation of a set of direction and topological relations, called the modelling space. Spatial indexes preserve the spatial relations of the modelling space and discard visual information and irrelevant spatial relations. Although we assumed distinct modelling spaces M_D and M_T in this paper, we do not argue that direction relations are completely independent from topological relations. For instance, the partitions denoted with * (number 5 in Figure 4a and numbers 13, 29, 53 and 69 in Figure 7) incorporate topological in addition to direction information. Furthermore, it is not allowed to have certain direction relations in conjunction with some topological relations among the same objects. For example, according to the previous definitions it is not permitted to have $p \ strong\_north \ q$ and $p \ overlap \ q$ in the same array.

The expressive power of spatial indexes can be extended or decreased according to the application domain. If, for instance, we do not need to express the relations *cover* and *overlap* we do not need the distinction between boundary and interior of an object and thus we do not have to use separate symbols in our implementation. On the other hand if, we want to represent topological relations among region, line and point objects we need the 9-intersection definitions [4]. If we want a further extension of the expressive power to include the dimension of the intersections, we should add symbols to distinguish line from point data etc.

Topics which can be considered for further investigation include the implementation of computational construction processes which detect representative points and construct spatial indexes from input images. In addition, compression techniques can be developed to minimise the size of spatial indexes. Several other theoretical topics that evolved during this work include the combinatoric study of spatial relations using representative points (e.g., the number of spatial relations that we can represent as a function of the representative points per object), the notion of image equivalence with respect to a modelling space, as well as information equivalence of relation based representations.

References

1. Buisson, L. (1989) Reasoning On Space With Object-Centered Knowledge Representations. In Buchmann, A., Gunther, O., Smith, T., Wang, T., (eds.) Proceedings of the First Symposium on the Design and Implementation of Large Spatial Databases (SSD 89). Springer-Verlag.

2. Chang, S.K, Jungert, E., Li, Y. (1989) The Design of Pictorial Database upon the Theory of Symbolic Projections. In Buchmann, A., Gunther, O., Smith, T., Wang, T., (eds.) Proceedings of the First Symposium on the Design and Implementation of Large Spatial Databases (SSD 89). Springer-Verlag.

3. Egenhofer, M., Herring, J. (1990) A Mathematical Framework For The Definitions Of Topological Relationships. In Brazzel, K., Kishimoto, H. (eds.) 4th International Symposium on Spatial Data Handling.

4. Egenhofer, M. (1991) Reasoning About Binary Topological Relations. In Gunther, O. and Schek, H.J. (eds.) Advances in Spatial Databases. Second Symposium, SSD 91. Springer-Verlag.

5. Freksa, C. (1991) Qualitative Spatial Reasoning. In Mark, D.M. and Frank, A.U (eds.) Cognitive and Linguistic Aspects of Geographic Space. Kluwer, Dordrecht 91.

6. Glasgow, J.I, Papadias, D. (1992) Computational Imagery. Cognitive Science, 16, pp 355-394.

7. Hernandez, D. (1990) Using Comparative Relations to Represent Spatial Knowledge. Workshop R.A.U.M., University Koblenz.

8. Levine, M. (1978) A Knowledge-Based Computer Vision System. In Hanson, A. and Riseman, E. (eds.) Computer Vision Systems. New York: Academic. pp 335-352.

9. Samet, H. (1989) The Design And Analysis Of Spatial Data Structures. Addison Wesley.

10. Sellis, T., Roussopoulos, N., Faloutsos,C. (1987) The R^+-tree: A Dynamic Index for Multi-Dimensional Objects. In P. Stocker and W. Kent (eds.) 13th VLDB conference, Brighton, England.

Computing Visibility Maps on a Digital Terrain Model

Leila De Floriani   Paola Magillo

Department of Computer and Information Sciences
University of Genova
Viale Benedetto XV, 3 - 16132 Genova (Italy)

Abstract. In this paper, we present a model based on a collection of nested horizons to describe the visibility of a terrain with respect to a viewpoint. We introduce first a formalization of mathematical and digital terrain models, and some background notions for visibility problems on terrains. Then, we define horizons and shadows on a polyhedral terrain, and introduce a few horizon-based visibility maps of a terrain. Finally, we present two algorithms for building the nested horizons on a polyhedral terrain. An application to the solution of point-to-point visibility queries are briefly discussed.

1   Introduction

Describing a terrain through visibility information has a variety of applications, such as geomorphology, navigation and terrain exploration. Problems which can be solved based on visibility are, for instance, the computation of the minimum number of observation points needed to view a region, the computation of paths with specified visibility characteristics (e.g., hidden paths with respect to a predefined set of observers, scenic paths from which large portions of the terrain can be viewed), the computation of optimal locations for television transmitters [2,12,14]. In terrain navigation problems the profile of the horizon is an ideal tool which can be used by an observer to locate his/her position on a map.

In this paper, we address the problem of representing the visibility of a terrain, from a given viewpoint, based on a nested horizon structure. A terrain can be mathematically described by a continuous function $z = f(x, y)$, defined over a connected domain D. Two points on a terrain are considered mutually visible when they can be joined by a straight-line segment lying above the terrain (and intersecting it only at its two extreme points). In practice, a mathematical terrain model is approximated through a digital model, usually built from a finite set of points belonging to the terrain. Visibility algorithms developed in the literature operate on digital terrain models (DTMs) and, usually, on so-called polyhedral models (i.e., planar-faced approximations of the terrain).

The horizon of a viewpoint, in the usual sense, provides, in each direction around the viewpoint, the minimum elevation which must have a visual ray emanating from the viewpoint in the given direction to pass above the terrain. Intuitively, the *i-th horizon* provides, for every radial direction, the i-th obstacle for the view of an observer located at the given viewpoint; the corresponding *i-th shadow* marks the point, behind the i-th horizon, where the terrain surface becomes visible again. In particular, the first horizon represents the first transition from visible to invisible; the first shadow is the first transition from invisible to visible. Successive horizons, thus, define a set of curves around the viewpoint.

In this paper, we define horizons and shadows, and introduce visibility maps, induced by horizons and shadows, on a polyhedral terrain. Such horizon-based visibility structures are defined by projecting the straight-line segments forming the horizons (and the shadows) on the $x - y$ plane and joining them through rays emanating from the viewpoint. We also describe algorithms for building horizons and shadows. We show that, in order to represent the visibility of the terrain, the information represented by the horizons (without taking into account the shadows) is sufficient; moreover, it is not necessary to store each single horizon, but it is sufficient to store the segments forming the horizons (without taking into account which order of horizon they belong to). For this purpose, a map, containing only the segments forming the horizons, called the *influence map*, can be used. Application of the influence map for determining whether a given query point is visible or not from the viewpoint is discussed in [8].

A definition of nested horizons has been given first in [17,18], where the authors focus on the problem of reconstructing a terrain from a set of horizons, with respect to a given viewpoint. In [16], the connection between horizons, shadows and visible and invisible areas on the terrain surface is explored; applications concerning the extraction of terrain features (e.g., ridges) from a collection of horizons are also suggested.

The reminder of this paper is organized as follows. In Section 2, we introduce the concepts of mathematical and digital terrain model, and we provide some background definitions for visibility problems on terrains. In Section 3, we define the nested horizons, first in the simplified case of a one-dimensional terrain, and then in the two-dimensional case. In Section 4, we define a number of horizon-based visibility maps for a polyhedral terrain. Finally, in Section 5, we propose two algorithms for computing nested horizons on a polyhedral terrain.

2 Background

A natural terrain can be described as a continuous function $z = f(x,y)$, defined over a simply connected subset D of the x-y plane. Thus, a *Mathematical Terrain Model* (MTM), that we simply call a *terrain*, can be defined as a pair $\mathcal{M} \equiv (D, f)$. The notion of digital terrain model characterizes a subclass of mathematical terrain models, which can be represented in a compact way through a finite number of information. A *Digital Terrain Model* (DTM) is defined on the basis of a plane subdivision $\sum$ of the domain D into a collection of plane regions $\mathcal{R} = \{R_1, R_2, \ldots R_m\}$, and of a family $\mathcal{F}$ of continuous functions $z = f_i(x, y)$, $i = 1, 2, \ldots m$, such that each function f_i is defined on R_i and any two functions f_i and f_j assume the same value on the common boundary of two adjacent regions R_i and R_j. Thus, a DTM can be expressed as a pair $\mathcal{D} \equiv (\sum, \mathcal{F})$. We call *face* of a DTM the graph of each function f_i, *edge* and *vertex* of a DTM the restriction of each function f_i to an edge and a vertex, respectively, of $\sum$. For simplicity, we will denote with $\bar{o}$ the projection on the $x - y$ plane of a generic geometric entity o in the 3D space.

Polyhedral Terrain Models (PTMs) are a special class of digital terrain models characterized by a domain subdivision consisting of a straight-line plane graph and by linear interpolating functions. A special class of PTMs is that formed by *Triangulated Irregular Networks* (TINs), which are characterized by a triangular subdivision of the domain. Often, a Delaunay triangulation is used as domain subdivision for a TIN because of its good behaviour in numerical interpolation. Given a set S of points in the plane, a triangulation $\sum$ of S is a *Delaunay triangulation* if and only if the circumcircle of each triangle t of $\sum$ does not contain any point of S inside. A Delaunay triangulation has belongs to the class of *acyclic subdivisions* of the plane [5].

Given a plane subdivision $\sum$ and a point O in the plane, we define a relation, called the *before/behind* relation (denoted $\prec$), on the edges of $\sum$ with respect to O as follows: an edge $e_1 \prec e_2$ if and only if there exists a ray r emanating from O, intersecting both e_1 and e_2 and such thate $\overline{OP_1} \leq \overline{OP_2}$, where $P_1 \equiv e_1 \cap r$ and $P_2 \equiv e_2 \cap r$. A subdivision $\sum$, on which the $\prec$ relation is a partial order relation, is called an *acyclic subdivision* with respect to O. In this case, we call *distance order* any total order relation extending relation $\prec$ [5,10,11,3].

An *acyclic digital terrain model* with respect to a viewpoint $V \equiv (x, y, z)$ is a digital terrain model $\mathcal{D} \equiv (\sum, \mathcal{F})$ where $\sum$ is an acyclic subdivision with respect to $\bar{V} \equiv (x, y)$. A digital model is simply *acyclic* if it is acyclic with respect to any observation point. Acyclic digital terrain models have a special interest related to visibility applications, because the existence of a distance order allows the computation of the visibility of an edge only taking into account the portion of terrain formed by those edges which come before it.

Given a mathematical terrain model $\mathcal{M} \equiv (D, f)$, we call *candidate point* any point $P \equiv (x, y, z)$ belonging to or above the terrain, i.e., such that $(x, y) \in D$ and $z \geq f(x, y)$. Two candidate points P_1 and P_2 are said to be *mutually visible* (or *intervisible*) if and only if, for every point $Q \equiv (x, y, z) = tP_1 + (1 - t)P_2$, with $0 < t < 1$, $z > f(x, y)$. In other words, two points are mutually visible when the straight-line segment joining them lies above the terrain (and it touches it at most at its two extremes). We call *observation point* (or *viewpoint*) any arbitrarily chosen candidate point, and *visual ray* any ray emanating from a viewpoint. Given a viewpoint V and a spherical coordinate system centered at V, a visual ray r is identified by the pair (θ, α), called *view direction*, where θ is the angle between the projection $\bar{r}$ of r on the $x - y$ plane and the positive x-axis, and α is the complementary of the angle between r and the positive z-axis (see Figure 1). We also call *view sphere* any sphere centered at the viewpoint and large enough to contain the whole terrain inside. Intuitively, the view sphere is used as a "projection screen" for the visible image of the terrain from the viewpoint.

The *horizon* of an observation point on a terrain is a well-defined problem in computational geometry, and several algorithms have been developed for its solution [1,13,7]. Given a terrain $\mathcal{M} \equiv (D, f)$ and a viewpoint V, the *horizon* of the terrain with respect to V is a function $\alpha = h(\theta)$, defined for $\theta \in [0, 2\pi)$, such that, for every radial direction θ, $h(\theta)$ is the maximum value α such that each ray emanating from V in a view direction (θ, β), with $\beta < \alpha$, does not intersect the terrain. Such definition corresponds to the intuitive notion that the horizon of the terrain provides, in each radial direction, the minimum elevation which must have a visual ray emanating from

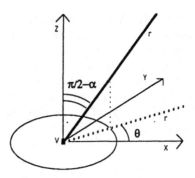

Fig. 1. Spherical coordinate system.

the point of view in the given direction to pass above the surface. On a polyhedral terrain the horizon is a radially sorted list of labeled intervals $[\theta_i, \theta_j]$. If an interval $[\theta_i, \theta_j]$ is labeled with a terrain edge e_j, then the visual ray defined by a direction $(\theta, h(\theta))$, with $\theta_i < \theta < \theta_j$, hits the terrain at a point belonging to an edge e_j. An example of a horizon on a polyhedral terrain is given in Figure 2.

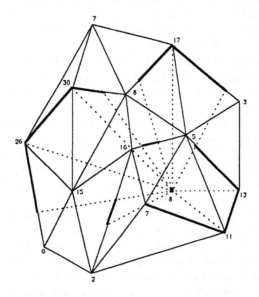

Fig. 2. Horizon of a viewpoint on a triangulated irregular network: the viewpoint is the marked point; the segments forming the horizon are drawn in thick lines.

The problem of computing the horizon of a viewpoint on a polyhedral terrain reduces to the problem of computing the upper envelope of a set of possibly intersecting segments in the plane. Given a set of p segments in the plane, i.e., p linear functions

$y = f_i(x)$, $i = 1 \ldots p$, each defined on an interval $[a_i, b_i]$, the *upper envelope* of such segments is a function $y = F(x)$, defined over the union of the intervals $[a_i, b_i]$, and such that $F(x) = max_{i|x \in [a_i, b_i]}(f_i(x))$. In other words, the upper envelope maps any x value into the segment having maximum y value over x (if such segment exists) (see Figure 3 for an example). The upper envelope of a set of segments can thus be

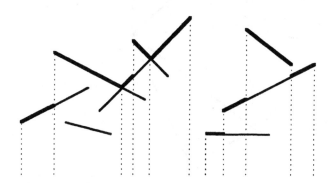

Fig. 3. A set of segments and its upper envelope (portions of segments taking part in upper envelope are drawn in thick lines).

represented as a sequence of labeled intervals. Each interval $I_i = [a_i, b_i]$ is labeled with the segment s_j of the given data set such that the upper envelope on I_i is the same as segment s_j.

In order to reduce the horizon on a polyhedral terrain to the upper envelope of a set of segments, we express the edges of the terrain in a spherical coordinate system centered at the viewpoint and consider only the two angular coordinates (as shown in Figure 1). This transformation produces a set of segments in the $\theta - \alpha$ plane, whose upper envelope corresponds to the horizon of the given terrain [2]. Figure 4 shows an example of a horizon on a polyhedral terrain and the corresponding upper envelope. It has been shown [2] that the complexity of the upper envelope of p segments in the plane is $O(p\alpha(p))$, and, thus, the complexity of the horizon of a polyhedral terrain with n vertices is $O(n\alpha(n))$.

The upper envelope of p segments in the plane can be computed either by a static divide-and-conquer approach with a $O(p \log p)$ worst-case time complexity [1,13], or by a dynamic incremental one, with a complexity equal to $O(p^2\alpha(p))$. A new randomized dynamic algorithm, with an expected time complexity equal to $O(p\alpha(p) \log p)$ in the worst case, has been proposed in [7].

3 Horizons and Shadows

In this Section, we define nested horizons for a terrain model, with respect to a predefined viewpoint V. We first examine the simplified case of the one-dimensional terrain model, and then extend the definition to a two-dimensional case, i.e., to a terrain model, as defined in Section 2.1.

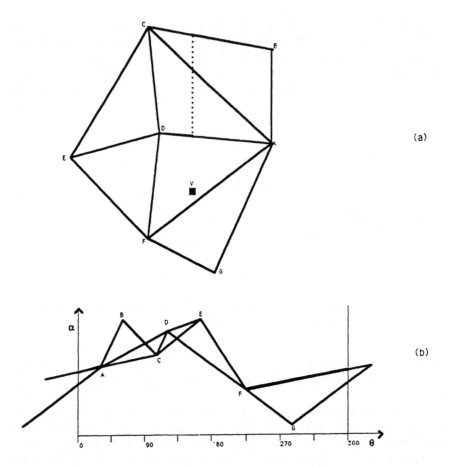

Fig. 4. Reduction of the horizon on a polyhedral terrain to the upper envelope of a set of segments: the horizon, with respect to viewpoint V (a); the set of segments in the $\theta - \alpha$ plane, obtained by projecting the terrain edges, and the corresponding envelope (b).

3.1 Horizons and Shadows on a One-Dimensional Terrain Model

A *one-dimensional terrain model* is a pair $\mathcal{M} \equiv (I, f)$, where $I = [a, b] \subseteq \Re$ and f is a continuous function defined over I. We represent the graph of a one-dimensional terrain model, called a *one-dimensional terrain*, for consistency with the notation used in Section 3.2. We call *left extreme* of the terrain the point $A \equiv (a, f(a))$, and *right extreme* the point $B \equiv (b, f(b))$. The definitions of *candidate point*, *intervisibility*, etc., for a one-dimensional terrain model are completely similar to those given for a two-dimensional terrain model in Section 2. Given a one-dimensional terrain model $\mathcal{M} \equiv (I, f)$, we put the *viewpoint* for $\mathcal{M}$ at an arbitrary candidate point $V \equiv (r_0, z_0)$. We term *vertical shadow* of V on $\mathcal{M}$, denoted V_0, the vertical projection $(r_0, f(r_0))$ of V on the terrain. In the following, we define *right* horizons and shadows on the terrain $\mathcal{M}$ with respect to viewpoint V, i.e., horizons and shadows concerning the portion of terrain that lies on the right of V_0. The definition of *left* horizons and shadows is clearly symmetric.

Informally, any right horizon of V on $\mathcal{M}$ is a point H on the terrain, that blocks the view of an observer located at V and looking rightwards. The i-th horizon of V will be the i-th such obstacle in a distance order from V. The corresponding shadow S of H is the point of the terrain, behind H, where the terrain becomes visible again from V. In other words, a horizon correspond to a transition from visible to invisible, and a shadow to a transition from invisible to visible.

A point $P \equiv (r_P, f(r_P))$ on a one-dimensional terrain, lying on the right of V_0, is a *transition from visible to invisible* if the points of the terrain immediately on the left of P are visible from V and those immediately on the right are invisible from V, i.e., if there exists a positive value ε such that points $(r, f(r))$ with $r \in (r_P - \varepsilon, r_P)$ are visible from V and points $(r, f(r))$ with $r \in (r_P, r_P + \varepsilon)$ are invisible. Conversely, a point P on the terrain is a *transition from invisible to visible* if every point of the terrain lying at the immediate left of P is invisible from V and every point at its immediate right is visible.

The *first (right) horizon* of a viewpoint V on a terrain $\mathcal{M}$ is the point $H_1 \equiv (r_1, f(r_1))$ defined as follows:

a) If every point $P \equiv (r, f(r))$ of the terrain on the right of V_0 (i.e., $r > r_0$) is visible from V, then $H_1 \equiv B$ (see Figure 5(a)). In this case nothing blocks the view of an observer located at V.

b) If V lies on the terrain ($V = V_0$) and the points $P \equiv (r, f(r))$ immediately to the right of V_0 are invisible from V (i.e., there exists a positive value ε such that, for every $r \in (r_0, r_0 + \varepsilon)$, $(r, f(r))$ is invisible from V), then $H_1 \equiv V$ (see Figure 5(b)).

c) Otherwise, H_1 is the first transition from visible to invisible immediately to the right of V_0, i.e., $r_1 = \min\{r > r_0 \mid (r, f(r))$ transition from visible to invisible$\}$ (see Figure 5(c)).

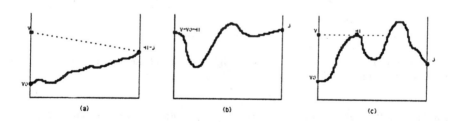

Fig. 5. First right horizon H_1: in (a) H_1 is the right extreme of the terrain, in (b) H_1 is the viewpoint and in (c) H_1 is the first transition from invisible to visible.

The *first (right) shadow* of V on $\mathcal{M}$, or *shadow of the first (right) horizon*, denoted S_1, is defined only if there are points of the terrain visible from V to the right of horizon H_1, and it is the first transition from invisible to visible on the right of H_1, i.e., $S_1 \equiv (s_1, f(s_1))$, where $s_1 = \min\{r > r_1 \mid (r, f(r))$ transition from invisible to visible$\}$ (see Figure 6).

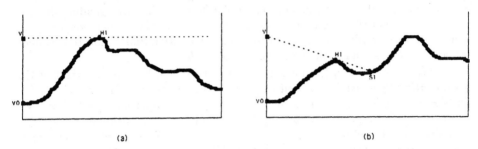

Fig. 6. First right shadow S_1: in (a) S_1 is not defined, in (b) it is defined.

The $(i + 1)$-th *(right) horizon* $H_{i+1} \equiv (r_{i+1}, f(r_{i+1}))$ of V on $\mathcal{M}$ is defined only if there exists shadow S_i, in the following way:

a) If every point $P \equiv (r, f(r))$ of the terrain on the right of S_i (i.e., $r > r_0$) is visible from V, then H_{i+1} is the same as the right extreme B of $\mathcal{M}$ (see Figure 7(a)).

b) Otherwise, H_1 is with the first transition from visible to invisible to the right of S_i, i.e., $r_{i+1} = \min\{r > s_i \mid (r, f(r))$ transition from visible to invisible$\}$ (see Figure 7(b)).

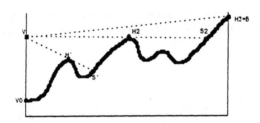

Fig. 7. $I + 1$-th right horizon: horizon H_3 is the right extreme of the terrain, horizon H_2 is the first transition from visible to invisible behind S_1.

The $(i + 1)$-th *(right) shadow* of V on $\mathcal{M}$, or *shadow of the $(i + 1)$-th (right) horizon*, denoted S_{i+1}, is defined only if horizon H_{i+1} is defined, and there are points of the terrain visible from V to the right of H_{i+1}. In this case, S_{i+1} is the first transition from invisible to visible on the right of H_{i+1}, i.e., $S_{i+1} \equiv (s_{i+1}, f(s_{i+1}))$, where $s_{i+1} = \min\{r > r_{i+1} \mid (r, f(r))$ transition from invisible to visible$\}$ (see Figure 8). The value i is called *order* of the i-th horizon and of the i-th shadow. We call *last (right) horizon* of V, denoted H_{last}, the horizon of maximal order, i.e., $last = \max\{i > 0 \mid H_i$ defined$\}$. H_{last} is always defined.

Moving on the terrain rightwards from V_0, horizons and shadow form a strictly alternating sequence. The first and last element (H_1 and H_{last}) of the sequence are

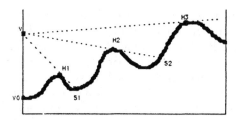

Fig. 8. $I + 1$-th right shadow: shadow S_2 is the first transition from invisible to visible behind H_2; shadow S_3 is undefined.

always horizons. Horizons and shadows mark the alternance between visible and invisible portions of the terrain. All the points of the terrain lying between V_0 and H_1, or between a shadow S_i and its subsequent horizon H_{i+1} are visible from the viewpoint V, and all the points of the terrain lying between a horizon H_i and its shadow S_i (or between H_i and the right extreme B of the terrain, if $i = last$) are invisible from V. An explicit representation of shadows is somehow redundant, being a shadow S_i the intersection point of the visual ray through H_i with the terrain behind. We define the *map of (right) nested horizons* the ordered sequence $HL \equiv [H_1, H_2, \ldots, H_{last}]$.

3.2 Horizons and Shadows on a Two-Dimensional Terrain Model

Let $\mathcal{M} \equiv (D, f)$ be a two-dimensional terrain model, and $V \equiv (x_0, y_0, z_0)$ a viewpoint on $\mathcal{M}$. Given a radial direction $\theta \in [0, 2\pi)$, the section of the terrain obtained by considering the vertical half-plane originating at the vertical line through V and extended in the θ direction can be represented as a one-dimensional terrain model, that we call the *section of $\mathcal{M}$ in direction θ*, and denote $\mathcal{M}_\theta$. $\mathcal{M}_\theta$ is defined as the pair (I_{theta}, f_θ), with $I_\theta \equiv [0, r_{max}]$, where $r_{max} \equiv \max\{r \geq 0 \mid (x_0 + r\cos\theta, y_0 + r\sin\theta) \in D\}$, and $f_{theta}(r) = f(x_0 + r\cos\theta, y_0 + r\sin\theta)$ for every $r \in I_\theta$. The viewpoint for such one-dimensional terrain model is the point $V_{1D} \equiv (0, z_0)$. The vertical shadow of V_{1D} is the left extreme of the terrain.

The *i-th horizon in direction θ* on two-dimensional terrain $\mathcal{M}$ with respect to viewpoint V is the point corresponding to the i-th right horizon of V_{1D} on the one-dimensional terrain section $\mathcal{M}_\theta$. Formally, if there exists the i-th right horizon $H_i \equiv (r_i, f_\theta(r_i))$ of V_{1D} on $\mathcal{M}_\theta$, then the i-the horizon in direction θ of V on $\mathcal{M}$ is defined and equal to $(x_i, y_i, f(x_i, y_i))$, where $x_i = x_0 + r_i\cos\theta$ and $y_i = y_0 + r_i\sin\theta$; otherwise, the i-the horizon in direction θ is undefined. The definition of *i-th shadow in the θ direction* is given in a similar way.

The *i-th horizon of V on $\mathcal{M}$* is a partial function $h_i : [0, 2\pi) \longrightarrow \Re^3$ that, for every radial direction $\theta \in [0, 2\pi)$, provides, if defined, the i-th horizon of V in direction θ. Similary, the *i-th shadow of V on $\mathcal{M}$*, is a partial function $s_i : [0, 2\pi) \longrightarrow \Re^3$ providing, for any radial direction $\theta \in [0, 2\pi)$, the i-th shadow of V in direction θ, if such shadow exists.

The *last horizon of V* is a function $h_{last} : [0, 2\pi) \longrightarrow \Re^3$ that associates, with every radial direction $\theta \in [0, 2\pi)$, the point $h_{k_\theta}(\theta)$, where $k_\theta = \max\{i \geq 1 \mid h_i(\theta)$

defined}. The order k_θ depends on the given direction θ, because in different radial directions a different number of horizons may be defined. Function h_{last} is total on $[0, 2\pi)$, since, in every radial direction, at least the first horizon is always defined. The last horizon h_{last}, defined above, corresponds to the horizon h, defined in Section 2, i.e., for every radial direction θ, the spherical coordinates of a point $h_{last}(\theta)$ are a triple (θ, r, α), where $\alpha = h(\theta)$. We can define an "enriched" version of the last horizon, by providing also the order of the last horizon in every direction. The *enriched last horizon* is thus a function $h_{last}^* : [0, 2\pi) \longrightarrow \Re^3 \times \mathcal{N}$ such that $h_{last}^*(\theta) = (P_\theta, k_\theta)$ where $k_\theta = \max\{i \mid k_i(\theta) \text{ defined}\}$ and $P_\theta = h_{k_\theta}(\theta)$.

3.3 Horizons and Shadows on a Polyhedral Terrain

In this Subsection, we give a characterization of horizons and shadows for a two-dimensional polyhedral terrain model $\mathcal{D} \equiv (\sum, \mathcal{F})$. Moreover, we estimate the space complexity of horizons and shadows, as a function of parameters related to the dimension of the polyhedral terrain $\mathcal{D}$.

A face f_i of $\mathcal{D}$, described by a linear function $z = a_i x + b_i y + c_i$, identifies two half-spaces: an *upper half-space*, defined by $z > a_i x + b_i y + c_i$, and a *lower half-space*, defined by $z < a_i x + b_i y + c_i$. Given a viewpoint V, a face f_i is said to be *face-up* with respect to V if V lies in upper half-space of f_i, and *face-down* if V lies in the lower half-space. An edge e_j of $\mathcal{D}$ is a *blocking edge* with respect to V if, denoted with f_{near} and f_{far} the faces adjacent to e_j on the same and on the opposite side of V, respectively, f_{near} is face-up and f_{far} is face-down. We call blocking edges also those edges of the boundary of the terrain, whose unique adjacent face is face-up with respect to V. The immediate projection of a blocking edge e_j from the viewpoint V on a face of the terrain lying behind e_j is called a *projection segment*.

Every visible portion b of a blocking edge defines an obstacle to the view of an observer located at the viewpoint V, by making a portion of terrain immediately behind b invisible. The distal boundary of the portion of terrain obscured by b is defined by the visible portions of the projection segments of b. In a polyhedral terrain, horizons are composed by visible portions of blocking edges, and shadows are formed by visible portions of projection segments. Thus, in the case of a polyhedral terrain model, we can collect, for every i-th horizon, the maximal radial intervals in which point $h_i(\theta)$ belongs to the same blocking edge e_j, and, for every i-th shadow, the maximal radial intervals in which point $s_i(\theta)$ belongs to the projection segment of the same blocking edge e_j on the same face f_i of the terrain. This leads to the following characterization of horizons and shadows as ordered lists of labeled intervals.

The i-th horizon can be represented as an ordered list $L = [I_1, \ldots, I_k]$ of maximal radial intervals, labeled with edges of the terrain, or with the special value *null*, in such a way that, if a radial interval I_l is labeled e_j, then for all $\theta \in I_l$, $h_i(\theta)$ is defined and belongs to a blocking edge e_j, and, if I_l is labeled *null*, then, for all radial directions $\theta \in I_l$, $h_i(\theta)$ is undefined. The i-th shadow can be represented as an ordered list $L = [I_1, \ldots, I_k]$ of maximal radial intervals, labeled with pairs (edge, face) or with the special value *null*, in such a way that, if a radial interval I_l is labeled (e_j, f_q), then for all radial directions $\theta \in I_l$, $s_i(\theta)$ is defined and belongs to the projection segment of blocking edge e_j on face f_q, and, if I_l is labeled *null*, then, for all radial directions $\theta \in I_l$, $s_i(\theta)$ is undefined.

Also the *last horizon* can be characterized as an ordered list of labeled intervals, by collecting the maximal radial intervals in which the last horizon is a point on the same blocking edge. The last horizon is the same as the horizon defined in Section 2. For a polyhedral terrain, the *enriched last horizon* is characterized as an ordered list of radial intervals whose labels are pairs (edge,horizon order), such that, if an interval I_l has a label (e_j, k_i), then for every radial direction $\theta \in I_l$, $h^*_{last}(\theta) = h_{k_i}(\theta)$ and is a point on edge e_j.

Upper bounds, in the worst case, to the spatial complexity of horizons and shadows, for a polyhedral terrain model, can be expressed as function of the number n of vertices of the terrain, and of three other parameters: the number d of maximal visible portions of blocking edges, the number m of maximal visible portions of projection segments, and the maximum number p of horizons defined in a radial direction θ. For complexity results related to the hidden line and surface elimination problem [15], we have that d and m are $O(n^2)$, since visible portions of projection segments and of blocking edges form the boundary of the visible and invisible regions on the terrain surface, as we will show in Section 4. The maximum number p of horizons in a single radial direction is $O(n)$ in the worst case, because in any direction there can be at most as many horizons as terrain edges.

In a polyhedral terrain horizons are formed by visible portions of blocking edges. Unfortunately, the same maximal visible portion of blocking edge may be subdivided, for different radial directions, in different orders of horizons (see Figure 9). Since the number of horizons in a single direction is p in the worst case, it follows that the total dimension of all the horizons if $O(dp)$, that is $O(n^3)$ in the worst case. Similarily, the total number of shadows is equal to $O(mp)$, i.e., $O(n^3)$ in the worst case.

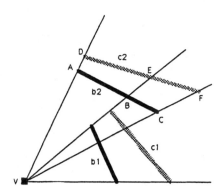

Fig. 9. The visible portion b_2 of blocking edge is split into a first horizon (fragment $A - B$) and a second one (fragment $B - C$). Its projection segment c_2 is split in the first shadow (fragment $D - E$) and the second one (fragment $E - F$).

The decomposition of a horizon into radial intervals is refined by the partition of $[0, 2\pi]$ defined by the radial directions corresponding to the extremes of the visible portions of blocking edges. The space complexity of a single horizon is thus $O(d) = O(n^2)$. The space bound of $O(n\alpha(n))$, derived from the upper envelope of n segments (see Section 2), holds for the last horizon h_{last}. The above bound does not apply to the enriched version h^*_{last} of the last horizon, since the extra information about the order of the last horizon in every direction may split every interval in several subparts. The dimension of h^*_{last} is still bounded from above by $O(n^2)$.

4 Visibility Structures Based on Horizons

In this Section, we consider a polyhedral terrain model $\mathcal{D} \equiv (\sum, \mathcal{F})$, and we define different visibility maps on $\mathcal{D}$, consisting of subdivisions of the domain D, induced by horizons and shadows defined in the previous Section.

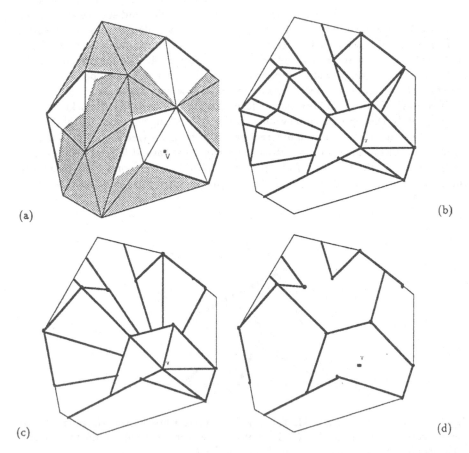

Fig. 10. Visible and invisible region (a) (the invisible region is shaded, horizons are drawn in thick lines); the visibility map (b), the horizon map (c), and the influence map (d).

The visibility structure of a terrain, with respect to a viewpoint V, is usually encoded as a partition of the terrain domain into a visible and an invisible region.

The *visible region* of V is the subset VR of D formed by the points $(x, y) \in D$, such that $(x, y, f(x, y))$ is visible from V. The *invisible region* of V, denoted IR, is the complement of VR in D (see Figure 10(a)). The projections of the visible portions of blocking edges and projection segments of $\mathcal{D}$ with respect to V on the $x - y$ plane form a set of semi-disjoint segments in the plane. For every extreme point P of such segments, we draw a ray through P and $\bar{V}$, to reach another segment (or $\bar{V}$ itself) inwards, and another segment (or a boundary edge of D) outwards. We call *visibility map* the subdivision of D obtained in this way.

The regions of the visibility map are *radial trapezoids*, i.e., quadrilaterals with two transversal edges (one corresponding to a fragment of blocking edge and the other one to a fragment of a projection segment) and two radial edges (both fragments of radial segments introduced during the construction). There are also *degenerate trapezoids*, i.e., triangles with one transversal edge (corresponding to a fragment of blocking edge) and two radial edges incident at $\bar{V}$ (see Figure 10(b)). Any radial trapezoid is completely contained either into the visible region VR or into the invisible region IR of V on $\mathcal{D}$. In a trapezoid contained in the visible region, the transversal edge nearer to $\bar{V}$ corresponds to a fragment of projection segment, and the one farther from $\bar{V}$ to a fragment of blocking edge. Conversely, in the trapezoids contained in the invisible region, the nearer transversal edge corresponds to a fragment of a blocking edge and the distal one to a fragment of projection segment. All degenerate trapezoids are contained into the visible region.

The visibility map represents the visibility of the terrain with respect to the given point of view. As in the one-dimensional case, a map based just on horizons, without considering shadows, is sufficient to provide a visibility structure of the terrain. We can thus repeat the above construction, but considering this time only the visible portions of blocking edges of the terrain. The resulting trapezoidal decomposition of D is called a *horizon map*. Any region of the horizon map is still a radial trapezoid, whose transversal edges are both visible fragments of blocking edges (see Figure 10(c)).

The horizon map is not a refinement of the partition of the terrain domain into visible and invisible regions, but it has an interesting visibility property. If R is a non-degenerate trapezoidal region of such map, and b denotes the blocking edge corresponding to the transversal edge of R lying towards $\bar{V}$, the invisibility of $R \cap IR$ is only due to the shadow cast on R by a vertical trapezoid hanging from b. Because of this property, it is possible to determine the visibility of a candidate point Q, whose projection belongs to R, just testing whether Q lies above or below the plane passing through b and the viewpoint V. All candidate points, whose projection lies in a degenerate trapezoid, are visible from V.

Still starting from the $x - y$ projections of the visible portions of the blocking edges, we can construct yet another subdivision of D by drawing, for every extreme point, a ray only outwards, to reach another segment or the external boundary of D. The resulting plane subdivision is called the *influence map* of V (see Figure 10(d)). The influence map is composed of regions with an arbitrary number of edges, and, in general, not convex. A region of such map is the union, along common radial edges, of those trapezoids of the horizon map whose nearer transversal edge corresponds to the same visible portion b of blocking edge; such region is called the *influence region* of b, denoted R_b. Exactly one region of the influence map is formed by the union of all the degenerate trapezoids of the horizon map, and it is called the *first visible region*. The influence region of b is completely contained in the radial sector defined by the $\bar{V}$ and $\bar{b}$. The same visibility property observed for the regions of the horizon map holds for the influence map as well, i.e., the invisibility of $R_b \cap IR$ is due only to b; the first visible region is totally contained into VR. In order to compute the visibility of a candidate point Q, it is sufficient to locate the region in the influence map containing $\bar{Q}$, i.e., to determine the visible portion of blocking edge which is immediately in front of Q.

The size of the visibility map, of the horizon map and of the influence map is equal to $O(d + m)$, $O(d)$, and $O(d)$, respectively, since a constant number of new vertices is introduced for every extreme point of the original set of segments considered. The dimension of these maps is thus $O(n^2)$ in the worst case.

5 An Incremental Algorithm

In this Section, we present an incremental algorithm for computing all the orders of the horizon on a polyhedral terrain, in all directions, with respect to a given viewpoint. The algorithms examines all the blocking edges of the terrain one at a time, in increasing distance order from the viewpoint. It requires a preprocessing step, in which the blocking edges of the terrain are sorted in a distance order from the viewpoint V. Thus, it works only for acyclic polyhedral terrain models.

The current enriched last horizon Hor^* (see Section 3.3) is maintained and updated at each step. For every blocking edge b examined, the algorithm locates the intervals of Hor^* in which b is above the current horizon; such intervals are called the *relevant* intervals for b. In each relevant interval I_i, the enriched last horizon is updated, and the subintervals of I_i in which the old horizon segment has been replaced by b are saved as belonging to previous horizons. The last enriched horizon Hor^* is represented as an ordered list of radial intervals, each labeled with a blocking edge b, or with the *null* value, plus the corresponding order of horizon of b. The horizon order is set to 0 for those intervals whose label is *null*. At the beginning, Hor^* is initialized as the list containing just interval $[0, 2\pi]$, labeled $(null, 0)$.

In more detail, a blocking edge b is processed as follows:

1) The *relevant* intervals for b are located. The relevant intervals are those intervals I_i on Hor^*, which properly intersect the radial interval $I \equiv [\theta_1, \theta_2]$ defined by the extreme points of b.

2) Each relevant interval I_i, labeled $[a_i, k_i]$, is processed. If the projection b' of b on the view sphere lies completely below the projection a_i' of a_i, then no update is required in I_i. Otherwise, I_i is split into at most three subintervals (by cutting I_i at the extreme points of b' which are above a_i', and at the intersection point between b' and a_i'). I_i is deleted from Hor^* and replaced by the new subintervals, each labeled with a_i or b, depending on whether a_i is above or below b inside the subinterval. Those subintervals having label a_i maintain the same horizon order k_i as I_i. The number k_{i+1} is assigned to all subintervals with label b; for each of such intervals, denoted J, the same interval J, labeled a_i, is saved as part of the k_i-th horizon (see Figure 11).

3) Adjacent intervals of Hor^*, labeled with the same pair (edge,horizon order), are merged together.

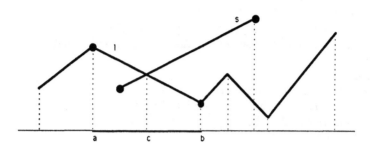

Fig. 11. Processing a new segment s in the incremental algorithm. The original interval $I_i = [a, b]$, labeled l, with horizon order k_i, is split into the two subintervals $[a, c]$, labeled l, with order k_i, and $[c, b]$, labeled s, with order $k_i + 1$. Interval $[c, b]$, labeled l, is saved as part of the k_i-th horizon.

After processing all blocking edges, for every interval I_i, labeled (a_i, k_i), of the final enriched horizon Hor^*, we save the interval I_i, labeled a_i, as a part of the k_i-th horizon. Finally, for each k_i-th horizon a re-arranging step is performed, in a similar way as performed at step 3) on Hor^*.

The pseudocode description of the algorithm is presented below.

Algorithm INCREMENTAL_HORIZONS

input

    $\mathcal{M} \equiv (\sum, \mathcal{F})$: acyclic polyhedral terrain model;

    V: viewpoint

output

    $Hor_1, \ldots, Hor_n$: horizons

begin

    determine which edges of $\mathcal{M}$ are blocking edges;

    sort the blocking edges by distance from V;

    for $i = 1, \ldots, n$ **do** $Hor_i \leftarrow$ *empty list* **end for** ;

    $Hor^* \leftarrow$ the only interval $[0, 2\pi]$, labeled *null*;

    for every blocking edge b, in increasing distance order **do**

        determine the intervals of Hor^* relevant for b;

        for every relevant interval I_i **do**

            $(a_i, k_i) \leftarrow$ segment and horizon order labelling I_i;

            split I_i in maximal subintervals such that in every subinterval b is either completely above a_i or completely below;

            delete I_i from Hor^*;

            for every generated subinterval J **do**

                if b above a_i in J

```
              then
                  insert J in Hor* labeled (b, kᵢ + 1);
                  insert J in Horₖᵢ labeled aᵢ
              else
                  insert J in Hor* labeled (aᵢ, kᵢ)
              end if
          end for ;
          for every pair of adjacent intervals I and J in Hor* do
              if I and J have same label then merge I and J end if
          end for
      end for ;
  for every interval I in Hor* do
      (a, k) ← segment and horizon order labelling I;
      insert I in Horₖ labeled a
  end for ;
  for every i = 1,..., n do
      for every pair of adjacent intervals I and J in Horᵢ do
          if I and J have same label then merge I and J end if
      end for
  end for
end algorithm
```

In the worst case, the main loop of the algorithm is executed $O(n)$ times (once for each blocking edge b), each time performing a number of operations proportional to the number of relevant intervals for b in Hor^*. The current enriched last horizon Hor^* is composed of at most $O(n^2)$ intervals, all relevant for b in the worst case. The worst case time complexity, thus, results to be equal to $O(n^3)$.

6 A Radial Sweep Algorithm

In this Section, we describe an algorithm for computing the maximal visible portions of blocking edges (i.e., the segments forming the horizons, without the information about the order of horizon they belong to) of a polyhedral terrain with respect to a given viewpoint V. The output of this algorithm is the information needed to build the horizon map or the influence map, defined in Section 4.1. The algorithm operates according to a radial sweep technique [19], by moving a vertical half-plane π, originating at the vertical straight-line l through V, radially around l. At any instant, the one-dimensional section of the terrain intersected by π is considered.

At any instant, the *status* of the sweep process is represented by the blocking edges which are currently intersected by π. Those blocking edges, which are horizons in the current radial direction of π, are called *active* edges. The status also maintains a horizontal and a vertical order of the blocking edges intersected by π at its current position. The horizontal order is defined according to the distance order of the blocking edges from V, while the vertical order is given by the height order among the projections of such intersection points on the view sphere. In more detail, the sweep status is represented by the lists L_1 and L_2 of the blocking edges intersected by half-plane π, sorted horizontally and vertically, respectively, and by the lists L'_1 and L'_2, obtained from L_1 and L_2 by considering only the active edges. A blocking edge b is active if and only if every blocking edge preceeding b horizontally (nearer to V than b) preceedes b also vertically (it is below b, and, thus, it does not obscure b).

The *events* of the sweep process, i.e., the radial directions θ at which the status changes, are represented by the radial directions corresponding to intersection points between projections of blocking edges on the view sphere and those corresponding to the extreme points of the blocking edges of the terrain (every terrain vertex is represented a number of times equal to the number of blocking edges incident at it; each time, information about its being a first or a second extreme is stored).

During the sweep process, events are processed in a radial (e.g., counterclockwise) order around V. At each event, the status of the sweep process is updated, depending on event type:

1) If the event is an intersection point between the projections of two blocking edges a and b (let a be before b in horizontal order), then we swap a and b in vertical order, and we update the active blocking edges. Three cases may occur:

1.a) a was below b in vertical order, and now a is above b:
b (if active) is disactivated, since b is now obscured by a, and cannot be a horizon anymore (see Figure 12(a)).

1.b) a was above b vertically, and now a is below b (in this case, b was inactive), and a was inactive:
no changes occur, since b cannot become a horizon as some active blocking edge c exists obscuring both b and a (see Figure 12(b)).

1.c) a was above b vertically, and now a is below b (b was inactive), and a was active:
b becomes active if and only if the active edge c which was immediately above a (and now is immediately above b) comes after b in horizontal order. In fact, b is obscured by c if and only if c is nearer to V than b. Moreover, b cannot be obscured by another blocking edge d different from c, otherwise d would obscure c (see Figure 12(c)).

2) If the event is a second extreme of a blocking edge b, then b is deleted from the status.

3) If the event is a first extreme of a blocking edge b, then we insert b in the horizontal and vertical order. Let b be between two edges a and a' ($a \prec b \prec a'$) in vertical order. We look for intersections between a, b and a', b, and (if they exist) we insert them among the events. Finally, b is made active if and only if the active edge c immediately above b follows b in horizontal order (see Figure 13).

The maximal visible portions of a blocking edge b are represented by the portions of b between any radial direction θ_1, corresponding to an event where b is made active, and the direction θ_2 of the next event where b is disactivated. For this purpose, we keep track, for every active edge, of the radial direction at which it became active.

The algorithm needs also a preprocessing step, in which the blocking edges of the terrain are determined, and every blocking edge crossing the sweep half-plane π at its initial position (e.g., $\theta = 0$) is split into two subsegments, lying on the right

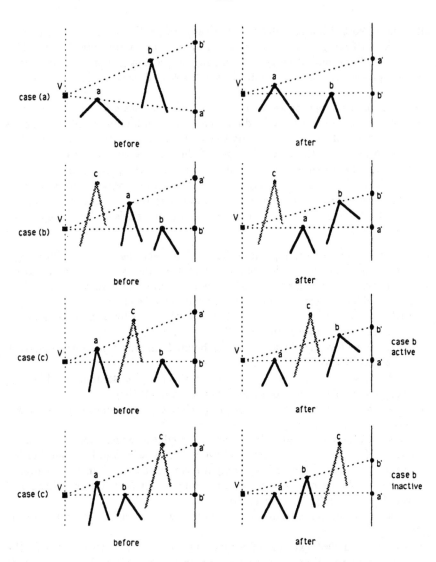

Fig. 12. Modification of the sweep status at an event represented by the intersection between two blocking edges a and b, projected on the view sphere.

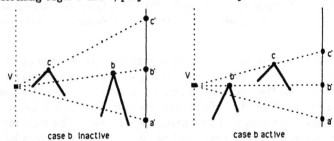

Fig. 13. Modification of the sweep status at an event represented by the first endpoint of a blocking edge b.

and on the left of π, respectively. The pseudocode description of the algorithm is presented below. The pseudocode descriptions of the procedures for activating and disactivating an edge are given separately.

Algorithm SWEEP_HORIZONS
    input
        $\mathcal{M} \equiv (\sum, \mathcal{F})$: polyhedral terrain model;
        $V \equiv (x_0, y_o, z_0)$: viewpoint
    output
        VP: list of maximal visible portions of blocking edges
begin
    compute the set BE of the blocking edges of $\mathcal{M}$ with respect to V;
    for every $b \in BE$ properly intersecting the positive x-axis **do**
        $b_+ \leftarrow$ portion of b above x-axis;
        $b_- \leftarrow$ portion of b below x-axis;
        replace b with b_+ and b_- in BE
    end for ;
    $EV \leftarrow$ *empty list*; {event queue}
    for every $b \in BE$ **do**
        $\theta_1, \theta_2 \leftarrow$ radial coordinates of the first and second extreme of b;
        insert θ_1 and θ_2 in EV
    end for ;
    $L_1, L_2, L'_1, L'_2 \leftarrow$ *empty list*;
    $VP \leftarrow \phi$;
    while $EV \neq$ *empty list* **do**
        pick first event θ from EV;
        case event type of θ **of**
        intersection point between two edges a and b: {$a \prec b$ horizontally}
            swap a and b in L_2 and in L'_2;
            if $a \prec b$ vertically **then**
                if b active **then** DISATCTIVATE(b) **end if**
            else
                if a active **then**
                    $c \leftarrow$ active edge in L'_2 immediately below b;
                    if $b \prec c$ horizontally **then** ACTIVATE(b) **end if**
                end if
            end if
        second extreme of an edge b:
            if b active **then** DISACTIVATE(b) **end if** ;
            delete b form L_1 and from L_2
        first extreme of an edge b:
            insert b in L_1 and in L_2;
            $a \leftarrow$ edge in L_2 immediately below b;
            $a' \leftarrow$ edge in L_2 immediately above b;
            if the projections of a and b on the view sphere properly intersect
            then
                $\theta \leftarrow$ radial direction of the intersection point; insert θ in EV
            end if ;

```
        if the projections of a' and b on the view sphere properly intersect
        then
            θ ← radial direction of the intersection point; insert θ in EV
        end if ;
        c ← active edge in L'₂ immediately above b;
        if b ≺ c horizontally then ACTIVATE(b) end if
    end case
  end while
end algorithm
```

```
Procedure ACTIVATE
    input  b: blocking edge
begin
    θ_b ← θ;
    mark b;
    insert b in L'₁ and in L'₂
end procedure
```

```
Procedure DISACTIVATE
    input  b: blocking edge
begin
    insert in VP the portion of b between θ_b and θ;
    unmark b;
    delete b from L'₁ and from L'₂
end procedure
```

During the sweep process, the size of the sweep status is bounded by $O(n)$, because, in the worst case, we have $O(n)$ blocking edges. The number of events is $O(n + d)$, where d denotes the number of intersections between pairs of blocking edges on the view sphere. Processing an event requires $O(\log n)$ time, regardless of its type, thus leading to a $O((n + d) \log n)$ time complexity for the entire sweep process. In the worst case, $d = O(n^2)$, and the complexity is equal to $O(n^2 \log n)$.

As pointed out at the beginning, the algorithm described computes the maximal visible portions of the blocking edges. The algorithm can be modified for determining all the horizons. For such purpose, it is necessary to take into account the events, where the same maximal portion of blocking edge is split into two different horizons.

At any time, the order of horizon of the current visible portion of a blocking edge b is given by the position of b in the horizontally ordered list $L'₁$ of the active edges. At any event, where an inactive edge becomes active, or an active edge is disactivated, the horizon order of all the active edges following it in horizontal order changes. More precisely, when a segment b becomes active (inactive), all the orders of horizon of the active edges horizontally after b are shifted forward (backwards) by 1. Activating or disactivating a blocking edge may be a operation needed at any type of event. A shifting operation costs $O(n)$, since it may involve all the edges in the current status. Such edges are $O(n)$ in the worst case. The computational complexity of the algorithm, thus, becomes $O((n + d)n)$, i.e., $O(n^3)$ in the worst case.

7 Concluding Remarks

In this paper, we have defined visibility structures based on *nested horizons* on a terrain model with respect to a viewpoint, and proposed algorithms to compute them. In particular, the *influence map* allows both an efficient computation of the sequence of horizons in a given direction and an answer to point visibility queries, when represented as a balanced tree, called the *influence tree* [8].

We consider the radial coordinates $\theta_1, \ldots, \theta_k$, in a spherical coordinate system centered at the viewpoint V, of the extreme points of the maximal visible portions of the blocking edges ($k = O(d)$, where d denotes the number of maximal visible blocking edge portions). The *influence tree* is a balanced binary tree, with $O(\log d)$ levels, in which the leaves correspond to the radial intervals $[\theta_i, \theta_{i+1}]$, for $i = 1, \ldots, k - 1$, and any internal node is associated with the union of the intervals corresponding to its two children. Every maximal visible portion b of a blocking edge is associated with a minimal number of nodes of T, in such a way that the corresponding radial intervals form a partition of the radial interval defined by the endpoints of b. The total size of the influence tree is equal to $O(d \log d)$, because it has $O(d)$ nodes and $O(d)$ segments, each of which can be associated with $O(\log d)$ nodes, in the worst case. The number of segments associated with nodes on the same root-to-leaf path is equal to $O(n)$.

The influence tree can be built by starting from the maximal visible portions of blocking edges (determined, e.g., by the radial sweep algorithm described in Section 4, or by any existent hidden line algorithm [9]). The time complexity of the construction algorithm, described in [8], is $O(d \log d)$.

Given a query point Q, the visible portion of a blocking edge, in whose influence region lies the vertical projection $\bar{Q}$ of Q, can be determined by using the influence tree. Let θ be the radial direction, with respect to viewpoint V, of point Q. We descend the influence tree to the leaf whose associated radial interval $[\theta_i, \theta_{i+1}]$ contains θ. While descending the tree, we maintain and update the segment immediately in front of Q (i.e., nearest to Q, on the same side as V), denoted b_Q, among the segments associated with the nodes encountered by now. At the end of the process, b_Q contains the visible portion b of a blocking edge, such that $\bar{Q}$ lies in the influence region of b. We can thus determine whether Q is visible from V just by testing whether Q is above or below the plane passing through b and V. The visibility from V of a query point Q is determined in $O(\log d \log n)$ time, i.e., in $O(\log^2 n)$ time, in the worst case.

Future developements of the research presented in this paper involve investigating the use of the influence map to answer other types of visibility queries, for example, the visibility of a line segment or of a polygon. Another field of research is represented by the study of a horizon based structure in the framework of hierarchical terrain models [4,6].

References

1. M.Atallah: Dynamic computational geometry. Proceedings 24th Symposium on Foundations of Computer Science, 1989, pp.92-99.
2. R.Cole, M.Sharir: Visibility problems for polyedral terrains. Technical Report 32, Courant Institute, New York University, 1986.
3. S.Cormack, C.M.Gold: Spatially ordered networks and topographic reconstruction. Int. J. GIS, vol.1, no.2, pp.137-148.
4. L.De Floriani: A pyramidal data structure for triangle-based surface description. IEEE Computer Graphics, March 1989, pp. 67-77.
5. L.De Floriani, B.Falcidieno, G.Nagy, C.Pienovi: On sorting triangles in a Delaunay tesselation. Algorithmica, Springer Verlag, 1991, N.6, pp.522-532.
6. L.De Floriani, E.Puppo: A hierarchical triangle-based model for terrain description. Theories and Methods of Spatio-Temporal Reasoning in Geometric Space, Springer Verlag, September 1992, pp. 236-251.
7. L.De Floriani, P.Magillo: Computing the horizon of a point on a polyhedral terrain. Submitted for publication, 1993.
8. L.De Floriani, P.Magillo: Computing point visibility on a terrain based on a nested horizon structure. Submitted for publication, 1993.
9. D.J.Foley, A.Van Dam, S.K.Feiner, J.F.Hughes: Computer Graphics: Principles and Practice. Addison Wesley, 1991.
10. C.M.Gold, U.M.Maydell: Triangulation and ordering in computer cartography. Proceedings, Canadian Cartographic Association's Third Annual Meeting, Vancouver, June 1978, pp. 69-81.
11. C.M.Gold: Spatial ordering of Voronoi networks and their use in terrain data base management. Proceedings, Auto-Carto 8, Baltimore, March 1987, pp. 185-194.
12. M.F.Goodchild, J.Lee: Coverage problems and visibility regions on topographic surfaces. Annals of Operation Research, 1989, 20, pp.175-186.
13. J.Hershberger: Finding the upper envelope of n line segments in $O(n \log n)$ time. Information Processing Letters, 33, 1989, pp.169-174.
14. J.Lee: Analyses of visibility sites on topographic surfaces. Int. J. of Geographical Information Systems, 1991, vol.5, n.4, pp.413-429.
15. M.McKenna: Worst case optimal hidden surface removal. ACM Trans. on Graphics, 1987, n.6, pp.19-28.
16. G. Nagy: Terrain visibility. Technical Report, Rensselaer Polytechnic Institute, 1990.
17. G.Nagy, N.C.Narendra, M.Sharir: Reconstruction of geographic terrains from horizon locations I: the 1-D case. In preparation, 1993.
18. G.Nagy, N.C.Narendra, M.Sharir: Reconstruction of geographic terrains from horizon locations II: the 2-D case. In preparation, 1993.
19. F.P.Preparata, M.I.Shamos: Computational Geometry: An Introduction. Springer Verlag, NY, 1985.

Toward a Theoretical Framework for Geographic Entity Types

David M. Mark

National Center for Geographic Information and Analysis
Department of Geography
State University of New York at Buffalo
Buffalo, NY 14261-0023

Telephone: (716) 645 3545
FAX: (716) 645 2329
geodmm@ubvms.cc.buffalo.edu

Abstract. This paper develops a theoretical framework for defining and representing kinds of geographic entities. The "Spatial Data Transfer Standard" (SDTS) of the United States is used as a starting point for discussion. SDTS defines entities in the world, objects as mathematical or computational constructs, and features as both entities and the objects that represent them. The paper reviews the central role played by categories in human cognition, and the relationships between categories and words of natural language. Although categories may appear to exist in an objective world, they are more properly (and conservatively) thought of as existing in human minds and cultures. This means that category definitions and boundaries can be expected to vary in cross-cultural, cross-linguistic, and cross-disciplinary comparisons, and even at an individual level. This in turn implies that development of 'universal' entity type schemes will be very difficult. Some of the difficulties are illustrated for the superficially-simple example of standing water bodies in English, French, and Spanish. Category boundaries appear to differ not only across languages but also geographically within languages. Human subjects testing will likely be required to explore the nature of geographical entity types as cognitive categories.

KEYWORDS: categories, cognition, GIS, data models, natural language, cross-linguistic, cross-cultural

1. Introduction

As part of the data transfer process, there is a need for common definitions of spatial features. Additionally, the spatial data community recognizes the intensifying need to share data. Agreement on a common format is not sufficient to ensure that the information transferred is meaningful to both the exporter and the importer. Part 2 of the SDTS is a formal attempt to develop a standardized list of entities. The lists in part 2 are the product of several years of effort which involved consideration of about 2,600 definitions of geographic features by different organizations with different requirements. The lists present 200 defined entity types, 244 defined attributes, and over 1,200 included terms. (Fegeas, Cascio, and Lazar, 1992, p. 286).

We have long known that classification is one of the fundamental tools of science. What may not be so obvious it that the way in which people assign individual observed objects or events to categories appears to be near to the core of human cognition itself. Without the ability to assign unfamiliar 'things' to categories, every new scene or view or other sensory input would have to be figured out from some sort of first principles. But with a set of categories, and default attributes for category members, we can learn a lot about a thing just by assigning it to some category. A person's set of cognitive categories provides a means for a 'top-down' interpretation of sensory inputs, mapping almost all novel stimuli onto existing concepts. Of course, there is a risk of mis-interpretation, but the alternative would be chaos. In fact, in a book on fuzzy logic, McNeill and Freiberger (1993) make an even stronger claim regarding the role of categories in cognition:

> "... Words and categories highlight the gist and dim down the details. They generalize and simplify, and without them, language would evanesce.
> It is almost impossible to over-estimate the importance of categories. Indeed, in 1970 theorist David Marr suggested that handling classes is the paramount role of the neocortex, the gray matter of the brain. It lies at the root of us all." (McNeill and Freiberger, 1993, p. 23)

A substantial complication is that categorizations are not necessarily the same for all individuals. Science asserts that nature can be 'cut at its joints', that is, there is a true ('God given') structure, which science attempts to discover. Even if that is true in some sense, it is not true of the ways different languages and cultures structure their worlds. Also, different scientific and engineering disciplines have different views of the world. And finally, differences that are truly 'individual' cannot be ignored.

This paper outlines strategies for investigating how geographical phenomena, features, and fields are categorized by humans in general and by GIS users or potential users in particular.

2. Cognitive Categories

Science has typically modeled categories as mathematical sets. Every possible object or event either is a member of some particular set, or it is not. And, there are procedures or rules available for determining set membership from the observable characteristics of the individual. Furthermore, every member of a set would be an equally good exemplar of that set. This familiar model also underlies multivariate mathematical models of classes, such as discriminant analysis, and most forms of cluster analysis.

Problems with this model, as it applies to the categories that people use in every-day perception, cognition, and thought, were identified long ago by Cassirer (1923), and perhaps earlier. For most categories and for most people, some members are better examples of the class than are others; furthermore, there is a great degree of agreement among human subjects as to what constitutes a good example. And sometimes, it is difficult to know whether or not some observed case is a member of a given class. Smith and Medin (1981) wrote a book that documents some of the evidence for this, and discussed at length two alternative models. Unfortunately, their "probabilistic view" shares many of the flaws of the classical model, and has some new problems as well. As an alternative, they more briefly describe an "exemplar view", which seems to account for departures from the classical view of categories. Briefly, they state:

"As its name suggests, the exemplar view holds that concepts are represented by their exemplars (at least in part) rather than by an abstract summary." (Smith and Medin, 1981, p. 141).

They go on:

"An exemplar of the concept clothing, for example, could be either 'your favorite pair of faded blue jeans' or the subset of clothing that corresponds to blue jeans in general. In the later case, the so-called 'exemplar' is of course an abstraction. Hence even the exemplar view permits abstraction." (Smith and Medin, 1981, p. 141).

However, as described by Smith and Medin, the exemplar view seems to have computational difficulties, because of the need to represent detailed exemplars, perhaps several or many, for each category.

A possible answer is contained in the works of Rosch (1973, 1978). Rosch and her co-workers found widespread examples that people consider categories to have prototypes or more central or typical members; that the same prototypes are given by many subjects. Lakoff and Johnson (1980) added the idea that many categories have a radial structure, with a core of the most basic or central meaning, and with other subclasses related to the core through similarities, metaphors, and family resemblances. This model was developed in great detail by Lakoff (1987), and seems to provide a fertile basis for work in cognitive models of geographic space (Couclelis, 1988; Mark, 1989; Mark and Frank, 1989).

Cognitive categories may seem like a complicated concept even for one individual mind, but the complication increases dramatically if we look at individual differences within a culture, and even more so if we look cross-culturally or cross-linguistically. More than half a century ago, the controversial linguist Benjamin Lee Whorf wrote this about categories, concepts, and words:

We dissect nature along lines laid down by our native languages. The categories and types that we isolate from the world of phenomena we do not find there because they stare us in the face; on the contrary, the world is presented in a kaleidoscopic flux of impressions which has to be organized by our minds—and this means largely by the linguistic systems in our minds. We cut nature up, organize it into concepts, and ascribe significance as we do, largely because we are parties to an agreement to organize it in this way—an agreement that holds throughout our speech community and is codified in the patterns of our language.[1] This agreement is, of course, an implicit and unstated one, BUT ITS TERMS ARE ABSOLUTELY OBLIGATORY; we cannot talk at all except by subscribing to the organization and classification of data which the agreement decrees. (Whorf, 1940, pp. 213-214.)

---

[1] The bold-faced italic type is my emphasis. The words in upper case are Whorf's emphasis as given in his original work.

In a scientific discipline, the agreement on the meanings of the words, on the links between words and categories, and on the category boundaries is much more explicit than for everyday language, and students are taught to accept a certain view of the phenomena addressed by that science or other formal field of study. However, as Whorf went on to point out, this more explicit agreement on meaning masks the fact that the categories themselves, and their definitions and boundaries, may be every bit as arbitrary as are the informal categories that underlie everyday speech:

> This fact [that we must subscribe to a common organization and classification of data in order to communicate verbally] is very significant for modern science, for it means that no individual is free to describe nature with absolute impartiality, but is constrained to certain modes of interpretation even while he thinks himself most free. The person most nearly free in such respects would be a linguist familiar with very many widely different linguistic systems. As yet no linguist is in any such position. We are thus introduced to a new principle of relativity, which holds that all observers are not led by the same physical evidence to the same picture of the universe, unless their linguistic backgrounds are similar, or can in some way be calibrated. (Whorf, 1940, pp. 213-214.)

Standards for the semantic aspects of digital data represent an even more explicit form of agreement on the relation between words and concepts. The cognitive theories of categories reviewed in this section may provide a theoretical framework for extending the semantic aspects of digital data standards across disciplinary, cultural, and linguistic boundaries.

3. The U. S. "Spatial Data Transfer Standard"

Data transfer standards seem essential if organization are going to share or transfer digital information. Such standards need to specify not only formats and data quality, but also the 'meanings' of the data elements. Data transfer standards are not necessarily equivalent to the schemas and data models for individual databases. For one thing, issues of efficient query support do not apply, since the transfer standard is only used between databases or other applications. However, if information loss during data transfer is to be avoided or at least minimized, the translation between the database contents and the transfer standard should be as complete as possible. If by analogy with Chomskian linguistics we think of information as having a surface structure (grammar, syntax) and a deep structure (semantics), a data transfer standard must be capable of representing as much of the deep structure as possible, and each system needs to be able to infer deep structure from its own surface structure (in order to produce output conforming to the transfer standard), and also create its own surface structure given the standard representation of deep structure.

In the United States, the "Spatial Data Transfer Standard" (SDTS) has recently (29 July 1992) been adopted as a Federal Information Processing Standard (FIPS) number 173 (Morrison, 1992, p. 277). Fegeas, Cascio, and Lazar (1992) provided an overview of SDTS. Much of Part 1 of SDTS "is concerned with logical specifications required for spatial data transfer and has three major components: a conceptual model of spatial data, data quality report specifications, and detailed logical transfer format specifications for SDTS data sets" (Fegeas et al., 1992, p. 278). They follow the distinction between entities (in the real world), objects (in the digital world) and features (in both) established in earlier Federal data standards (NCDCDS, 1988).

> "... Briefly, the SDTS model describes the real world as consisting of entities characterized by attributes which are assigned attribute values. "Entity" is the term chosen to describe real-world phenomena because of confusion invoked by the more common term "feature." ... The term "feature" has in the past been applied to either a real-world phenomenon or a digital representation, with resulting ambiguity. Reserving "entity" for the real-world, and "object" for the digital world, the SDTS defines "feature" as both a real-world entity and its object representation. ... Spatial objects are used to digitally represent real-world entities. The SDTS defines a set of 13 simple zero-, one-, and two-dimensional spatial objects ... . The object definitions are oriented toward surface representations (i.e. two-dimensional data), but are also valid for nonplanar geometry, and coordinates associated with the objects may include z values." (Fegeas et al., 1992, p. 279).

Part 2 of SDTS "provides a model for the definition of real-world spatial features, attributes, and attribute values and includes a standard but working and expandable list with definitions" (Fegeas et al., 1992).

3.1 Entity Types
Most cartographically-based geographic data bases use some kinds of 'feature codes' to denote 'kinds' of things in the world that the cartographic objects or symbols represent. Often, these codes, which represent categories, merely encode all and only the distinctions that were made on maps through symbols. Sometimes, they are arranged hierarchically (such as USGS's DLGs), but in other cases they are a list of categories all at the same level (USGS's GNIS files). SDTS defines an entity type as "a set into which similar entity instances are classified (e.g., bridge)." An assumption of a set-theoretic model of categories is commonly made in feature coding schemes, either explicitly (as in SDTS) or implicitly.

4. A Case Study: Water Entities
The ideas presented thus far in this paper may seem rather abstract, and it seems appropriate to present a relatively-simple example. Even though the categorization of standing water bodies turns out to be very complicated, it still is very likely simple, relative to many other broad classes of geographic information.

4.1 Water Entities in SDTS
The "Normative Annex A: Entity Types" in SDTS includes 200 defined entity types. These might be considered to be a first approximation to basic-level for geocartographic categories, as seen by the SDTS authors, although they explicitly avoid defining any hierarchy among or above these entity types. The list contains the following terms for 'standing' water features:

BACKWATER An area of calm water unaffected by the current of a stream.
FISHING_GROUND A water area in which fishing is frequently carried on.

HARBOR An area of water where ships, planes or other watercraft can anchor or dock. Also spelled HARBOUR.

LAGOON A sheet of salt water separated from the open sea by sand or shingle banks. The sheet of water between an offshore reef, esp. of coral and mainland. The sheet of water within a ring or horseshoe shaped atoll.

LAKE Any standard body of inland water.

LOCK An enclosure in a water body with gates at each end to raise or lower vessels as they pass from one level to another.

MARINA A harbor facility for recreational craft where supplies, repairs, and various services are available.

PILOT_WATERS Areas in which the services of a marine pilot are essential.

PLUNGE_POOL A hollow eroded by the force of the falling water at the base of a waterfall, particularly by the eddying effect.

POLYNA Any enclosed water area in pack ice other than a lead, not large enough to be called open water. When frozen over, a polyna becomes an ice skylight from the point of view of the submariner. Also called BIG CLEARING, CLEARING, GLADE, ICE CLEARING, POOL, REGIONAL CLEARING.

PORT A landing place provided with terminal and transfer facilities for loading and discharging cargo or passengers, usually located in a harbor.

SEA The great body of salt water of the oceans.

SWASH The mass of broken foaming water which rushes bodily up a beach as a wave breaks. Synonymous with SEND.

TANK A structure used for the storage of fluids.

TURNING_BASIN A water area used for turning vessels.

Many of the above apply to parts of larger water bodies. In this paper, however, I will concentrate on the three of these features that refer to entire bodies of standing water:

SEA The great body of salt water of the oceans.

LAKE Any standard body of inland water.

LAGOON A sheet of salt water separated from the open sea by sand or shingle banks. The sheet of water between an offshore reef, esp. of coral and mainland. The sheet of water within a ring or horseshoe shaped atoll.

Whereas the sea is clearly distinctive, it is not nearly so clear whether the distinction between lakes and the first type of lagoon is fundamental, especially if one examines these entity types from a cross-linguistic perspective, as may become clearer below.

In the SDTS documentation, "Normative Annex C: Included Terms" lists approximate synonyms and/or subsets of the basic terms. The Annex lists only two terms with LAGOON as their basic reference type: "barrier lagoon" and "laguna". In contrast, LAKE forms the reference type for a long list of other water features: bayou, catch basin, inland sea, millpond, mortlake, open water, oxbow, pasteuer lake, paternoster lake, pond, pool, proglacial lake, reservoir, salina, salt lake, sound, and (even) swimming pool. Included terms associated with SEA are archipelago, closed sea, marginal sea, ocean, open sea, open sound, and open water.

4.2 Cross-language Comparisons for Standing Water Bodies

In attempts to relate word means to concepts, it is almost always useful to examine terminology in a cross-linguistic framework. This is because, no matter how objective and obvious the match between words and concepts may seem in one's native language, other languages may organize things just as apparently-obviously, but in very different ways. In rare cases where meanings appear to be similar in genetically-unrelated languages, one can consider the associated concepts to be very fundamental indeed; almost always, however, the mis-match of terms across languages reveals the arbitrary nature of the match between cognitive concepts at the 'real world'. In the spirit of this approach, the following sections examine meanings of terms for standing water bodies in three languages that are rather closely related: English, French, and Spanish. In addition, regional variants of the last two languages also are examined.

Before this analysis, it is interesting to examine briefly a standardized set of entity types across these and some other languages. The Digital Geographic Information Working Group (DGIWG) is composed of the military mapping branches for Belgium, Canada, Denmark, France, Germany, Italy, the Netherlands, Norway, Spain, the United Kingdom, and the United States. This working group has developed standards for digital data exchange. Part 4 of the DGIWG documentation provides a catalog of feature codes, which are arranged hierarchically (DGIWG, 1992). Lakes are included in code BH080. The leading "B" refers to "Hydrography", one of 10 top-level classes. Within the hydrography theme, 10 subclasses are distinguished, including "BH" for "Inland Water." Lastly, there are 28 different kinds of inland water features. Each of these has a brief definitional phrase in English, followed by terms in six languages, plus an additional version of English, such as in the following example (DGIWG, 1992, p. A-66):

BH080: A body of water surrounded by land

US	Lake / Pond
FR	Lac / Étang
Œ	See / Teich
IT	Lago / Stagno
NL	Meer / Plas / Vijver
SP	Lago / Laguna
UK	Lake / Pond

The DGIWG documentation does not explain why two terms are supplied (three for Dutch) for each language for this feature, but in those cases for which the author could determine relative meanings, the sense of difference generally went from a term for larger water bodies first, with smaller following. Another examples of interest is BH190 (DGIWG, 1992, p. A-71):

BH190: Open body of water separated from the sea by sand bank or coral reef.

US	Lagoon / Reef Pool
FR	Lagon / Lagune
Œ	Lagune
IT	Laguna
NL	Lagune / Strandmeer
SP	Albufera
UK	Lagoon / Reef Pool

As presented in the DGIWG standard, the provision of terms in each language matched to a single definition seems straightforward. The following sections will provide more detailed analysis that raises some doubts about such a conclusion.

4.2.1 English

In English, the term 'lake' is used is used for many standing water bodies, from the Great Lakes to rather small features; 'lake' is probably the most generic term for inland bodies of standing water, and is treated as such in the US Geological Survey's GNIS data sets. Most of the terms listed in the SDTS section (4.1) above as being included (subordinate) under 'lake' would indeed probably be considered to be synonyms or subsets (subordinate classes, kinds) of 'lake' by most English speakers. In fact, I suspect that the coastal type of 'lagoon', which is given in SDTS as a basic entity type parallel to 'lake', would actually be considered to be a subclass of 'lake' by many (most?) English speakers. The term 'pond' is normally used in English to refer to much smaller water bodies, although many small bodies that would be considered to be ponds have names including the generic part 'Lake', and some large lakes in some English-speaking areas, especially western Newfoundland, have names ending in 'Pond' (for example, 'Western Brook Pond'). Figure 1 presents a schematic representation of distinctions among lakes, lagoons, and ponds as the terms commonly are used in both British and American English.

4.2.2 French

French-English dictionaries typically give lac as the French translation of 'lake', and indeed most lakes in Quebec have names including Lac.. This translation also fits the first terms in DGIWG's feature class BH080, as given above. Those same dictionaries often list étang as equivalent to 'pond', matching the second terms in class BH080. In France, however, many large water bodies that would be called lakes in English are called étangs; in fact, the second largest 'lake' in France is the Etang de Berre, near Marseilles Etangs in the central part of France (say, in the Dombes region northeast of Lyon) are shallow, with marshy edges, but often are rather large. Even larger are the coastal lagoons between the Rhône and the Spanish border, also called étangs, separated from the Mediterranean by barrier beaches. These would typically be called 'lagoons' in English, or perhaps 'lakes' or 'bays', but in no way match the normal meaning of the English word 'pond'.

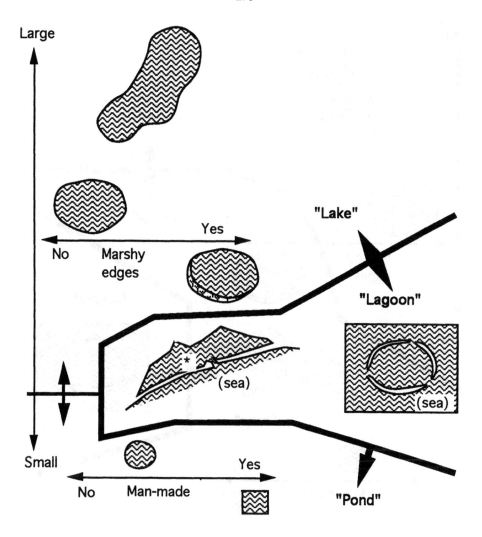

Figure 1: Schematic representation of some basic entity types and included terms for standing water bodies in English.

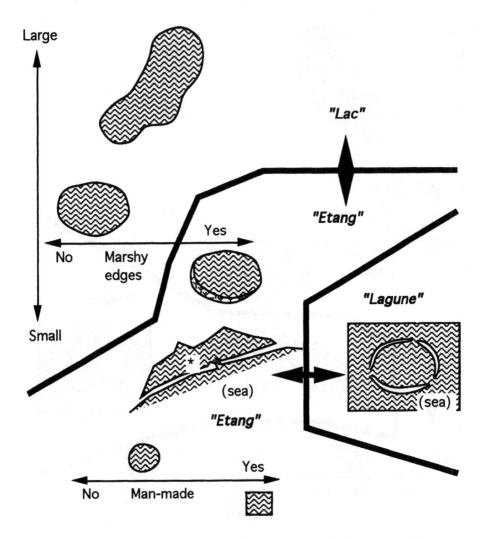

Figure 2: Schematic representation of some basic entity types and included terms for standing water bodies in Standard French; compare with the English terms in Figure 1.

Interestingly, in Quebec, the categorization of water bodies as lac or étang seems to parallel the lake-pond distinction in English much more closely. The "Glossary of Generic Terms in Canada's Geographic Names" (Canada, 1987, p. 94) confirms this. The English definition of the French Canadian étang is "body of water of shallow depth and limited size," and the English equivalent term listed is 'pond'. Thus étang seems to have a much more narrow meaning in Quebec than in France, applying mostly to small bodies of water that would be called 'pond' in English.

The Cajun version of French is even more different in its terminology for standing water bodies. According to Daigle (1984), the common French term, étang, is not a part of Cajun at all. Daigle lists marais as the Cajun translation of the English term 'pond', although in Standard French, marais is usually translated to 'marsh', but in Cajun French, marsh translates to mèche. Mèche is not given as a water-body term in a dictionary of Standard French, nor is it listed in the Canadian glossary cited above.

4.2.3 Spanish. In a standard Spanish-English dictionary, lago is translated to 'lake', whereas laguna has a broader range of meaning, including the following that apply to water bodies and wetlands: "1. Pond, lake, a large diffusion of stagnant water, marsh. 2. An uneven country, full of marshes." (Velázquez de la Cadena, 1973, p. 402). The term estanque is used in Spanish to refer to things that would be man-made ponds in English, a subset of étangs in Quebec, and a very small subset of étangs in France. Estanque and étang are etymologically related to each other and to the English word 'stagnant', but the extents of their meanings are quite different. Ponds dug for water supply for cattle in the southwestern United States are often called 'cattle tanks', perhaps derived from the Spanish term.

SDTS lists 'laguna' as an alternative of subset term under lagoon; such land forms are indeed termed lagunas in Mexico and Texas, but as noted above, in Spain, the term laguna refers mainly to small lakes, and seems not to be used for the coastal water bodies, which in Spain are termed albufera. The DGIWG terminology is consistent with this usage from Spain, with albufera listed under 'lagoon' (feature code BH190). Velázquez de la Cadena (1973, p. 24) translates albufera as "a large lake formed by the sea," explicitly making an albufera a subclass of lake, and not necessarily salty. However, equivalent landforms to the albufera of Spain are called étangs in southern France, and étang is listed DGIWG feature class BH080 as an equivalent of 'pond'! Clearly, water bodies in France whose proper names include the word étang fall under two distinct DGIWG feature classes (BH080, BH190), and furthermore span both included terms (lake, pond) under BH080. Clearly, the terms 'lake', 'pond', and 'lagoon' in English "cut nature up, organize it into concepts, and ascribe significance as [they] do," (Whorf, 1940, pp. 213) differently than do lac, étang, and lagune in the French spoken in France, even if the central meanings of the terms match up well.

5. Discussion

The case study presented in section 4 of this paper found that entity types (feature codes) for standing water bodies would group entity instances (individual water bodies) differently for English, French, and Spanish. If a three by three matrix between English and French were constructed, with 'lake', 'lagoon', and 'pond' along one axis and lac, lagune, and étang on the other, features falling into at least 5 of the 9 cells could be found, and probably more. This means that a common set of entity types cannot exist at a 'basic' level in both languages. Rather, if common entity types are required, they will have to be at a subordinate level within each language, with in this example the French having to place étangs into at least three entity types (étang-pond, étang-lake, and étang-lagoon). For genetically-related languages, the combinatorial growth of numbers of entity types will probably be less than the product of the number of terms per language, but if genetically-unrelated languages are included, the number of cross-language entity types might grow at a rate approaching this upper limit.

Considering how difficult it may be to come up with a suite of cross-linguistically valid entity types, even for a domain as simple as standing water bodies, it may be appropriate to ask a much deeper question: are discrete sets of non-overlapping entity types ('feature codes') necessary, or even desirable, in geographic data transfer standards? Perhaps it would be better to have just one 'kind' of geographic entity, and then attach attributes and values to each instance that would allow entity types to be determined according to various schemes of interest in various languages. For water entities, it seems that we could assign individual cases to the major entity type terms of English, French, and Spanish if we knew the size (surface area) of the water body, its spatial relation to the ocean, whether the water is fresh or salty, whether the edges are marshy, and whether the feature is man-made. However, as Lakoff (1987) has argued quite persuasively, there are many cognitive categories for which membership cannot be uniquely determined by some function of the attribute-values of entities. A basic research question, then, is whether a significant number of geographic entity categories would fail attempts at 'objective' definition; perhaps the exceptions Lakoff talks about would not come up in a geographical context.

If geographical entity types could defined in such an objective fashion, entity types would almost certainly be very useful within any particular database, for the same reasons that they appear to be so important to human cognition—computational efficiency. The computational load of always having to compute the category whenever a query involving a category specification is made would be unreasonable in many applications. But this sort of reasoning does not apply to data transfer standards, where computational efficiency is not so important, since transformations between the transfer standard and a database format should only need to be done occasionally, and not during query mode where system response times become crucial.

If we decide that entity-type codes are useful, or even necessary, for transfer standards, then how do we construct entity typologies that will stand up across individuals and cultures? What do we do if we want to merge data sets that were developed using different entity-type schemes? This even applies within organizations, since for example at the USGS, the problems of linking the geographic names in GNIS with their geometric features from DLG is non-trivial, and the difference in the entity-type schemes contributes to the problem. The problems will be much greater as we face the cross-linguistic and cross-cultural challenges that will have to be faced to create geographic databases of continental or global scales, since a single database to support users from different languages will need either duplicate entity type schemes, or primitives from an entity type cross-tabulation, or will have to have in the common database only the 'objective' attributes from which categories can be determined. Human subjects testing, literature review, and formal type definitions will all be needed in the agenda to achieve a theoretical framework for geographic entity types.

6. Acknowledgments

This paper is a part of Research Initiative 10, "Spatio-Temporal Reasoning in GIS," of the U.S. National Center for Geographic Information and Analysis (NCGIA), supported by a grant from the National Science Foundation (SES-88-10917); support by NSF is gratefully acknowledged. Cyrille Simard provided me with a useful extract from the "Glossary of Generic Terms in Canada's Geographic Names," and Catherine Dibble provided valuable discussion of an earlier version of this paper. Lastly, the author's participation in the Federal Geographic Data Committee's "Spatial Features Forum" at the US Geological Survey in March 1993 provided him with many valuable insights on this problem.

7. References

Abler, Ronald F., 1987. The National Science Foundation National Center for Geographic Information and Analysis. International Journal of Geographical Information Systems, vol. 1, no. 4, pp. 303-326.

Berlin, B., and P. Kay, 1969, Basic Color Terms: Their Universality and Evolution. Berkeley: University of California Press.

Canada, 1987. Glossary of generic terms in Canada's geographic names (Génériques en usage dans les noms géographiques du Canada). Terminology Bulletin 176, Ottawa : Secrétariat d'Etat du Canada (Department of the Secretary of State of Canada).

Cassirer, E., 1923. Substance and Function. New York: Dover.

Couclelis, Helen, 1988. The truth seekers: Geographers in search of the human world. In Golledge, R., Couclelis, H., and Gould, P, editors, A Ground for Common Search. Santa Barbara, CA: The Santa Barbara Geographical Press, pp. 148-155.

Daigle, J. O., 1984. A Dictionary of the Cajun Language. Ann Arbor, Michigan: Edwards Brothers, Inc.

DGIWG, 1992. The Digital Geographic Information Standard (DIGEST): Part 4, Feature and Attribute Coding Catalog (FACC). Washington, DC: U.S. Defense Mapping Agency.

Fegeas, R. G., Cascio, J. L., and Lazar, R. A., 1992. An Overview of FIPS 173, the Spatial Data Transfer Standard. Cartography and Geographic Information Systems 19 (5), 278-293.

Frank, A. and D. M. Mark, 1991. Language Issues for GIS, in: D. Maguire, M. Goodchild, and D. Rhind, editors, Geographical Information Systems: Principles and Applications, Longman, London, vol. 1, pp. 147-163.

Lakoff, George, 1987. Women, Fire, and Dangerous Things: What Categories Reveal About the Mind. Chicago: University of Chicago Press.

Lakoff, George, and Johnson, Mark, 1980. Metaphors We Live By. Chicago: University of Chicago Press.

Mark, D. M., 1989. Cognitive image-schemata for geographic information: Relations to user views and GIS interfaces. Proceedings, GIS/LIS'89, Orlando, Florida, in press.

Mark, D. M., and Frank, A. U., 1989. Concepts of space and spatial language. Proceedings, Ninth International Symposium on Computer-Assisted Cartography, Baltimore, Maryland, 538-556.

McNeill, Daniel, and Freiberger, Paul, 1993, Fuzzy Logic. New York: Simon & Shuster

Morrison, Joel L., 1992. Introduction. Cartography and Geographic Information Systems 19 (5), 277.

National Committee for Digital Cartographic Data Standards, 1988. The Proposed Standard for Digital Cartographic Data. The American Cartographer, 15[1], 9-140.

NCGIA (National Center for Geographic Information and Analysis), 1989. The research plan of the National Center for Geographic Information and Analysis. International Journal of Geographical Information Systems, vol. 3, no. 2, pp. 117-136.

Rosch, E., 1973. On the internal structure of perceptual and semantic categories. in T. E. Moore (editor), Cognitive Development and the Acquisition of Language, New York, Academic Press.

Rosch, E., 1978. Principles of categorization. In E. Rosch and B.B. Lloyd (editors) Cognition and Categorization. Hillsdale, NJ: Erlbaum.

Smith, E. E., and Medin, D. L., 1981. Categories and Concepts. Cambridge, Massachusetts: Harvard University Press.

Velázquez de la Cadena, Mariano, 1973. A New Pronouncing Dictionary of the Spanish and English Languages (compiled with Edward Gray and Juan L. Iribas). Englewood Cliffs, New Jersey: Prentice-Hall, Inc.

Whorf, B. L., 1940. Science and linguistics. Technology Review (M.I.T.), 42(6). Reprinted in Carroll, John B., , editor, 1956. Language, Thought, and Reality: Selected Writings of Benjamin Lee Whorf. Cambridge, Massachusetts: The M.I.T. Press, pp. 207-219.

Land, Space and Spatial Planning
In Three Time Regions

Albert Z. Guttenberg

Department of Urban and Regional Planning, University of Illinois
Urbana, IL 61801

Abstract. What is space? The term denotes primarily an ubiquitous natural resource. It also denotes a cultural resource, an artifact fashioned out of natural space by means of legal enactments and physical design. This paper identifies three varieties of cultural space which have their source in three different models for the nation: the family, the neighborhood, and the city. All three models are present and effective in every time and place. In certain places, however, and certain times, one is dominant. In the late nineteenth century the city (or civic) model predominated. In the first quarter of the present century the neighborhood model was ascendant. From the Great depression to the end of World War II it was the family model. Currently, the three models are very much in contention with each other for primacy. They present a challenge to both the planner and the GIS specialist. The planner is challenged to invent new settlement forms that adjust the rival spatial regimes to one another. The GIS specialist confronts the task of developing new ways of conceptualizing and capturing geographic data that are equal to the present day complexity and multi-dimensionality of present day cultural space.

1 Introduction

No inquiry into the cultural origins of the concept of space should ignore its debt to Walter Firey. Firey characterized and criticized the dominant theories of ecological space of our day and his. These theories hold that space is an external quantity given by nature to which social systems must adapt according to a principle of competition for all metropolitan sites. One of the most influential theories of this type was that of Robert Murray Haig who argued that the principal quality of an urban site is the access it affords to all other sites in the metropolitan system. Some sites, e.g, geographically central sites, give more access, others less. Which agent gets which site? Haig's answer is that each site goes to that adapting agent which can best use it in the sense of realizing from the specific quantity and quality of access it affords the highest profit. This is not only a plausible explanation of how certain business firms relate to one another spatially within a metropolitan framework, it is also a prescription for an ideal city. When every site is dedicated to its "highest and best" (most profitable) use, the whole community is at its land use optimum. The good city is the city of efficient land use.

Haig's theory was later added to by William Alonso to account for the behavior of residential locators in this system but it did not change the defining attribute of this type of theory which is to view space as no more than an impediment, an obstacle to access, a friction to be overcome.

The "friction of space" is a powerful concept for helping to explain the urban land use pattern. Its sufficiency as an explanatory concept is another matter. Firey's most important contribution to regional science was to ask an unsettling question: If land use is determined by firms and households adapting to space on a highest and best use basis, how account for those phenomena which appear to fly in the face of this principle? The most striking example he adduced is that of the Boston Common, a tract of land of almost fifty acres in the economic heart of Boston. Not only does the Common preempt prime land, but it is dysfunctional in its effect, distorting the overall land use pattern, forcing vehicles to go around it, causing traffic jams, spreading out establishments that could benefit from closer proximity to each other and- not the least of its disadvantages -- removing potentially valuable ratables from the tax rolls. And yet Boston has successfully maintained its Common intact in the face of every kind of challenge raised against it for the past 300 years. How can this be so if efficiency in the allocation of sites is the principle that determines land use? With this question and the answer he gave, Firey opened the way to an entirely different understanding of the nature of space: Space as a quantity *internal* to social systems and differing as they themselves differ. The Boston Common, he avers, is a symbol. It stands for Boston, itself, and Boston, with its revolutionary past stands for the nation. The city retains that symbol because it represents an interest which underlies and ultimately sustains the long term interests of all of Boston's inhabitants however much it may be in conflict with their short term particular interests.

Although the foregoing is an overly simple rendition of Firey's argument, our interest here is in the argument's general implications, not in its fine points: Space denotes a natural resource, but that does not exhaust its meaning. Space also denotes a cultural resource, an artifact fashioned out of natural space. The present paper draws on that pregnant insight of Firey. Three varieties of cultural space are identified which, although they are to some extent ideal types, are also real geographical and historical phenomena. Viewed geographically they are part of that constellation of cultural characteristics that differentiate one region of the world from another. Viewed in an historical framework, they are among the distinguishing aspects of the three time-regions which are the principal subjects of this paper.

2 Time Regions

The term time-region has its source in a definition by the economist Kenneth Boulding. A region, he says, is any quantity of space or time treated as a unit (Boulding, 1985). As with many of Boulding's observations, this one, too, has the power of breaking down, or at least of rendering transparent, the walls between established disciplines, in this case between history and geography.

What historians call an historical era can easily be seen as a region in time, a time-region. Historical eras are structured entities just as are space regions. They have centers (climaxes, points in time when they are most intense) and peripheries (points in time where they die out as other eras are beginning). They even have subregions (currents which are changing faster or slower than the prevailing rate).They are inhabited, and their populations have traits, goals, problems, and life styles that distinguish them from the inhabitants of other time-regions. In short, they have their own cultures. Most pertinent for this study, they have their own conceptions of space. But why the special terminology? Why not just call them historic eras? Only because the term era, though it snatches a segment from the flow of time and frames it as though it were a self-sufficient unit, still carries with it too much of a sense of fluidity and change. Region, on the other hand, at least to my mind, better captures the discontinuity between two temporal entities. It is precisely the discreteness of the two eras which I wish to stress here.

Any question asked about two space regions can also be asked about two time regions, such as how a particular word or concept is used, what it means, what it refers to in different regions. A well-know example is the meaning of democracy in the west and the meaning of the same word in the former eastern European block of nations. Louis Wirth, the sociologist, once observed that today's United States is not the United States of the founding fathers, that what the two have in common is no more than a pair of words. This is hyperbole, but it contains more than a grain of truth.

"Planning" is a word. "Land" and "space" are words which have been constant companions of "planning" giving it scope and focus as in "spatial planning" and "land planning", etc. What happens to these words, to what do they refer, as they move from one time region to another?. Do they remain constant or do their meanings change? Behind these questions others lurk: Can a profession or discipline remain self-identical if its key terms and concepts change their meaning across either a time or space boundary? What then of planning as it crosses boundaries? Does it have a self, a continuing self? Does it have a history? Or should planning be pluralized? Is there one planning or several plannings?

Similar questions can be asked of space and land, planning's companion words- What is space? What is land? Are they one thing or many things? To deal with questions such as these, it is necessary to consider planning in relation to space as well as to time.

3 Space

Terrestrial space comes in two varieties- wild and domestic. Wild or natural space is space which has not yet come under the woodsman's ax, the farmers's plow or the builder's hammer and saw. This is the space which Aldo Leopold envisioned as a distinct land use The complement of wild space is the space structured and shaped in accord with human purpose. Two forces shape man-made space: physical design and legal enactments. Design shapes space in the most elementary possible way by controlling our movements- allowing or denying us physical movement in this or that direction. Law does the same thing, it governs our movements. Although the legal enactments that restrain our movements are not visible they are hardly less substantial than the physical restraints - as anyone who habitually lets his parking meter expire soon learns. (It would be illuminating to be able to see the legal landscape with the eye as we see the physical landscape).

Since the influences of design and law are reciprocal, physical shape and legal shape generally coincide and reinforce each other but not in every case. A "no-trespassing" sign, for example, is not always accompanied by a fence surrounding the protected property. A park may be labeled "open space" on the physical planners map and appear open to the eye, but it is by no means legally open.

Law and design collaborate to create man-made structured space. They define paths and boundaries, the two structural elements of the built environment that determine where and whether we can move. These are also the major components of individual buildings. What are buildings other than tightly wound paths and boundaries -walls? (cf. Louis Kahn: "a street wants to be a building"). However, those who lay down streets and raise buildings are not in themselves the ultimate crafters of that environment. Working through them and using them to create human settlements are human ideals. Every human settlement is some concretized vision of what essentially we are and how we ought to be related to each other. Such visions are not revelations from on high. They are idealizations of actual and quite commonplace human institutions. Thus, terms, such as "family", "farm", "factory", "neighborhood", "nation", "city". denote familiar social forms. At the same time each of them epitomizes a unique design for living, working, and governing that at some point in time has been held up as a model for the nation itself. Each encodes a specific net of social and physico-spatial relationships.

In this paper I identify three planning models which have their sources in three different models for national society: neighborhood, family, city. All are present in every time and place. Yet,in certain places and certain times, one or another of them predominates. In the late nineteenth century it was the city model. In the first quarter of the present century it was the neighborhood model. From 1930 to the end of World War II, it was the family model. Since then, the three models have been very much in contention with each other for primacy.

4 The Nation as Neighborhood

City, family, neighborhood. Each model says something essential about what the people of a nation are, how they relate to each other, what they owe one another, how they should be governed, what land is, what space is and, above all for our purposes, what planning is (Table 1).

As closely related as they are in everyday life, family and neighborhood are profoundly diverse as societal models. The basis of the family is kinship; the basis of the neighborhood is friendship, friendly association through physical propinquity. Consequently, in the neighborhood model, space is a social field divided between those who qualify as friends, that is, as social equals, and those who do not. In the first quarter of the present century this was the ruling model. It was the intent of influential social reformers to make of the American society a great neighborhood where comity should prevail, which meant excluding the "uncongenials", those with different tastes, values and habits, characteristics for which race, religion and ethnicity served as the crude indicators. The restrictive immigration laws enacted in 1924 were crafted to implement this program for the nation as a whole vis-a-vis the rest of the world. Identical ends were pursued at the local level through zoning ordinances. Despite the difference in scale both were forms of population control, both were measures to keep out the unwanted, Americanization was an essential aspect of this policy. Insofar as outsiders were already breaching the borders, they were to be melted down and recast as worthy neighbors. We are describing (not judging) the motive and meaning of planning in the U.S. as it was in the Progressive and post-Progressive era, say, 1890-1920. These motives shaped space itself. This was the era of "Garden Cities", "Neighborhood Units" parks and playgrounds and the beginning of the era of the one-class suburb- all of them structures for fostering community. The term "land" also had distinctive overtones in those days. It connoted the people's soil, the ground of their communal life. Land was a synonym of the term country (*patria*). Small wonder that the conservationist attempt under Theodore Roosevelt to redefine land as a resource made little headway.

5 The Nation as Family

Although it was always latent, the notion of land as a resource to be husbanded came into its own with the Great depression. In a time of the severest economic stress the social characteristics of people became less important than their functions as producers and consumers, functions that characterize a family. When I say family, what I am referring to is not the post-modern American family, but the family as it used to be, and as it still is in most parts of the world. In the traditional family the individual members do not belong to themselves. Nor are they really whole personalities. They are the functionally differentiated parts of a single organic unit. They are defined by their biological roles (husband, wife, parent, child) and by their economic functions. Some are earners, others by dint of age or circumstance are dependents. So to the extent that actual American families were forced by hard times in the 1930s to hand over some of their functions to the state, to the nation, the nation itself began to acquire the characteristics of a family, a Great Family, a family in the classic mold. In the New Deal days the Great Family model had a traditional Dad (breadwinning) side- that is, state-supported industry. It also had a traditional Mom side that distributed what Dad provided not according to just deserts but according to need. We are talking now of the 1930s and half of the 1940s. This is the era of the Agricultural Adjustment Act (AAA) and the National Recovery Act (NRA), of public housing, public industry, public employment. It is the era of 12 million men under arms living in close, family-like interdependency.

As the Great Family replaced the Great Neighborhood as dominant model, space tended to lose its quality as a field for expressing social class distinctions. In an actual family the principle that governs the relation of one person from another is function, not status. Likewise, the organization of household space is economic, not social. It consists of functionally differentiated areas - kitchen, bath, bedroom, etc.-which are complementary to each other just as the family members are complementary. Analogously, under the family image, the national territory is conceived as the family patrimony, a set of differentiated resource regions equivalent to rooms, to be managed and maintained, i.e. conserved as productive factors in a national household. In fact the term economy means household management.

Lewis Mumford once observed that one might search the literature of the New Deal from end to end and scarcely find a single reference to his work. If that is true, at least part of the reason may lie in Mumford's philosophy of planning. No one was more closely identified with the Great Neighborhood ideal than he. He was its most gifted and eloquent proponent. However, when the Great Family replaced the Great Neighborhood, land as a noble setting for human life, as a stage for neighborly association yielded to land as a resource, as a factor in production. Mumford's talk of "Regions-To-Live-In" was drowned out by louder talk of resource regions - agricultural regions, mineral regions, industrial regions, forest region, and their conservation.

The move from the neighborhood to the family model transformed American planning thought by substituting economics for politics. Planning on the neighborhood models allows for jousting, inventiveness, even a certain playfulness among titular equals. Neighbors vote their interests and bargain their way out of an impasse. Planning on the family model is driven by economic (and biologic) necessity, not by interests. Its aim is mainly the adjustment by experts of the production and consumption functions of specialized family members. This is a technical not a voting matter. Voting has no place in such a system. One might as well expect the parts of a machine to vote.

6 The Nation as City

Finally, there is the civic model of the nation as opposed to the community (neighborhood) and family models, and city planning as opposed to neighborhood planning and to planning on the family model.

The distinction between city planning and community planning (that is, planning on the neighborhood model) is by no means a new one, although it has been half-forgotten. As long ago as the 1920s the Regional Planning Association of America (RPAA) distinguished between community planning and city planning. The city, in the Association's view, was not just a larger community containing neighborhoods. It was the antithesis of a community. It was also more than a place. It was what Louis Wirth said it was- a way of life. Moreover, the RPAA members made no bones about where their preference lay: The city was bad. The neighborhood was good; correspondingly, community planning was good and city planning, if not bad, was at best ineffectual. (American Institute of Architects, 1925, pp.113-114).

Their preference has its basis now in the circumstances of the time, but that is no reason why we should not give the city and city planning their due. After almost seventy years let us make a case for city planning as opposed to community planning. The term neighborliness suggests kindliness. It has, however, an obverse side, a negative side. It smacks of cronyism, clubiness, exclusionism, nepotism, clientism, special interestism, exceptionism of all kinds. But the United States is *par excellence* the place where the bonds of blood and other special relationships are not supposed to count. And the American model is essentially the civic model -- despite the traditional agrarian imagery. In a city the basic unit is not a cultural group nor the family. It is the free-standing, sovereign individual- the freeman, the citizen. In a city (ideally) to live you do not have to be somebody's friend or relative or neighbor. All you need is money. In the city it is money and ballots that regulates the relations of millions of people most of whom do not know each other. They do not need to, may not want to, know each other. From this standpoint, to propose otherwise, to propose as a matter of principle that public affairs be conducted on the basis of kinship or friendship, is considered the highest form of political immorality. It is for this reason that feeling runs so high at the notion of group rights, a notion that obliterates the individual citizen as the basic unit of society. (In fact, the current agitation signals the emergence of a new and assertive principle of affinity -- "multiculturalism" -- the exalting of membership in a group defined by a single characteristic, such as gender, generation, language, race, what have you. It would not be surprising if this latest social paradigm were to produce its own sense of space).

The significance of "land" also changes as we pass from the neighborhood and family models to the civic model. In the city one may find a sentimental attachment to land (as with the Boston Common) but this is a vestige, a survival of a neighborhood phase, not an essential quality of the civic model. Land is not a symbol of the community as it is in the neighborhood model. Nor is it a resource to be used as in the family. Its value, instead, is as a location and as a vantage point for access to other locations. Here is where Haig's theory of space fits in. It shows itself as a theory generated by the civic as opposed to the neighborhood and family view of society. Correspondingly, in this model the definitive works of planning are not garden cities or "regions-to-live-in". In their place is the planning of the great civilizing, i.e. "citifying", as opposed to tribalizing, physical plans and projects that give access within the nation and beyond - the structures that transcend, traverse, open up, and undermine local, particularistic formations. In fact, to break up such self-enclosed or enclosed-by-others communities is precisely their function. I refer here to the ports and airports, canals and railroads, bridges and highways, telecommunication networks, downtowns, etc., all structures which are the enemies of enclosedness and exclusiveness. (The latest example is Vice President Gore's proposed "telecommunications superhighway"). There are also their counterparts, the non-physical civilizing works, such as those laws that did away with

Jim Crowism in the American south. I count these civil rights acts as city planning acts, city creating acts of the non-physical variety.

7 Conclusion

This paper is now reaching its allotted limit. I will summarize briefly some of its implications for GIS and planning.

The planner and the GIS specialist alike must recognize the cultural character of space. Only in a geodetic context perhaps can we speak of a point in space as though it were an objective datum. A far more elusive entity is the spatial datum in a cultural framework. What such a datum is, where it is, and, indeed, whether it exists at all are questions that can be answered not by pointing to an object and specifying its objective attributes and distance from other objects, but solely with reference to its function in a system whose meaning can be known but not directly observed. It is no easy matter to know, simply by observing, in what kind of space and in whose space we stand, and in whose sight even the most commonplace object may have an unsuspected significance. An example is the by now trite case of the Indian burial ground beneath the new subdivision.

Every spot on earth is the point of intersection of an indeterminate number of spatial realms, past and present, such as the three which we have featured in this paper. These nodes reveal themselves when they become the foci of intense conflicts among the diverse spatial systems. We see this in the bitter land use battles that break out all over the United States, especially at the edges of metropolitan regions: Values transposed from the metropolitan center by high-speed transportation facilities (accessibility values) clash with conservationist and preservationist interests emanating from the cultivation of the soil or from some other environmental feature (resource values) and with neighborhood values that sustain and defend residential amenities and tranquillity.

The planner's role is to adjust the claims of the various cross-cutting spatial realms so that they can co-exist peaceably in the same general territory and not engage in fruitless conflict or simply cancel one another out. One of the most interesting means for achieving this is called Transfer Development Rights (TDR). It concerns specifically the resolution of conflicts between civic (accessibility) space and family-type (resource) space, and neighborhood (social) space. The following example illustrates how it works: Two land use districts are set up, a preservation district and a receiving district. Imagine a retailer in the receiving district who sees an opportunity to extend his reach into the surrounding region by constructing a new higher density building but is prevented from doing so by the current zoning ordinance. Imagine in the other district (preservation) a family farmer who would like to keep his land in cultivation and pass it on as a farm to his son rather than convert it to urban uses if only he could afford to do so. Under TDR the dilemma for each of them can be resolved reciprocally. The farmer is simply authorized to sell (transfer) his development rights to the retailer. Result: The retailer can pursue his plan without hindrance from the zoning ordinance and the farmer now has the wherewithal to maintain his land in its current agricultural use. In short, a market in development rights has been established under public jurisdiction and to the benefit of the public.

This example throws some light on questions raised earlier: Does planning have a self, does it have a history or is it simply a word tossed about on the winds of history? The answer is that as long as planning remains content to remain within the confines of the dominant model of the day, be it civic , social or economic it will always be subject to catastrophic changes in meaning, such as those we have described. History will see to that. In this case there can be no such thing as planning, only plannings. On the other hand, if planners are willing and able to see and stay with their higher calling as adjusters of the various societal models in their spatial realms, then they can indeed rise above the vicissitudes of history and claim a continuing and self-consistent identity.

As for the GIS specialists, they must, like the planner, become students of the hidden meanings of the spatial field they explore and whose information they record and manage. Lacking access to those meanings they will be as blind men groping in the dark. On the other hand, by penetrating to the deep structure of the spatial fields, to its intangible but determining philosophical underpinnings, they will sharpen their sense of what constitutes relevant geographic information and devise more perfect instruments for the capture and analysis of that information.

References

1. William Alonso: A theory of the urban land market. In Papers of the Regional Science Association, 6. Philadelphia: Wharton School of Finance, pp. 149-158 (1960)

2. American Institute of Architects, the Committee on Community Planning: Proceedings of the 58th Annual Convention of the American Institute of Architects, pp. 113-122 (1925)

3. Kenneth E. Boulding: Regions of time. Papers of the Regional Science Association, 57, pp. 19-32 (1985)

4. Walter Firey: Land use in central Boston. Cambridge Mass: Harvard University Press (1947)

5. Albert Guttenberg: "A Note on the Idea of Cycles in American Planning History". Proceedings: Third National Conference on American Planning History. Cincinnati, Ohio. November 30-December 2 (1989)

6. R.M. Haig: Major economic factors in metropolitan growth and arrangement. In R.M. Haig and R. C. Mc Crae, eds. Regional survey of New York and its environs (1927)

Table 1: Three Models of the Nation

	CITY	NEIGHBORHOOD	FAMILY
ORGANIZING PRINCIPLE	Individualism Money Nexus	Social Likeness/ Propinquity	Kinship/ Economic Interdependence, Functional Complimentarity
RELATION TO OUTSIDE	Openness Easy Access Porous Boundaries	Closeness Difficult Access Guarded Boundaries	Closure No Access Sealed Boundaries
SPACE	A field of vantage points of access	A stage, a setting for neighborly association among social equals	The family lifespace- "lebensraum"
LAND	A commodity	"Sacred soil", a symbol of community	A resource base
PLANNING OBJECTIVES	To maintain access/ openness To overcome insidedness	To achieve recess, protection from outside	To achieve self-sufficiency, To eliminate outsidedness
SUBJECTS OF PLANS	Ports, Airports, Railroads, Highways, Telecommunication Networks, Civil Rights Legislation	"Garden Cities," "Neighborhood Units," Exclusionary covenants, Zoning, Growth Controls, Immigration Laws	Public Housing, Public Employment, Public Industry

Geographic and manipulable space in two Tamil linguistic systems[1]

Eric Pederson

Max Planck Research Group for Cognitive Anthropology

Abstract. This paper concerns the linguistic and conceptual contrast between 1) egocentric or speaker-relative spatial reference (such as left/right systems) and 2) absolute spatial reference (such as the use of cardinal directions). Urban Tamils, like European culture, use NSEW exclusively with large-scale or geographic space. In stark contrast with this, rural Tamils use absolute NSEW to depict their manipulable space, e.g. objects located on a table, as well as their geographic space. Urban and rural Tamil speakers were asked to match photographs by verbal description. The director and matcher had identical sets to select from, but they could not see one another's choices. The photos in these sets varied either in the horizontal relations of the depicted objects or along some other non-targeted relationship. Matches involving horizontal plane relationships were relatively more difficult for speakers using a relative system than for speakers using NSEW. The nature of these errors suggests that fundamental methods of manipulating conceptual representations of space vary according to the basic linguistic system used by each community.

TYPOLOGY OF SPATIAL SYSTEMS

Contrary to the literature focusing on familiar Western systems, there is a tremendous amount of cross-linguistic variation among linguistically encoded spatial reference. This variation across languages is of two basic types: 1) variation among similar systems which are grounded on the same basic terms and 2) variation in contrasting types of systems which are organised with significantly different basic terms.

As an example of the first type, we can compare th different uses of Left/Right/Front/Back (*LRFB*) in Hausa (cf. Hill 1982) and French (cf. Vandeloise 1991). A Hausa speaker will conceive of the side of a tree which is closer to the speaker as the tree's *back* by imputing an orientation of the tree facing the same direction as the speaker. An English or French speaker will typically conceive of the back of the tree as the side further from the speaker -- assuming an orientation of the tree facing the towards the speaker.

Both uses of front/back are grounded in the same basic egocentric or speaker-relative system. This type of system determines reference by dividing the world into regions which are extensions from the body of the speaker, e.g. the left and right sides of the body extend to zones of associated space. These regions necessarily shift as the

[1]Thanks to V. Krishnaswami and P. Velraj of Madurai for their invaluable assistance with data collection and transcription. Additional thanks to Christy Bowerman for help with the graphics and Misja Peters and Bernadette Schmitt for help with the statistics. Considerable thanks to David Wilkins and two anonymous reviewers for their comments on the earlier draft of this paper. The remaining problems are mine alone.

This work is part of a larger project of about a dozen people (in the Cognitive Anthropology Research Group at the Max Planck Institute for Psycholinguistics) using such techniques as described here to investigate various linguistic systems of representing spatial relationships from around the world. Thanks to the entire group for their contributions to the design of the project and for comparing their results in their languages of research with the Tamil results.

speaker shifts the alignment or location of his or her body. That is, they are *perspective dependent.*

Alternatively, many languages and cultures use an absolute or *perspective-independent* system of spatial reference, but vary in the exact details of how it is used. Human navigation systems are well known for sharing certain basic features (e.g. the use of *North* or *Northerly*) which nonetheless are used differently in specific instances. There are many different types of absolute systems which all describe the relationship of two locations in such a way that the position of the speaker and hearer is irrelevant. The systems make reference to conventionally agreed upon "directions" which will always remain constant throughout the world commonly encountered by the speakers. The most common subtypes:

A: *NSEW*; monsoonwards; towards sunset/sunrise; etc. ...
B: *Uphill/downhill* systems (common to hill cultures with typically limited geographical range, eg. on a single slope of a mountain range's rise from the plains); similarly in plains river cultures: an upstream/downstream system. The reference of these systems may shift with large-scale movement.
C: *Inland/seaward* (common to island cultures). Clearly, in terms of a NSEW system, the direction of seaward will change depending on ones' location on the island.
D: *Conventional landmark,* e.g. towards the bathing place, towards the headman's house. "Conventional" means that there is a linguistic convention for the use of these landmarks to designate directions.
Each of these systems may be either *symmetric* with a pair of equivalent axes (NS vs. EW) or *asymetric* with a dominant or developed axis (e.g. Uphill/Downhill) with a less specified axis (traverse, 90 vs. 270 degrees not specified).[2]
E: *situationally based local landmark (LLM)* system can be contrasted with an absolute system: each ascribes directionality to relatively distant points. In the case of a situationally based LLM system, the points are stable only for a select set of universes of discourse. This contrasts with a *conventional* landmark system which has become a conventional part of the linguistic system and which typically involves landmarks at or beyond the horizon of common places of communication. An absolute system remains viable regardless of the universe of discourse.

This paper is more concerned with the second type of variation (between different basic types). In discussing the same topical domain, speakers of some languages will typically use a speaker-relative system, while speakers of other languages will typically use an absolute system.

For example, English speakers will typically use an absolute NSEW system while referring to topics the scale of whole nations (*Holland is North of Belgium*), but will use a speaker relative *LRFB* for describing locations on the scale of objects one commonly handles or interacts with (*The phone is left of the computer terminal (as you face it)*). That is, some cultures have a distinction in the grammatical/linguistic system between *geographical* (large-scale) space and *manipulable* (local) space. Other cultures/languages do not make such a distinction. Many Australian languages are known for having grammatical markers which refer to motion towards/away from

---

[2]Cf. Brown 1991:38-43 for a discussion of the priveledged uphill/downhill axis in Tzeltal and the conventionalized strategies for differentiating the other axis.

a reference point with respect to abstracted regions in the horizon (which correspond roughly to the European NSEW). They use these cardinal direction markers for organising information about both geographic space and their extremely local space (i.e., the space immediately surrounding their bodies and the objects which they can directly manipulate). That is, they do not use an egocentric relative reference system at all for describing their environment on any scale. In some groups, a NSEW system is even used for describing some of their own body parts. ("There's a beetle on your East shoulder.") This is in obvious contrast with cultures such as mainstream European cultures which virtually never use NSEW with regards to their immediate physical environment.[3]

Each system has it's advantages and disadvantages in various communicative contexts. One obvious difference between the two systems is the sort of information which must always be readily retrievable. In a relative system, a speaker must always be able to quickly decide on the difference between his or her left and right side and their associated spaces. The world is divided, as it were, into a left hemisphere and a right hemisphere which shift with each movement or change in perspective. Typically, children in educated "Western" cultures are carefully taught to make the distinction and the ability to learn that this distinction is indeed relative (*on my left* may be *on your right*) develops quite late.

In an absolute system, on the other hand, the speaker must have constant reference to the absolute direction (e.g. which way is North) at the time of spatial utterances if the utterance is to be communicatively successful.[4]

Since in circumstances involving regions of the manipulable space, some cultures use a speaker relative system (LRFB) and others an absolute system, and these two systems require different mental calculations, the question arises as to whether or not these distinct cultures have fundamentally distinct conceptualisations regarding their manipulable space. Do speakers who switch systems on the basis of scale or topic have distinct categorizations for these two scales? Conversely, do speakers who invariably use an absolute system conceptualise their manipulable space as essentially the same sort of space as geographic space?

The challenge is to compare the use of the two systems in a way which is not completely subjective. The most formidable problem is to be certain that any determined difference between populations is the result of the linguistic system used when there is also considerable non-linguistic cultural variation between the populations. This difficulty is especially acute given the tremendous cultural variation typically found between the cultures which use these absolute vs. relative systems. One might find a systematic difference in the construal of space or in the methods of

---

[3] Perhaps the earliest reference to this phenomena for Australian langauges is Laughren 1978 for Warlpiri. Cf. Haviland 1986, 1992; Levinson 1986, 1992 for a more elaborated discussion of the absolute reference system in the language and gesture of Guugu Yimithirr.

[4] Levinson (1992) found that elder Guugu Yimithirr (Australia) speakers when led into the bush could dead reckon the direction of distant non-visible, but known sites to within 13 degrees of accuracy. My own informal tests with 5 Dutch amateur naturalists in similar cicumstances gave a mean accuracy of dead reckoning estimates at 75 degrees across subjects or somewhat better than random. Some subjects were better than others, but none exceeded a 40 degree mean accuracy.

manipulating objects in space if one compares, e.g., urban Western European cultures with hunter-gatherer cultures of Australia. However, how can one clearly attribute the difference to the linguistic reference systems when there are also many cultural differences which may contribute to different performances on any given task?

There are two possible investigative solutions to this dilemma: 1) the researcher may examine groupings of languages which exemplify each reference system and which are presumably balanced for cross-cultural factors; 2) the researcher may search for a single cultural area which exhibits each system in two distinguishable linguistic sub-groups. This paper explores an initial investigation of the second type.

ABSOLUTE AND RELATIVE SYSTEMS IN TAMIL

Tamil (a South Indian language) provides an unusual test case for this question of the cognitive impact of spatial reference systems. The Tamil language contains a rich vocabulary of directional terms. It has NSEW (having essentially the same referential qualities as the European NSEW system) as well as a complex system of LRFB terms which can vary as to their deictic center (i.e. the position from which the LRFB values are calculated).[5]

In the Madurai district of Tamil Nadu, the urban-dwelling Tamil speakers use the LRFB speaker-relative system exclusively for descriptions of objects in manipulable space. NSEW terms are only used in describing distant features such as major cities one could travel to.

In contrast to this, many rural agricultural communities, which speak the same dialect of Tamil, seldom, if ever, use "Left" and "Right" for spatial regions despite using the terms for body parts. Instead, they rely primarily on a NSEW system of reference for describing objects in their manipulable space with occasional supplementation with such terms as "in front of", "closer" etc.

These populations identify themselves as essentially part of the same cultural group -- though there will necessarily be some differences between them which result from urban versus rural ways of life and to the varying degrees of literacy and education. By studying the urban and rural Tamil populations, we maximize the chances that any systematic differences in the way they work within their manipulable space is directly linked to their contrasting and habitual linguistic encoding.[6] The two sub-groups are

---

[5]For the purposes of this paper, I will simply use the reliable English translations of the Tamil terms. Most briefly, the stem forms are vaTa *north*, teeR *south*, kiz *east*, meeR *west*, vala *right*, iTa *left*, mun *front*, piN *back*. See appendix 2 for an example of glossed Tamil text.

The overall structure of Tamil allows occurance of NSEW or LRFB morphology in the same syntactic slot of locative constructions. Accordingly, we should not speak of two distinct *grammatical* strategies, but rather two distinct *lexical* sets, the uses of which are determined by community and which require distinct cognitive operations.

[6]The question remains why the difference exists between these two populations. This cannot yet be answered, but on the basis of a preliminary survey of absolute-speaking populaions known to the author, living in a rural environment appears to be a necessary, but not a sufficient condition for this habitual speech pattern. However, this may also be only an epiphenomenon, not a true causal correlation. I would speculate that the use of absolute vs.

determined on the basis of which linguistic encoding they used on a range of tasks involving manipulation and description of various small objects. This sub-grouping correlated perfectly with the linguistic use of NSEW/LRFB in the task discussed in more detail below. There is a high, though imperfect, correlation between urban/rural and LRFB/NSEW use respectively. Accordingly, the terms *urban* and *rural* will be simply short-hand for LRFB-speaking and NSEW-speaking samples. There are isolated instances of individuals shifting from the absolute-speaking to the relative-speaking population sub-groups. Further study is needed here.

METHODOLOGY

To take advantage of these contrasting populations, I conducted a series of "language games" in both the city of Madurai and several adjacent rural communities.[7] While these games are not true psychological experiments, interesting trends nonetheless emerge when comparing the two populations.

All of the games were between two players: a director and a matcher sitting side by side. These roles were intermittently exchanged. The director had access to a model or photograph of a particular arrangement of objects and had to give a verbal description of this model to the matcher who could not see the model. On the basis of this description, the matcher was to select from a set of models or to build a similar model which would match the director's model. Both could ask questions of the other but a screen prevented communication by gesture as well as preventing the matcher from seeing the model and the director from seeing the matcher's attempts. When both the director and matcher were reasonably confident that a match had been successful, the screen was removed, the model and match were compared and discussed, and then the screen was replaced and a new model begun. This method elicits rich linguistic data and allows the researcher to determine how linguistic encoding is interpreted in (an admittedly unusual) context. Further, it allows reasonably precise determination of the nature of any communicative or cognitive problems which may arise during the games.

While similar results obtain in each of the games, I focus here on the results from the photo to photo matching game. In this game, the matcher and director are each given identical sets of 12 code numbered photographs in a random array of 3 by 4 which cannot be seen by the other player. The number is only printed on the director's set of photographs to prevent identification by number for the players. This game is highly efficient in its prompting for exact discussion of preselected contrasts. Responses are easy to evaluate (mismatched or correctly matched photos) and it is relatively simple

---

relative spatial systems has less to do with the living environment than with habitual contact (or non-contact) with artifacts such as written language and movable pictures/photographs. For such artifacts, the spatial referential qualities change drastically for an absolute speaker from instance to instance.

[7]The first of these language games and the inspiration for the subsequent games is owed primarily to Lourdes DeLeón (cf. DeLeón 1991). Subsequent development was by various members of the Cognitive Anthropology Research Group with various members working together on various games. The photo-to-photo matching game and its coding described in detail below was primarily the development of Eve Danziger and myself -- with contributions from Penny Brown. I use the term "game" rather than "experiment" because there are too many extraneous variables involved to be considered true psychological experiments.

to determine the cause of the mismatch. All games were video-taped and mismatches were examined.

The director selects a photograph and describes it to the matcher who must select the same photograph from his or her set. After questioning and deciding verbally that they have both selected the same photo, the director selects another photograph and so on through each set of 12. There are four sets in all. Each set contains groups of photos (from 2 to 8) which vary from one another by one or two parameters in the arrangement of objects in the photos. The major differences within the larger contrast groups in sets 2-4 are all variations of directionality and location within the horizontal plane. These groups will be refered to as the "target" photographs. In each target photo, there is a toy man and either another toy man or a tree. Each photo in the set varies the direction the man/men is/are facing by 90 or 180 degrees and/or varies the position where the man/men/tree is/are standing by 90 or 180 degrees around the center of the photograph. That is, the target photographs vary both position and facing (orientation) details. Three examples from one contrast set of target photos is reproduced below.[8]

Photo 2.4   Photo 2.5

Photo 2.6

The first set and many of the photographs in the other sets probe for different contrasts: Full vs. partial containment, full vs. partial covering, single vs. dual piles of material, in the middle of a line vs. at the end of a line, canonical locative relationship vs. atypical locative relationship, etc.. While the strategies used to match these photographs are often interesting, they are not of immediate concern here, so these photographs will be termed the *non-target* photographs, e.g., 1.5 and 1.6 below.

---

[8]The numbering for the photographs is **n.m** where n= the set the photograph belongs to and m= the particular photograph of the set.

Photo 1.5   Photo 1.6

Even when describing the horizontal plane arrangements of objects represented within photographs, rural Tamil speakers in the Madurai district use NSEW extensively (e.g., "the man is North of the tree") and they seldom, if ever, use LRFB systems. That is, a rural Tamil speaker describing photo 2.5 (shown above) would say something equivalent to "There's a man North of a tree looking at the tree" if the photograph were held with the top of the photograph towards the West. If the same photograph were held rotated 90 degrees to the South, then the description would be "There's a man West of a tree looking at the tree". Should the director and matcher be facing different directions, the photographs would have different orientations with respect to absolute North. In such cases, it can be quite difficult to select the same photograph on the basis of verbal description.[9]

In stark contrast to these rural speakers, urban Tamils (similar to English speakers) use varying LRFB and deictic (*this side/that side*) systems to depict the same spatial arrangements which are coded with NSEW by rural speakers. Because of the variety of relative systems available in urban Tamil, there is often contradicting use of seemingly similar relative systems between speakers. An urban Tamil speaker might describe the same photo 2.5 using LRFB or deixis. Because of the interchangeability of the systems for projecting LRFB onto a ground object, an Urban speaker might say *The boy is {R of / L of / behind / on that side of} the tree.*

In none of the video tapes of the games do the rural Tamils negotiate between themselves which direction is North, etc.. Again, this is in contrast to the urban Tamil speakers who typically don't know the compass directions relative to their current position (like many Europeans and Americans). Further, discussion between urban Tamil players about the exact sense of the terms "left", "right", etc. is common. This correlates with the more ambiguous range of uses available for these terms.

---

[9]I attempted one trial with the speakers holding the photographs in different orientations (sitting shoulder to shoulder with 90 degrees of rotation), but this had to be abandoned as almost every photograph was "mismatched" in terms of being matched with a non-like photograph. In the error count of this set, I corrected for a 90 degree rotation between speakers (who did not allow for this difference in their description): they matched almost perfectly (one error). Sabine Neumann, who recently conducted the same game with Sengologa (Kalahari Bantu) speakers who use a partial absolute cardinal direction system, reports (p.c.) similar results.

FUNCTIONAL EQUIVALENCE OF SYSTEM TYPES

Since otherwise essentially the same dialects speak completely intelligibly with one another, but one group uses the NSEW lexical set and the other the LRFB lexical set, we have an argument that the two systems are essentially interchangeable. Further, both groups of speakers have the ability to shift between absolute and relative terminology -- though the rural Tamils tend to not use *left/right* and the urban Tamils do not use absolute cardinal directions. For example, rural Tamils who do not use left/right (speaker-relative terms) nonetheless frequently augment their use of absolute orientation co-ordinates (NSEW) with purely relative terms (e.g. *neeraa* "straight ahead [for speaker or oriented figure]"). Conversely, urban Tamil speakers very occassionally invoke the use of situationally based local landmarks (e.g., "towards the video camera"), which introduces (temporary) earth-based orientation to disambiguate confusions between the different LRFB projections onto figures facilitated by Tamil.

Note that while each system (NSEW, local landmark, LRFB, deixis) can be used autonomously of the other, individual terms within each system can only be used with (implicit) reference to the other terms in that system -- to wit, there can be no left without a right. All systems are effectively equivalent in that they describe relative positions in terms of approximate right angles.

Despite the ability to create appropriate terms (eg. "slight angle from __"), all speakers had difficulty communicating about angular relationships which were not aligned on the x-y co-ordinates established by the use of NSEW or LRFB when playing other games which involved placement according to non-right angles. Speakers who used local landmark techniques did not use such landmarks to fix non-right angles despite the availability of such landmarks in the immediate environment. This suggests that the absence of a non-right angle reference system (and the conceptualizations necessary for its use) may inhibit spontaneous creation of alternatives dedicated to non-right angles. It remains to examine some of the languages with extensive non-right angle systems used in navigation (cf. Lewis 1972), to see whether any of these angles are ever incorporated into the regular locative constructions of the language.

While both an observer-relative and absolute system must require different types of mental calculations, there is nothing contradictory about the two systems. Since both LRFB and NSEW systems create x/y axes on the horizontal plane with which to locate objects, the two systems are essentially functional equivalents. However, if the different ways people habitually speak imply different conceptions of manipulable space, then it seems reasonable that speakers would tend to avoid use of both an absolute and a speaker-projective system to avoid different underlying mental operations. Use of deixis, such as *near* or *far*, seems compatible with both systems.

Certainly, some systems seem better suited to certain tasks. In all the games, speakers using NSEW had *relatively* less difficulty in matching photos or objects to descriptions of horizontal plane relations. However, LRFB speakers, despite generally finding the tasks less difficult, had relatively more difficulty with matches critically involving horizontal plane distinctions.

ERROR CODING

I made an error count of the mismatched photos from the photo to photo matching games I conducted. The results are errors per photo selected, not per linguistic token of LRFB, etc. The basic coding assumption is that when the target photos (2.3-2.8, 3.1-3.8, 4.5-4.12) are mismatched with one another, this constitutes an error of horizontal spatial arrangement -- either of the position of the figures, or of their orientation, or both. While it is possible that two target photos could be mismatched on the basis of other miscalculations, the assumption is that overall, the errors of matching photographs are the result of the particular photographs and the communication/thinking about them.

These games were run with the director selecting the order of the photographs played. Accordingly, in virtually all selection cases, the match for a photo will be selected in one game from one set of alternatives while the match for the same photo in other games will be selected from a different set of alternatives. The only exceptions to this are games which share the same sequence of photos to be matched. This is extremely unlikely beyond the first photo or two. Thus mismatches are seldom completely identical. If a photo is mismatched because there was no choice (the correct match had already been mismatched earlier) this was not counted as an error.

Mismatches on target photos may be the result of errors varying along two parameters: they may be *Facing errors* (for human figures) or they may be *Positional errors* (the position of the figure in the picture, e.g. standing to the North or the East). Both types of errors may involve either *90 or 180 degree errors of rotation* from the correct match. A 270 degree error is the same as a 90 degree error. Some cases (like 2.4 matched with 2.6, shown above) are tagged as an instance of each type of error and two errors are recorded for that single mismatch. In many cases, what will be coded as a facing error will seem completely derivative of the positional error of the photograph, cf. 4.5 matched with 4.8 (below) where the positional error is a 90 degree (P90) rotation and the men are in both cases "facing one another". This is coded as a facing error (F90) as well despite being derivative from the positional error (P90) and the linguistic devices ("looking at one another") used to select the match. Note that there may be more than one error per mismatched photo pair in the target set, so total errors may not equal total mismatches.

Differences in the directionality of the mismatch was ignored -- that is, whether 4.5 was described and matched with 4.8 or the reverse. Appendix 1 gives the error coding tables used.

Photo 4.5   Photo 4.8

DISCUSSION OF ERRORS

The subjects are divided into two samples: LRFB-speaking (or primarily urban) and NSEW-speaking (or primarily rural). This categorization was determined by the nearly exclusive use of NSEW or LRFB in both this game and in other independent games involving manipulation of small toys on a table or floor. There were six trials (of two people and four games) for the LRFB sample and six trials for the NSEW sample.[10] I then calculated the errors from mismatched target photos according to the above charts. These errors are summarised below.

Error type	LRFB Tamils (6 trials)	NSEW Tamils (6 trials)	All Tamils (12 trials)
90 deg. Facing	3 (15%)	3 (10%)	6 (12%)
90 deg. Position	8 (40%)	7 (23%)	15 (30%)
Total 90 deg.	11 (55%)	10 (33%)	21 (42%)
180 deg. Position	3 (15%)	3 (10%)	6 (12%)
left/right axis	3 (15%)	2 (7%)	5 (10%)
front/back axis	0 (0%)	1 (3%)	1 (2%)
180 deg. Standing	4 (20%)	4 (13%)	8 (16%)
left/right axis	3 (15%)	4 (13%)	7 (14%)
front/back axis	1 (5%)	0 (0%)	1 (2%)
Total 180 deg.	7 (35%)	7 (23%)	14 (28%)
Total posit. errors	12 (60%)	11 (37%)	23 (46%)
Total facing errors	6 (30%)	6 (20%)	12 (24%)
Total LR errors	6 (30%)	6 (20%)	12 (24%)
Total FB errors	1 (5%)	1 (3%)	2 (4%)
Target photos	18 (90%)	17 (57%)	35 (70%)
Non-target photos	2 (10%)	13 (43%)	15 (30%)
Total errors	20 (100%)	30 (100%)	50 (100%)

Tamil errors on photograph matching task (field work 6-7/1992; 4/1993)

Note that the overall error rate for the two populations is similar. This is striking given that the urban (LRFB speaking) Tamils were generally more educated and more comfortable with the peculiarity of the task. It is remarkable how similar the error patterns are for target errors across the two samples. For example, both groups had equal error rates for positional and for facing errors. Within the category of 180 degree errors, we find that further specification of whether or not the error was on the left-of-the-photo or on the right-of-the-photo axis reveals again the same pattern for both populations -- though the numbers are too small to be considered significant. This indicates that for both populations, there are certain types of errors within the

---

[10]There was a seventh trial with a rural population, but the subjects used LRFB primarily in the begining of the task and then shifted to a predominantly NSEW system later on. Accordingly, this pair is not included in the current report. It is interesting to speculate that the players themselves found it preferable to use the NSEW system after initially trying the task with the LRFB. This could be because this is their more customary system (they are rural Tamils, who had received some education in town), or because they found it to be a more advantageous system for the task.

target errors which are more likely to be made than others. That is, there is probably something inherent to the task itself which produces this pattern among Tamils regardless of the lexical set of lacative terms which they use. Thus the task inherently appears more difficult for Tamils on the left-right axis than on the front-back axis even for those speakers who use a NSEW system which does not privilege either axis over the other.[11]

However, looking at the ratio of target errors to non-target errors, we see a significant difference in the error rates in the two samples. The deviation between the expected and observed frequencies is reasonably low, so the probability (p)-value should be fairly reliable. We see that there is considerable difference between the two populations in the *relative* difficulty in avoiding errors between target and non-target photographs, despite the similarity in the types of errors made within just the target photographs.

Error ratio of target to non-target	LRFB Tamils	NSEW Tamils
Target photos (horizontal plane)	18	17
Non-target photos (misc. configuration)	2	13

Chi-squared = 6.35, p= 0.012 (Pearson)
With Fisher-Yates 2x2 continuity correction: chi-squared = 4.86, p=0.027

To wit, the rural Tamils appear to have roughly equal difficulty with the photos which crucially contrast horizontal plane relationships as with the photos which contrast other relationships. The urban Tamils, on the other hand, find the horizontal relationship photos proportionally more difficult. In a sense, the Urban (LRFB) Tamils have difficulty in proportion to the task. This is because the contrast sets of target photos are larger than the contrast sets for the non-target photos, and hence there are simply more possible likely errors available in the target photos. Thus we may conclude that the rural Tamils -- despite finding the task overall somewhat more difficult, found the horizontal plane relationships relatively easy, while the urban Tamils did not.

The question, of course, is why? Essentially, the LRFB system is inherently more confusing in these tasks. In addition to having to consistently differentiate between left and right sides, there is the problem that many Tamil terms like *pinnale* "behind" are multiply ambiguous in describing the position of one object (the Figure in Talmy's (1978) terminology) relative to another (the Ground). Thus *The horse is behind the tree* in Tamil may refer to either the first or the second of the situations pictured below.

---

[11]One can easily argue that the notion of front/back (or closer/nearer to speaker) is inherently simpler cognitively than left/right contrasts.

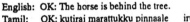

English: OK: The horse is behind the tree.   * The horse is behind the tree.
Tamil: OK: kutirai marattukku pinnaale   irukku.   OK: kutirai marattukku pinnaale   irukku.
           horse tree-Dat behind-Loc be-Pr-3sn   horse tree-Dat   behind-Loc be-Pr-3sn

English: * The tree is behind the horse.
Tamil: OK: maram kutiraikki pinnaale   irukku.
           tree   horse-Dat behind-Loc be-Pr-3sn

It should be noted that this greater range of ambiguity is not just simple lexical ambiguity. Rather, it corresponds to a greater range of permitted cognitive operations (manipulation of deictic orientation) associated with the use of these morphemes. Thus the higher error rate reflects the cost of greater conceptual freedom. The absolute orientation system constrains the speakers to a single mental operation ("Where is North?") This inflexibility serves perfectly in fixed orientation situations, but breaks down with (the culturally uncommon) situations involving rotation of graphically represented landscapes. The relative orientation system is inherently flexible and serves well in situations involving rotation, but this flexibility increases the likelihood of misunderstanding.

CONCLUSION

This study has shown how -- despite the seemingly complete interchangeability of the two lexical sets discussed here -- the habitual language of a speaker strongly affects the ability to perform on relevant problem solving tasks. While these tasks were largely linguistic, they do require contrasting non-linguistic mental operations. This suggests that the linguistic system of encoding spatial relationships may ultimately prove to have implications for the way speakers conceptualise their environment. On the basis of this and the other language games of the study, it would appear that speakers who use an absolute system of orientation for their immediate environment do not contrast immediate manipulable space with large-scale geographic space. Speakers with orientation-relative systems necessarily will treat their manipulable environment in a way distinct from any geographic environment which is encoded with an absolute NSEW lexical set.

Of course, this study only examines behavior in an extremely limited and highly linguistically influenced domain. However, this is leading to the design of further non-linguistic tests, e.g. sorting, memory, and inference tasks. Thus we may obtain more precise cognitive information about how these two contrastive populations fundamentally conceive their surrounding space: both their manipulable space, as well as their geographically-scaled space.

REFERENCES

Brown, P. 1991. Spatial conceptualization in Tzeltal. Working paper no. 6. Cognitive Anthropology Research Group, Nijmegen, The Netherlands.

DeLeón, L. 1991. Space games in Tzotzil: Creating a context for spatial reference. Working paper no. 4. Cognitive Anthropology Research Group, Nijmegen, The Netherlands.

Haviland, J. 1986. Complex referential gestures. Ms. Center for Advanced Study in The Behavioral Sciences, Palo Alto, California.

-----. 1992 Anchoring, iconicity and orientation in Guugu Yimithirr pointing gestures. Working paper no. 8. Cognitive Anthropology Research Group, Nijmegen, The Netherlands.

Hill, C. 1982. Up/Down, Front/Back, Left/Right: A contrastive study of Hausa and English. Pragmatics and Beyond 3:13-42/

Levinson, S. 1986. The semantics/pragmatics/kinesics of space in Guugu Yimidhirr. Unpublished paper presented at the University of Bamberg.

-----. 1992. Langauge and cognition: The cognitive consequences of spatial description in Guugu Yimithirr. Working paper no. 13. Cognitive Anthropology Research Group, Nijmegen, The Netherlands.

Laughren, M. 1978. Directional Terminology in Warlpiri. Working paprers in language and linguistics, no. 8. Launceston: Tasmanian College of Advanced Education.

Lewis, D. 1972. We, the navigators: The ancient art of landfinding in the Pacific. Honolulu: University of Hawaii Press.

Talmy, L. 1978. Figure and Ground in complex sentences. In J. Greenberg, et al. (Eds.) Universals of human language. Stanford, California: Stanford University Press.

Vandeloise, C. 1991. Spatial prepositions: A case study from French. Chicago: University of Chicago Press. (Originally published under the title: L'espace en francais: semantique des prepositions spatiales, 1986.)

APPENDIX I

Coding charts for photo matching tasks -- only for target contrast sets

	2.4	2.5	2.6	2.7	2.8
2.3	F180	P180 F180	P180 F90	F90	P180 F90
2.4	XXXXX	P180	P180 F90	F90	P180 F90
2.5	XXXXX	XXXXX	F90	P180 F90	F90
2.6	XXXXX	XXXXX	XXXXX	P180 F180	F180
2.7	XXXXX	XXXXX	XXXXX	XXXXX	P180

NB: no P90 errors possible in this set; all positional errors are "left/right" inversion types

	3.2	3.3	3.4	3.5	3.6	3.7	3.8
3.1	P90 F180	P90 F90	F180	P90 F90	F90	F90	P90
3.2	XXXXX	F90	P90	F90	P90 F90	P90 F90	F180
3.3	XXXXX	XXXXX	P90 F90	F180	P90	P90 F180	F90
3.4	XXXXX	XXXXX	XXXXX	P90 F90	F90	F90	P90 F180
3.5	XXXXX	XXXXX	XXXXX	XXXXX	P90 F180	P90	F90
3.6	XXXXX	XXXXX	XXXXX	XXXXX	XXXXX	F180	P90 F90
3.7	XXXXX	XXXXX	XXXXX	XXXXX	XXXXX	XXXXX	P90 F90

NB: no P180 errors possible in this set

	4.6	4.7	4.8	4.9	4.10	4.11	4.12
4.5	P180	P90 F90	P90 F90	F90	F90	P90	P90
4.6	XXXXX	P90	P90 F90	F90	F90	P90	P90
4.7	XXXXX	XXXXX	F90	P90	P90	F90	F90
4.8	XXXXX	XXXXX	XXXXX	P90	P90	F90	F90
4.9	XXXXX	XXXXX	XXXXX	XXXXX	P180 F180	P90 F90	P90 F90
4.10	XXXXX	XXXXX	XXXXX	XXXXX	XXXXX	P90 F90	P90 F90
4.11	XXXXX	XXXXX	XXXXX	XXXXX	XXXXX	XXXXX	P180

NB: no F180 errors possible in this set

APPENDIX II Sample text of rural Tamil mismatch
(non-target photos: 2.2 selected by director; 2.1 selected by matcher)
Extract from Murugan and Balu 9 August, 1992.

Photo 2.2   Photo 2.1

\sc Ki: aTutta   nampar colluppaa.
\mr   aTu-tt-a nampar col-u-ppaa
\gle   next-Prt-ps number tell-2s.Imp-Dim
\fr Assistant: "Tell the next number."

\sc Mu: nampar reNTu puLLi reNTu
\mr nampar reNTu puLLi reNTu
\gle number two   point two
\fr M: "Number 2.2"

\sc Mu: oru paya
\mr oru paya
\gle a   boy
\fr M: "A boy..."

\sc Ba: ong
\gle B: mm

\sc Mu: vaTakku   mokamaa   paatta   maanikki
\mr vaTa-ku mukam-aa paar-tt-a maanikki
\gle North-Dat face -Adv look_at-Prt-ps man
\fr M: "is facing North."

\sc Ba: ong
\gle B: mm

\sc Mu: valatu   kaiyila kampe   puTicca   maanikki
\mr vala-tu kai-la kampu-e puTi-tt-a maanikki
\gle : right-Nmr hand-Loc stick-Acc grasp-Prt-Adj person
\fr M: "A person who's holding a stick in (his) right hand."

\sc Ba: ong
\gle B: mm

\sc Mu: reNTu pannikkuTTi munnaala pookutu.
\mr reNTu panni-kuTTi munnaala poo-kiRa-tu
\gle two pig-baby ahead go-Pr 3sn
\fr M: "Two pigs are going in front."

\sc Ba: ong
\gle B: mm

\sc Mu: ingkiTTu, ingkiTTum puucceTi
\mr i-kiTTu i-kiTTu-um puu-ceTi
\gle Prox-Assoc Prox-Assoc-Incl flower-plant
\fr M: "There's a flowering plant this way and this way."

\sc Ba: ong
\\gle B: mm

\sc Mu: irukkaa?
\mr iru-ka-aa
\gle be-Inf-Adv
\fr M: "Is there?"

\sc Ba: irukku ...
\mr iru-k-u
\gle be -Pr-2s.Imp
\fr B: "Yes .."

\sc meeRku paRam onnu kizakku paRam onnu.
\gle west side one east side one
\fr "One on the west side and one on the east side."

\sc ceTi irukkaa?
\mr ceTi iru-k -aa
\gle plant be -Pr ?
\fr "Is there a plant?"

\sc Mu: kizakku paRam onnu.
\gle east side one
\fr M: "One on the East side."

\sc 1: nii collu murukaa anta aNNan keekkanum.
\mr nii col-u murukan-oo anta aNNan keeL-ka -num
\gle 2s tell-2s.Imp M. -Voc that elder_brother ask-Inf-need
\fr Bystander: "You say it. Murugan, you need to ask your older brother."

\sc Ki: illa cari cari
\gle Assistant: no right right

\sc Mu: ong irukku.
\mr ong iru-k-u
\gle mm be -Pr 3sn
\fr M: "Mm, there is."

\sc Ba: irukku   m cari.
\mr iru-k-u   m cari
\gle be -Pr-3sn mm right
\fr B: "There is .. mm .. right."

\sc Mu: paya vaTakku mokamaa tirumpi paatta   maanikki
\mr paya vaTa -ku mukam-aa tirumpi   paar-tt -a maanikki
\gle boy North-Dat face-Adv back_again see Prt-Adj person
\fr M: "The boy is turned facing to the North."

\sc Ba: ong
\gle B: mm

\sc Mu: kampe   puTicca   maanikki
\mr kampu-e puTi-tt-a maanikki
\gle stick-Acc grasp-Prt-Adj person
\fr M: "A person holding a stick"

\sc Ba: reTi
\mr reTi
\gle B: "ready"

\sc Mu: reNTu pannikkuTTi munnaaTi varutu.
\mr reNTu panni-kuTTi munnaaTi vaa-kiRa-tu
\gle two   pig-baby in.front come-Pr-3sn
\fr M: "Two piglets are coming in front (of him)."

\sc Ba: cari ...
\gle B: right

\sc munnaaTi onnu
\gle in.front one
\fr "Right, one in front"

\sc Mu: ingkiTTu   kizakku pakkam oru maram meeRkku pakkam oru maram.
\gle Prox-Assoc east   side   one tree west   side   one tree
\fr M: "There's one tree on the East side here and one tree on the West side."

\sc Ba: ong ceTiyaa irukku.
\mr ong ceTi-aa iru-k-u
\gle mm plant-? be-Pr-3sn
\fr B: "Ah, it's like with a plant."

\sc Mu: raiTTu ....nampar reNTu puLLi ... ezu
\gle right .... number two   point ... seven
\gle M: "right   Number 2.7...."

Scale and Multiple Psychologies of Space

Daniel R. Montello

Department of Geography, University of California
Santa Barbara, CA 93106 USA

Abstract. The importance of scale to the psychology of space (perception, thinking, memory, behavior) is discussed. It is maintained that scale has an important influence on how humans treat spatial information and that several qualitatively distinct scale classes of space exist. Past systems of classification are reviewed and some novel terms and distinctions are introduced. Empirical evidence for the need to distinguish between spatial scales is presented. Some implications of these scale distinctions are briefly considered and research needs identified.

1 Introduction

> *...the Schools of Cartography sketched a Map of the Empire which was of the size of the Empire and coincided at every point with it. Less Addicted to the Study of Cartography, the Following Generations comprehended that this dilated Map was Useless and, not without Impiety, delivered it to the Inclemencies of the Sun and of the Winters."
>
> Suárez Miranda, *On Exactitude in Science*
> [2, p.123]

Of what import is spatial scale? As a problem for formal analysis, the absolute spatial scale of a geometric object is largely irrelevant. The relationships between sides, angles, etc., are scale-independent; the properties of an isosceles right-triangle hold no matter its size. As a problem for geography, the "science of space", scale has always been a concern of cartographic coding and decoding. But once the scale of the cartographic representation is fixed, all of the decisions made with the map become largely scale-independent. A clustered pattern is a clustered pattern. It is this scale-independence of maps, of course, that gives them their great power and utility. They represent spatial relations and patterns of any size at whatever convenient scale fits on our laps, in the glove compartments of our cars, or on the computer monitors of our geographic information systems.

So properties of space, or relations between objects in space, are typically treated as scale-independent when studied as formal problems and, for the most part, as geographic problems. But when studied as a problem for human perception, thought, and behavior (i.e., when studied as a *psychological* problem), it is the thesis of the present essay that space is not scale-independent. "Space is not space is not space" when it comes to human psychology. Instead, multiple spatial psychologies are called for. And these will not differ in a merely quantitative way, but as I hope to show, in a qualitative way. This thesis is not particularly novel, as such. Below I review previous conceptual distinctions about scales of psychological space but also introduce some terms and distinctions of my own. Furthermore, I add some empirical justification to the conceptual justifications for classifying psychological space into multiple scale classes. Finally, the importance of the issue is considered, and some pressing questions for future research are outlined.

A terminological confusion must be addressed at the outset. My dictionary defines *scale* as (among other things) the "ratio between the dimensions of a representation and those of the thing that it represents", in other words, the *relative* size of a representation. Thus, a *large-scale map* is large compared to the space it represents, at least relative to a *small-scale map*. Of course, even very large-scale maps are commonly much smaller than the spaces they represent. Behavioral scientists, on the other hand, speak of *large-scale spaces*, spaces that are relatively large compared to *small-scale spaces*. But more than a statement about the relative sizes of spaces, as I discuss below, this terminology reflects a concern for the size of a space relative to a person, more precisely, to a person's body and action (e.g., looking, walking). Thus large-scale spaces are large relative to a person. Large-scale spaces might be less ambiguously termed "large-size spaces", but such a terminology ignores the critical role that scale relative to the organism plays in the psychological properties of spaces differing in size [cf. 25]. The classificatory terms I introduce below will avoid this confusion because the word *scale* is not used in them.

2 Previous Scale Distinctions

Researchers studying orientation and navigation have long made a distinction between "proximate" or "near" space and "distal" or "far" space [see 23, 24, and the earlier references cited there]. The former can be apprehended from one spot; the latter require locomotion for their apprehension. In the literature on distance estimation, such a distinction is made between "perceptual" and "cognitive" distance [e.g., 3]. Piaget did not explicitly consider the importance of size or scale per se in his extensive body of work on the development of spatial knowledge [e.g., 19], other than the problem of translating between different scales. He did imply that scale was important, however, in so far as a mature "Euclidean" knowledge of spatial layout, involving externally-based co-ordinate systems or reference frames, depends in his theory on experience with multiple perspectives. Knowledge of smaller objects in themselves does not require this.

The most influential work on this topic is certainly the chapter by Ittelson [11]. In it, he compared "environmental" space with the object space of traditional space perception research. He discussed several special aspects of environments as spaces and urged the need to study spatial perception within the context of environments. Unlike objects, the environment "surrounds, enfolds, engulfs" [p.13]. The role of scale is critical: "Environments are necessarily larger than that which they surround...large in relation to man....large enough to permit, and indeed require, movement in order to encounter all aspects of the situation....large-scale environments, extending from rooms through houses, neighborhoods, cities, countrysides, to the whole universe in size....require[s] some process of spatial and temporal summation" [p.13]. In addition, Ittelson noted that information about environments is acquired multimodally, not through a single sense, and that more information is provided than can be processed in any reasonably finite period of time. These distinctions clearly depend to a large extent on the relative size of environmental and object spaces.

Many writers have cited Ittelson's distinctions [1, 13, 25], typically in order to characterize "large-scale" and "small-scale" spaces. Especially important have been Ittelson's distinctions about the opportunity to view large-scale space from multiple vantage points, and the need for locomotion and information integration over time to apprehend large-scale space. The role of different motor systems in the exploration of spaces differing in scale has also frequently been emphasized [e.g., 20].

Mandler [16] distinguished three classes of psychological spaces: small-, medium-, and large-scale. Small-scale spaces are apprehended from a single perspective, outside of the space itself (e.g., table-top models). Medium-scale spaces are apprehended through locomotion about the space, but spatial relationships within them can still be directly observed from one point (e.g., rooms). The spatial relationships within large-scale spaces cannot be directly observed but must be "constructed" over time from locomotion within the space (e.g., houses, towns). Though not formally part of her classification, Mandler acknowledged that very large-scale spaces such as states and countries probably constitute a special case because they are normally apprehended via maps.

As did Mandler [16], Gärling and Golledge [6] further subdivided Ittelson's concept of environments. They distinguished between small-, medium-, and large-scale "environments". A room would be an example of a small-scale environment, a building or neighborhood would be medium-scale environments, and spaces at the scale of cities and beyond are large-scale environments. Gärling and Golledge did not specify exactly upon what these distinctions were based. One important consequence of the distinction, however, is that knowledge of large-scale environments is learned in a very "piecemeal" fashion and is hierarchically organized.

Zubin has discussed an interesting classification of scales relevant to spatial language [reported in 17]. He identified four categories of spaces: Types A, B, C, and D. Type A spaces are smaller than or equal in size to the human body; examples include pens and other hand-sized objects. They are visible within a static visual field and can be manipulated. Type B spaces are larger than the human body and can be viewed from a single point; examples include trees, the outside of a house, and a mountain. They can also be viewed, at least from one perspective, within a static visual field, but they cannot be manipulated. Type C spaces (he also calls them "scenes") are quite a bit larger than the human body and must be scanned to be perceived; examples include a large room, a field, and a small valley. Their apprehension as single spaces is a constructive process. Finally, Type D spaces (he also calls these "territories") are very much larger than the human body; examples include a forest, a city, a state, and an ocean. These spaces cannot be perceived as a unit, according to Zubin.

3 A (Slightly) Novel Classification

The distinctions made by previous authors provide insight into the issues surrounding a classification of psychological spaces (and of course the impetus to consider such a classification in the first place). My own classification largely integrates the distinctions made by earlier writers and provides some novel terms for the classes that focus on their functional properties. I distinguish four major classes of psychological spaces: *figural*, *vista*, *environmental*, and *geographical*. I distinguish these on the basis of the *projective* size of the space relative to the human body, not its actual or apparent absolute size. Large spaces viewed from a distance effectively become small spaces (more on this in my section below on research directions). The most critical functional consequences of the relative projective size of a space are the means by which it may be apprehended and its cognitive treatment by the mind.

Figural space is projectively smaller than the body; its properties may be directly perceived from one place without appreciable locomotion. It may be usefully subdivided into *pictorial* and *object* spaces, the former referring to small flat spaces and the latter to small 3-D spaces. Figural space is the space of pictures, small objects, distant landmarks, and the like. Although one may sometimes haptically manipulate (touch) objects to apprehend their spatial properties, no appreciable movement of the entire body is required.

Vista space is projectively as large or larger than the body but can be visually apprehended from a single place without appreciable locomotion. It is the space of single rooms, town squares, small valleys, and horizons.

Environmental space is projectively larger than the body and surrounds it. It is in fact too large and otherwise obscured to apprehend directly without considerable locomotion, and is thus usually thought to require the integration of information over significant periods of time. It is the space of buildings, neighborhoods, and cities. Although environmental spaces cannot be apprehended in brief time periods, I maintain that their spatial properties can be apprehended from direct experience alone, given enough exposure to them.

Geographical space is projectively much larger than the body and cannot be apprehended directly through locomotion; rather, it must be learned via symbolic representations such as maps or models that essentially reduce the geographical space to figural space. This bears repeating: Maps represent environmental and geographic spaces, but are themselves instances of pictorial space! I therefore expect the psychological study of map use to draw directly on the psychology of pictorial space rather than on the psychology of environmental space. States, countries, and the solar system are good examples of geographical spaces (not withstanding the earth-bound reference in the word *geographic*). The surface of the earth as seen from an airplane, however, would constitute a vista space because of its small projective size and our consequent ability to apprehend it directly from our seat in the plane. One point of clarification: Geographers have very commonly studied spaces at what I have defined as a geographic scale. But by choosing this label, I do not mean in the least to dictate the "appropriate" spatial scale for geographers. I have no problem with the view that geographers can profitably apply their spatio- temporal tool kit to whatever scale they wish. Indeed, as I have said, the maps and computer images studied so often by geographers are instances of figural spaces.

4 Some Empirical Evidence

I now review some empirical evidence that justifies classifying psychological space into several classes on the basis of scale. Very little research has been done explicitly for the purpose of establishing classes of psychological scales; the issue is a potentially fruitful one for generating researchable questions. Given this paucity of research and the limited space of the present format, I will only briefly discuss three bodies of evidence: (1) the effects of learning from maps vs. from direct environmental experience, (2) differences in the frames-of-reference used to organize and manipulate spatial knowledge at different scales, and (3) attempts to measure individual differences in spatial ability at different scales.

Some of the earliest work to look at maps vs. direct environmental experience was done by Evans and Pezdek [4; see also, 26]. They had subjects judge the accuracy of relative positions of triads of US cities. Time to respond was a linearly increasing function of the degree of rotation of the triad from 0° (north to the top, as in standard cartographic convention). This pattern was also found when subjects judged triads of campus buildings that were learned from a map. However, it was not found when subjects learned the campus layout from direct experience; response-time was roughly equal no matter in what orientation the stimuli were presented. Evans and Pezdek suggested that the multiple perspectives of direct experience was responsible for the lack of an *alignment effect*.

However, an important paper by Presson et al. [21] presents strong evidence that multiplicity of perspectives does not, in itself, explain how scale influences the way we treat spatial information. They had subjects learn simple paths marked on plastic sheets, imagine they were standing at one of the places on the path, and then make directional judgments between places on the paths. Thus, the orientations of the learned space and the judgment space could be aligned or misaligned. Several aspects of the paths and procedures were systematically and independently varied: the absolute size of the space from which the path was learned (.36 m^2 to 13.7 m^2, i.e., ranging from "map size" to "room size"), whether the space was described as a "map" or a "path", and the amount of scale translation necessary to match the space to another path. In all cases, the spaces were learned by inspection from a single vantage point. Only the absolute size of the path space (the plastic sheet) influenced the occurrence of alignment effects. That is, subjects retrieved spatial knowledge in an *orientation-specific* manner when they learned it from a small spatial representation; they made large errors when asked to indicate directions as if they were viewing the space from a misaligned perspective (e.g., they learned the space with point 3 in front of point 2, but when judging directions, they were asked to imagine they were standing so that 2 was in front of 3). When subjects learned the space from a large representation, on the other hand, they retrieved spatial knowledge in an *orientation-free* manner; their errors did not differ in size as a function of the perspective from which they were asked to imagine the space. The fact that a switch from orientation-specific to orientation-free coding occurred somewhere around 5-6 m^2 is consistent with the importance of body size in scaling spaces.

Empirical work on *frames-of-reference* provides another body of evidence for the importance of distinguishing psychological spaces according to scale. While spatial perception and knowledge at any scale involves reference frames (the points, objects, or axes relative to which spatial locations are defined), the particular type of frame differs at different scales. Models of spatial perception and knowledge commonly acknowledge this [e.g., 5]. Acredolo [1] reviewed evidence from her lab showing that the tendency to code space *egocentrically* (with reference to one's own location) depends in part on scale. Huttenlocher and Presson [10] asked subjects to answer questions about object locations in a small table-top space vs. a larger room space. The questions could be answered by mentally moving either the object array or one's body; the former was more common with the small space, the latter with the room space.

Another property of reference frames that has been investigated is the way they explain patterns of distortions in locational judgments. Huttenlocher et al. [9] found that subjects distorted their memories for the location of dots within a small circle towards the center axes of the four quadrants of the circle (i.e., towards the diagonals) [but see 27]. A very different pattern was found by Sadalla and Montello [22], however, who had vision-restricted subjects estimate headings after walking a pathway in a large room. Their subjects distorted path headings towards the orthogonal quadrant boundaries themselves (defined by orthogonal axes emanating from the subjects' bodies).

A third body of evidence suggests that ways that individuals differ in their abilities to solve spatial problems may be different at different spatial scales. There has been surprisingly little work using traditional psychometric paper-and-pencil tests (pictorial spaces) of spatial ability to assess performance in environmental spaces, on tasks such as navigation or landmark location. Perhaps an uncritical assumption of the irrelevance of scale has operated here. One extensive attempt to do this revealed only a very weak prediction of environmental abilities with pictorial measures [15]. This would be expected if environmental space is in fact psychologically distinct from pictorial space. Similarly, Loomis et al. [14] performed a principal-components analysis on performance of several table-top and room spatial tasks. In the resulting factor structure, the table-top tasks grouped together in one component, and the room tasks grouped together in two other components. More work on individual differences in environmental spatial abilities is clearly called for.

5 Importance of the Issue

In this section, I briefly suggest some answers to the question: Why is the topic of scale and psychological spaces important? I provide three reasons; others could undoubtedly be given.

5.1 Effective Spatial Communication: Clearly, differences between spatial scales should have implications for spatial communication and the optimal design of a variety of spatial information systems. Examples include maps, other navigation guides such as personal navigation systems, and effective user-interfaces for GIS. A concern for effective communication involves both linguistic and nonlinguistic information [see 17, 12].

5.2 Validity of Simulations: This reason partially overlaps with the first, though the emphasis here is primarily methodological. What are the implications for the validity of spatial simulations if space at different scales is processed differently? To give a common example, research that uses maps, models, or other figural spaces to study the psychology of environmental space may be questionable. For instance, Hanley and Levine [8] used table-top spaces to make general statements about spatial learning and "cognitive maps". McClurg and Chaillé [18] provocatively entitled their paper "Computer games: Environments for developing spatial cognition?". Considerations of scale suggest that computer screens are not environments. Of course, this issue is not only a methodological one; simulations also have practical applications such as flight simulation training. Research comparing results using simulation and real environments is quite germane to the question of scale in spatial psychology.

5.3 Scale and Spatial Choice: Models of spatial choice [e.g., 7] incorporate factors such as time and effort that are scale-dependent. Thus, we wouldn't expect choice patterns and decision processes underlying choice to be identical at different scales.

6 Research Directions

I conclude by offering some comments on the important directions for future research. The most critical is certainly to obtain more evidence of the validity and utility of a qualitative scale classification. Both new findings and replications of existing findings are needed. How many classes are necessary and what should they be? Of particular importance to theories of spatial learning are questions pertaining to what humans can learn about environmental and geographic spaces from direct experience. To what extent can the layout of a city be learned without maps? When does a space become too large to learn directly (thus becoming a geographic space)? Finally, what is the relevant way to treat size in such a scheme? I proposed above that projective size is the determining factor in how spaces are psychologically treated. It is also plausible, however, that apparent or *visual size* is the crucial factor (the size the space appears to be, regardless of its projective size). Is a city seen from an airplane treated as a vista space or an environmental space?

References

1. L.P. Acredolo: Small- and large-scale spatial concepts in infancy and childhood. In: L.S. Liben, A.H. Patterson, N. Newcombe (eds.): Spatial representation and behavior across the life span. New York: Academic 1981, pp. 63-81

2. J.L. Borges, A.B. Casares (eds.): Extraordinary tales. New York: Herder and Herder (1971)

3. D. Canter, S. Tagg: Distance estimation in cities. Environment and Behavior 7, 59-80 (1975)

4. G.W. Evans, K. Pezdek: Cognitive mapping: Knowledge of real- world distance and location information. Journal of Experimental Psychology: Human Learning and Memory 6, 13-24 (1980)

5. J.A. Feldman: Four frames suffice: A provisional model of vision and space. The Behavioral and Brain Sciences 8, 265- 289 (1985)

6. T. Gärling, R.G. Golledge: Environmental perception and cognition. In: E.H. Zube, G.T. Moore (eds.): Advances in environment, behavior, and design (Vol. 2). New York: Plenum 1987, pp. 203-236

7. R.G. Golledge, G. Rushton (eds.): Spatial choice and spatial behavior. Columbus: Ohio State University (1976)

8. G.L. Hanley, M. Levine: Spatial problem solving: The integration of independently learned cognitive maps. Memory & Cognition 11, 415-422 (1983)

9. J. Huttenlocher, L.V. Hedges, S. Duncan: Categories and particulars: Prototype effects in estimating spatial location. Psychological Review 98, 352-376 (1991)

10. J. Huttenlocher, C.C. Presson: The coding and transformation of spatial information. Cognitive Psychology 11, 375-394 (1979)

11. W.H. Ittelson: Environment perception and contemporary perceptual theory. In: W.H. Ittelson (ed.): Environment and cognition. New York: Seminar 1973, pp. 1-19

12. W. Kuhn, M.J. Egenhofer: Visual interfaces to geometry (NCGIA Technical Paper 91-18). National Center for Geographic Information and Analysis (1991)

13. B. Kuipers: The "map in the head" metaphor. Environment and Behavior 14, 202-220 (1982)

14. J.M. Loomis, R.L. Klatzky, R.G. Golledge, J.G. Cicinelli, J.W. Pellegrino, P.A. Fry: Nonvisual navigation by blind and sighted: Assessment of path integration ability. Journal of Experimental Psychology: General 122, 73-91 (1993)

15. C.A. Lorenz, U. Neisser: Ecological and psychometric dimensions of spatial ability (Rep. No. 10). Atlanta, GE: Emory University, Emory Cognition Project (1986)

16. J.M. Mandler: Representation. In: P. Mussen (ed.): Handbook of child psychology, Vol. III (4th ed.). New York: John Wiley & Sons 1983, pp. 420-494

17. D.M. Mark, A.U. Frank, M.J. Egenhofer, S.M. Freundschuh, M. McGranaghan, R.M. White: Languages of spatial relations: Initiative Two specialist meeting report (NCGIA Technical Paper 89-2). National Center for Geographic Information and Analysis (1989)

18. P.A. McClurg, C. Chaillé: Computer games: Environments for developing spatial cognition? Journal of Educational Computing Research 3, 95-111 (1987)

19. J. Piaget, B. Inhelder: The child's conception of space (F.J. Langdon & J.L. Lunzer, Trans.). New York: W.W. Norton & Company. (Original work published 1948) (1967)

20. H.L. Pick, J.J. Lockman: From frames of reference to spatial representations. In: L.S. Liben, A.H. Patterson, N. Newcombe (eds.): Spatial representation and behavior across the life span. New York: Academic 1981, pp. 39-61

21. C.C. Presson, N. DeLange, M.D. Hazelrigg: Orientation specificity in spatial memory: What makes a path different from a map of the path? Journal of Experimental Psychology: Learning, Memory, and Cognition 15, 887-897 (1989)

22. E.K. Sadalla, D.R. Montello: Remembering changes in direction. Environment and Behavior 21, 346-363 (1989)

23. C.I. Sandström: Orientation in the present space. Stockholm: Almqvist & Wiksell (1951)

24. F.N. Shemyakin: Orientation in space. In: B.G. Anan'yev et al. (eds.): Psychological science in the USSR (Vol. 1). Washington: U.S. Off. of Tech. Services (No. 11466) 1961, pp. 186-255

25. A.W. Siegel: The externalization of cognitive maps by children and adults: In search of ways to ask better questions. In: L.S. Liben, A.H. Patterson, N. Newcombe (eds.): Spatial representation and behavior across the life span. New York: Academic 1981, pp. 167-194

26. P.W. Thorndyke, B. Hayes-Roth: Differences in spatial knowledge acquired from maps and navigation. Cognitive Psychology 14, 560-589 (1982)

27. B. Tversky, D.J. Schiano: Perceptual and conceptual factors in distortions in memory for graphs and maps. Journal of Experimental Psychology: General 118, 387-398 (1989)

GIS and Modeling Prerequisites

Arthur Getis

Department of Geography
San Diego State University
San Diego, CA 92182

Abstract. In this paper structural elements are identified for the preparation of GIS-based data for use in econometric or statistical modeling. These elements include the need to know about the special characteristics of spatial data, such as map scale, spatial dependence, spatial variance heterogeneity and spatial trend heterogeneity, and the usual problems faced by modelers, such as nonspherical disturbances, stationarity of data, heteroscedasticity, and temporally and spatially autocorrelated disturbances. Detective work proceeds on the basis of the varying structures implied by the cross product statistic. These include measures of spatial differences, covariance, and interaction, and the exploratory data analysis functions included in the S-Plus statistical package.

1 Introduction

Because of technological advances, researchers seeking empirical verification of their models are faced with larger data sets made up of more detailed spatial information than has been the case in the recent past. Among the types of expanding data sets are: detailed consumer surveys (household data), detailed land cover information (remotely sensed data), environmental indicators (air and water quality data), selling price of houses (real estate data), detailed transportation flows (origination-destination data), consumer demand (individual choice data), industrial location (plant data), and migration (individual household movement data). As the new technology evolves, larger and larger data sets will contain information on smaller and smaller spatial units. Smaller unit data, such as census block data, are currently available in easy to use graphic form (for example, Claritas Corporation data).

Geographic information systems (GIS) represent the technology that is used to store the data in spatially meaningful, logical formats. In addition to the storage function, increasingly GIS is being used to perform analytical operations on the data. This capability together with a well-developed graphics display technology ranks GIS as the major spatial analytical tool of the future.

Coming to grips with the new technology is not a straightforward undertaking for scientists who are used to modest sample sizes, aggregate data, and models that require assumptions about data. Nonetheless, the new technology should be welcome for the following reasons:

- The need to test theory at the scale that the theory requires tends to favor larger scales of analysis over smaller scales. Most social, economic, and environmental theory is based on individual rather than group behavior.

- Scientists usually opt for larger rather than smaller data sets, not only so that they may obtain more reliable estimates of model parameters, but to reserve data for sensitivity analyses.

- Increasingly, scientists favor analytical schemes in which the strict rules of statistical theory can be relaxed. Fast computers capable of handling large sets of data are ideal for devising Monte Carlo schemes that can be statistically distribution free.

- Exploratory data analysis can become a highly productive prelude to confirmatory data analysis.

In this paper, I outline the problems social and environmental scientists face when dealing with the types of data mentioned above. The main concern is with the preparation of data for use in model building and eventually in the confirmatory analysis that model building implies. Data preparation represents, in a formal way, a new step that must be added to traditional empirical analyses (see Figure 1). In the past, models were devised, data collected, tests made, and results reported. Small spatial unit data suffer from a variety of inherent problems that make them generally unfit for confirmatory analysis. These problems revolve around the assumptions made about the distribution of data that are to be used to test our models. We now must face such issues as:

- the appropriate scale of analysis

- the intrinsic stationarity of spatially distributed variables

- the degree and characteristics of any spatial association in the data.

Before outlining the steps that must be taken to ready data for confirmatory analysis, I should like to point to an opportunity modern researchers now have that allows for efficient data preparation. This is exploratory data analysis (EDA). Although exploratory techniques have been in existence for a long time, such technology as that embodied in GIS allows for the manipulation of large amounts of data quickly, resulting in handsome and helpful displays of data. Models can be developed by means of identifying particularly constructive aspects brought forward in EDA. In the process of EDA, the model becomes clear, data are prepared, and confirmatory analysis can be carried out effectively. This sequence does not obviate the need for looping backwards to EDA to stimulate further thought on the nature of the proposed model. In the next section of this paper, I will outline in more detail why EDA, including data preparation, is necessary when working with the new technology. In section three, I will briefly outline the data issues within a model framework. This is followed by a section on EDA where I explore remotely sensed, raster data showing land cover in Oceans ide, California.

STAGES IN ANALYSIS

TRADITIONAL APPROACH	GIS AND SPATIAL ANALYSIS APPROACH
Identify issues, problems	Identify issues, problems
Collect relevant data	Collect relevant and potentially relevant data
Exploratory data analysis histograms, charts, maps, graphs	Exploratory data analysis 1. histograms, charts, maps, computer graphics displays, overlays, interaction with data base 2. Preparation for tests data manipulation data reduction map scale selection
Build model	Build model
Run model tests on parameters	Run model(s) simulations tests on parameters sensitivity analysis
Report results	Report results

Fig. 1.

2 Preparing Spatial Data for Model Building and Testing

There are two special considerations that require that steps be taken before engaging in any confirmatory analysis using spatial data. These pertain to the spatial scale of analysis and the nature of spatial stationarity or non-stationarity in the data. If not addressed successfully, the first of these can cause a major theoretical problem (recall the ecological fallacy) and a major technical problem (recall the modifiable areal unit problem). In the following paragraphs we shall discuss only the technical issues of spatial scale and spatial stationarity.

2.1 Problems of Scale *[As discussed in Anselin and Getis (1992)]*

Openshaw and Taylor (1979) summarize the results of extensive experimentation in which scale changes radically alter the correlative and autocorrelative relationships among variables. Arbia (1989) claims that it is the spatial autocorrelation, or the dependence of nearby spatial units on one another, that is responsible for changes in summary measures as scale is changed. If units are summed into larger units, the mean increases, the covariance increases, and the correlation decreases in absolute value in proportion to the change in the size of the units. In all but a few circumstances, however, the variance increases in relation to the changed size of units and to the correlation between specified neighboring units. Immediately it becomes clear that statistical tests will be affected by the chosen scale (Haining, 1991). This being the case, the selection of an appropriate scale is a crucial decision to which a great deal of attention must be given. This is particularly important since it often is not clear whether the modifiable areal unit problem is indeed an artifact of a particular data set, as is typically assumed, or instead should be attributed to the use of an improper model and/or technique, as argued by Tobler (1989).

It is difficult to predict how the moments of a spatial sample will change with changing scale in all but the simplest circumstances, that is, when the specification of the relationship between spatial units is simple, and therefore, not particularly interesting. In addition, when spatial units are of unequal size, weighting schemes to "equalize" them must be arbitrary and, as a consequence, one must settle for a range of test results rather than a specific value.

Most likely there is no exact solution to the problem. In any case, data preparation techniques must allow for an assessment of changing variable parameters as scale is changed. Clustering and aggregation algorithms are useful in this. Technically, the optimum scale of analysis is most likely just small enough to avoid having to employ the concept of intrinsic stationarity (see next section), and where the classical concept of stationarity can be assumed to hold.

2.2 Assumptions About Spatially Distributed Variables

The classical stationarity assumptions of constant mean and variance used in econometrics and time-series modeling cannot be employed for models describing correlated, geo-referenced data. The assumptions needed for correlated spatial data come from the field of geo-statistics, and are defined in terms of first differences (Cressie, p.40, 1991). The concept is called *intrinsic stationarity*. Intrinsic stationarity is represented by the semivariogram. This means that the covariance at spatial lag s is a function of s, the distance apart in space of the observations (first difference), but not of the absolute locations of the data points in question. As distance increases between data points the variance increases until the variance of the population is reached. The word intrinsic is used to signify that one should expect that in any fine-mesh spatial data set nearby units are correlated. Further, it is assumed that the correlation and variance are constant throughout the region under consideration for the separation distance in question (unless a further non-isotropic assumption is employed). The classical stationarity assumptions can hold for spatial data only if the distance separating spatial units is greater than the range required for making intrinsic stationarity assumptions or if the fact that the data are spatial is incidental to the problem at hand. Nonstationarity, in the nonspatial case means that moments of the distribution function of the variable in question vary from sample to sample. Nonstationary spatial data implies that moments observed in one window (or kernel) are either systematically or unsystematically different from those in another window. Intrinsic stationarity implies that the data for particular spatial units are related and similar to data for other nearby spatial units in a spatially identifiable way. The resulting correlation in the data under intrinsic stationarity signifies that there is dependence, interaction, or some such relationship between spatial units. Proper specification of models dealing with spatially correlated data must explicitly take into account the "spatial effects."

2.3 Finding Heterogeneity and Dependence in Practice

With regard to spatial heterogeneity and dependence, a variety of tools is designed to help describe or model variance structure. The typical measures include the semivariogram, parameters of variables or error terms in spatially autoregressive systems, Moran's I, Geary's c, and filter parameters (see Getis 1991 for a review). In addition, various econometric tests for heteroscedasticity (heterogeneous residuals from regression models), such as the Breusch-Pagan (1979) test, can be adapted for spatial situations (Anselin 1988). Although not straightforward, all of these measures can be entered into a GIS via kernel routines.

For any variable, values of the variance greater than that which would be expected in a sample drawn from a population with mean μ and variance σ^2 are identified as representative of heterogeneous circumstances and are thus nonstationary. Statistically significant consistency in the variance (little variance), that is, less variance than that which would be expected in a set of data which has been sampled from a population having mean μ and variance σ^2, represents a spatially dependent situation (intrinsic stationarity).

Trend may be identified in any of a number of ways. Getis (1993 forthcoming) suggests that trend be identified by the parameters of a function that decribes the changes in spatial association for a variable as distance increases from each datum (Getis and Ord 1992). If the parameters vary with distance monotonically, then trend exists. Other techniques such as universal kriging (Cressie 1991), trend surface analysis (Davis 1983), smoothing spline-equations (Wahba and Wendelberger 1980), spatial adaptive filtering (Foster and Gorr 1986), and spectral analysis (Rayner 1971) are often used for this purpose.

3 Issues Requiring Action in a Model Framework

In the preceding section, we dealt with the inherent problems of spatial data. In addition, before making any tests on models, it is necessary to make assumptions about the population from which we draw our sample spatial data. I shall list rather than discuss the problems of which research workers must be aware. The questions under each heading must be considered and, if needed, acted upon so that the problem is either avoided or is explicitly taken into account in the modeling process. Although not discussed here, the term that best represents the pitfalls that may result from overlooking these questions is "specification error." Simply put, the researcher must avoid incompleteness and unidentified systems.

- Sample Errors
  Do the data fairly represent the phenomena under study?
  What steps must be taken to aggregate or weight observations?

- Nonspherical Disturbances
  If data or residuals are nonspherical, that is, they do not have uniform variance and/or are correlated with one another, how should data be organized?

- Heteroscedasticity
  If residuals of an exploratory model are heteroscedastistic, what corrective steps should be taken? The same question can be asked of heterogeneous data. Should the heteroscedasticity have been included in the model proper or in the disturbance term?

- Temporal Autocorrelation
  What ARIMA-type model is appropriate?

- Spatial Autocorrelation
  How should the data be sampled in order to avoid spatial autocorrelation? Should the autocorrelation have been included in the model proper or in the disturbance term?

The point of this paper is that these questions should be answered before the actual confirmatory model is developed by engaging in EDA. The processes of data preparation and model building will be unsuccessful unless data characteristics are taken into account.

4 The Detective Work: Exploratory Data Analysis

On the following pages a number of illustrations are given of exploratory data analysis. Each of the figures is based on a single set of univariate data. For this paper we use the S-Plus statistical package for our graphic analysis. In the future, it is our understanding that S-Plus will be linked to the GIS called ARC/INFO. Currently, the package lends itself to data ordered in a raster format.

For this paper, we explore a small data set containing gray-scale values of a remotely sensed image of land cover in the Oceanside, California area. Each of the 256 (16 by 16) pixels represents an area 30 meters on a side (see Figure 2). A series of exploratory displays is given of these data in Figures 3 to 11.

First, we consider the distribution of the entire set of data. The histogram, box plot, and density plot in Figure 3 indicates that the data are clearly not normally distributed. The probability distributions shown in Figure 4 indicate that the data are distributed most closely to a uniform probability distribution (all values equally represented). The box plots in Figure 5 tell us that the variation in the data differs markedly from column to column and row to row with the highest median values associated with greatest variation. There is a great deal of similarity, however, in the structure of the data from column to column. This is further demonstrated by the column plots in Figure 6. All possible scatterplots shown in Figure 7 allow for more detailed correlation work. The Chernoff faces in Figure 8 give us a view of the data according to criteria that we choose. In this case, the happier, less troubled faces correspond to normal, higher variance data by rows. The star plots of Figure 9 emphasize the heterogeneity of the rows. The data can be represented as a surface with a perspective of our choosing (Figure 10). In preparation for the statistical EDA, the lagged plots in Figure 11 show us the strong association among pixels that are contiguous to one another but that the association deteriorates quickly by the second spatial lag. Perfect spatial dependence would be represented by a series of observations on a 45° angle.

In addition to the exploratory graphs, we now look at exploratory data (see Table 1). Knowing that the data are best approximated by a uniform distibution, it is not surprising to have a median close to the mean and a mode over one standard deviation from the mean. There appear to be few, if any, outliers; the coefficient of variation is low. The data would have to be classified as heterogeneous within a narrow range. There does appear to be a great deal of autocorrelation in the data. Both the Z values of the Moran's and the Getis and Ord statistics indicate a very strong nearest pixel association. The Getis and Ord Z values, however, begin a downward trajectory immediately, indicating that the greatest clustering of values is within the one pixel distance. Moran's values, show a peak of autocorrelation in the third ring of pixels around each pixel. The data therefore are highly dependent and most likely should be aggregated to a width and length of from two to three pixels on a side.

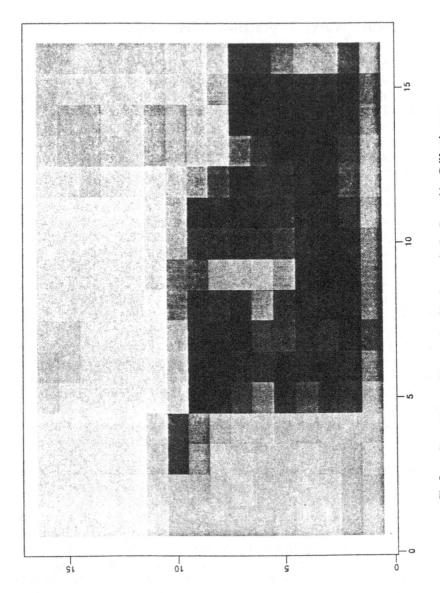

Fig 2.   Remotely sensed image plot of land cover in the Oceanside, California area.

330

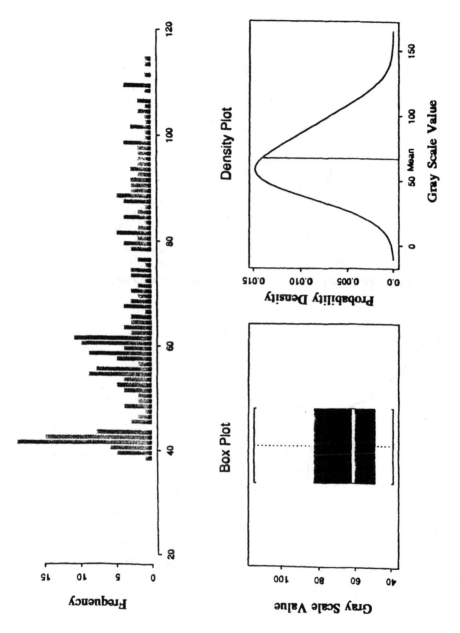

Fig. 3. Histogram, box plot, and density plot of gray scale values.

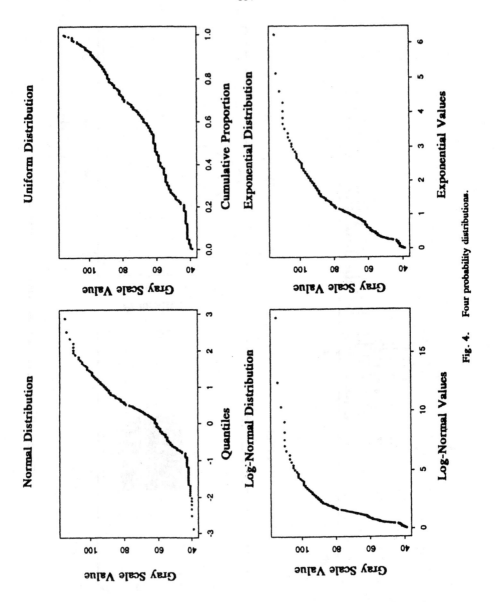

Fig. 4. Four probability distributions.

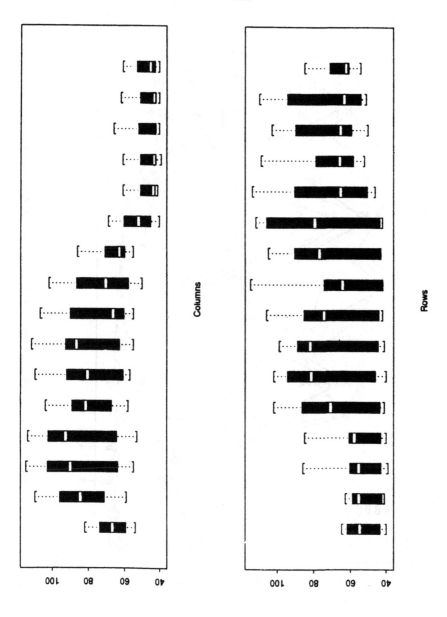

Fig. 5. Box plots of gray scale values by columns and rows.

333

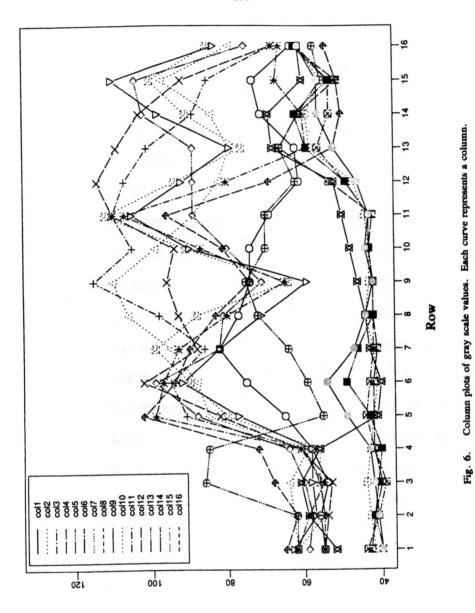

Fig. 6. Column plots of gray scale values. Each curve represents a column.

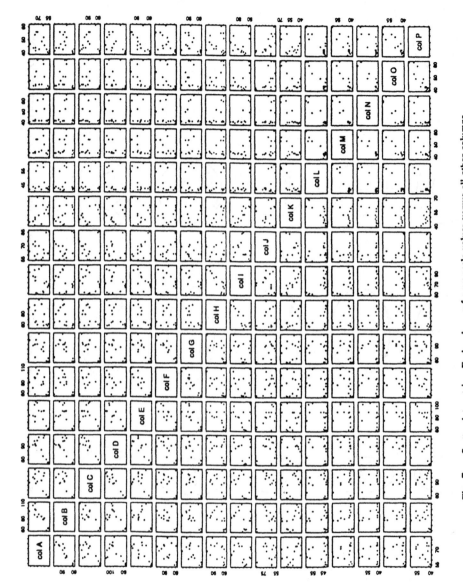

Fig. 7. Scatterplot matrix. Each column of gray scale values versus all other columns.

335

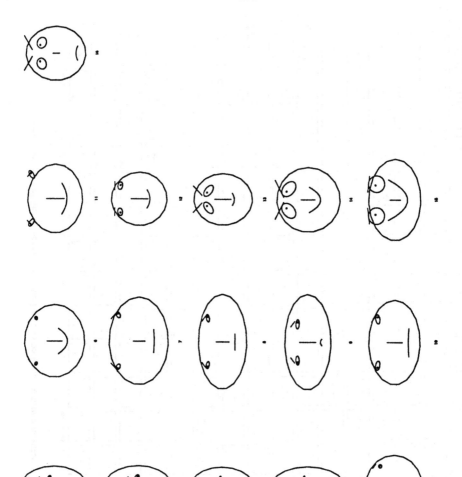

Fig. 8. Chernoff faces for gray scale values. Rows represent observations, columns represent variables.

336

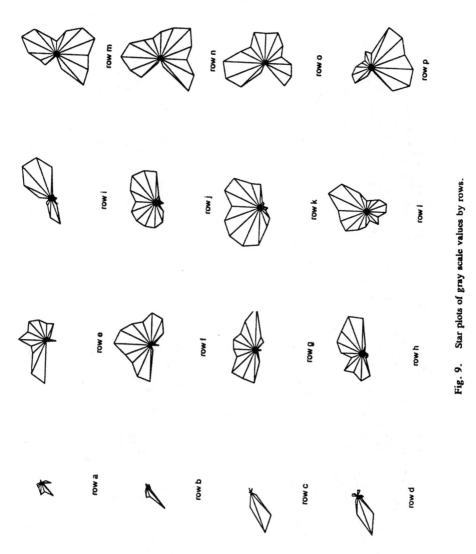

Fig. 9. Star plots of gray scale values by rows.

337

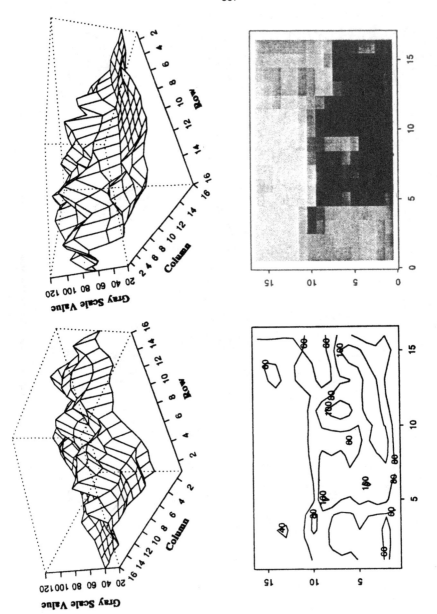

Fig. 10. Perspective and contour plots of gray scale values.

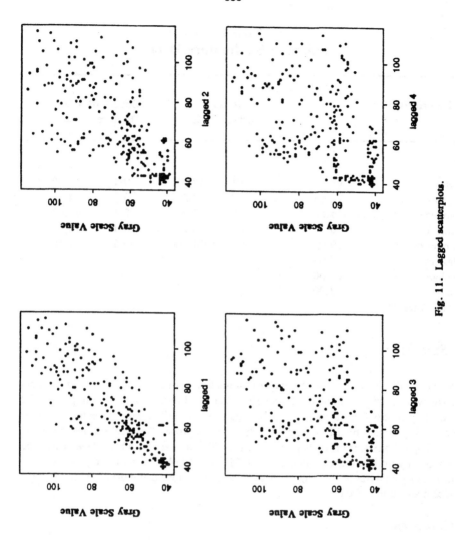

Fig. 11. Lagged scatterplots.

Table 1
Summary Exploratory Data

Descriptives		Moran's I		Getis/Ord's G	
		Distance	Z	Distance	Z
Mean	66.61	0-1	19.71	0-1	15.59
St. Dev.	20.53	1-2	28.22	0-2	14.37
Variance	421.56	2-3	29.72	0-3	11.62
Coef. Var.	0.31	3-4	26.26	0-4	10.33
Minimum	39.00	4-5	24.16	0-5	9.12
Maximum	115.00				
Median	61.00				
Mode	42.00				
Sample Size 256					

5 Summary

The graphic displays shown in this paper represent only a sample of the information that can easily be obtained by use of various EDA statistical packages. For example, we did not display results from a package that computes semi-variograms or spectra. There is much to be learned by EDA, and as technology improves, researchers are becoming increasingly dependent on it. Especially in spatial data analysis, where dependence and heterogeneity problems are particularly in evidence, ESDA allows for the preparation of confirmatory analyses by helping to identify the assumptions that must be made about the spatial distribution of data.

References

1. L. Anselin: Spatial Econometrics, Methods and Models. Dordrecht: Kluwer Academic (1988)

2. L. Anselin, A. Getis: Spatial statistical analysis and geographic information systems, Annals of Regional Science, 26, 19-33 (1992)

3. G. Arbia: Spatial Data Configuration in Statistical Analysis of Regional Economic and Related Problems. Dordrecht: Kluwer Academic (1989)

4. T. Breusch, A. Pagan: A simple test for heteroscedasticity and random coefficient variation, Econometrica, 47, 203-7 (1979)

5. N. Cressie: Statistics for Spatial Data, Wiley (1991)

6. J.C. Davis: Statistics and Data Analysis in Geology. New York: Wiley (1986)

7. S.A. Foster, W.L. Gorr: An adaptive filter for estimating spatially-varying parameters: Application to modeling police hours in response to calls for service, Management Science, 32, 878-889 (1986)

8. A. Getis: Spatial dependence and heterogenity in geographic information systems, in NCGIA publication edited by S. Fotheringham, P. Rogerson consisting of papers of I-14 Conference (1993 forthcoming)

9. A. Getis: Spatial interaction and spatial autocorrelation: a cross-product approach, Environment and Planning, A, 23, 1269-1277 (1991)

10. A. Getis, J.K. Ord: The analysis of spatial association by use of distance statistics, Geographical Analysis, 24, 189-206 (1992)

11. R. Haining: Spatial Data Analysis in the Social and Environmental Sciences. Cambridge: Cambridge University Press (1990)

12. S. Openshaw, P. Taylor: A million or so correlation coefficients: Three experiments on the modifiable areal unit problem. In N. Wrigley, R.J. Bennett (eds), Statistical Applications in the Spatial Sciences, pp. 127-144, London: Pion (1979)

13. J.N. Rayner: An Introduction to Spectral Analysis. London: Pion (1971)

14. W. Tobler: Frame independent spatial analysis, in M.G. Goodchild and S. Gopal (Eds.), The Accuracy of Spatial Databases, pp. 115-122, London: Taylor & Francis (1989)

15. G. Wahba, J. Wendelberger: Some new mathematical methods for variational objective analysis using splines and cross validation, Monthly Weather Review, 108, 1122-1143 (1980)

Acknowledgement

The author would like to thank Long Gen Ying, San Diego State University, for the preparation of the illustrations.

A Map Editing Kernel Implementation: Application to Multiple Scale Display

Philippe Rigaux*°   Michel Scholl*°   Agnès Voisard•

rigaux@cnam.cnam.fr
scholl@cnam.cnam.fr
Agnes.Voisard@inria.fr

Abstract. User interfaces are a fundamental component of applications using as a support Geographic Information Systems (GIS) and geographic DataBase Management Systems (DBMS). We propose in this paper a model and an architecture for designing Geographic Database User Interfaces (GDUI). A GDUI kernel based on this architecture has been implemented using, at the interface programming level, X Windows, Motif and C++, and, at the database level, the GeO_2 data model [DRS93b] implemented itself on top of the O_2 DBMS. From the starting model, we make a first proposal for multiple scale map display.

1 Introduction

User interfaces are a fundamental component of applications using as a support Geographic Information Systems (GIS) and Geographic Database Management Systems (DBMS). Maps are the basic objects handled by such DBMS. There are two classes of functions in Geographic Database User Interfaces (GDUI): (i) the display of maps (geometry plus textual and structured description of the geographic objects displayed on the screen) and (ii) interactive queries on these maps.

[EF88, Voi91] give some user requirements for GDUI's. From a first experience with the implementation of a geographic database prototype [SV92, Voi92] with the O_2 object-oriented DBMS [BDK91], it was clear that these user requirements cannot be handled by regular interfaces or interface tools for database systems. In particular, GDUI's present specific features such as the following:

---

*Cedric/CNAM, 292 rue St Martin, F-75141 Paris Cedex 03, France.

°INRIA, BP 105, F-78153 Le Chesnay Cedex, France.

Work partially supported by the French *CNRS GDR Cassini*, and the ESPRIT BRA *AMUSING*.

•Institut für Informatik, Universität München, Leopoldstr. 11b, D-8000 München 40, Germany (on an INRIA grant).

1. These queries are fully or partially graphically expressed. In particular some parameters correspond to the displayed geometry and are identified with the mouse. This identification is complex and requires computational geometry code not provided by "off the shelf" low level interface tools,

2. several maps corresponding to the same geographical area are overlaid in the same window on the screen (e.g., administrative layout and road network),

3. browsing and navigation through data is peculiar; it takes for example the form of scrolling or zooming.

4. the geometry is displayed at various scales and may have various representations (encodings) according to the scale (multiple scale representation: e.g., a town is either an icon or a polygon or a point, or...),

5. with the geometry of the map is associated a legend that includes graphical features such as line texture, color inside a polygon,..., as well as the display of alphanumeric descriptions (e.g., name of the city at the center of the polygon),

6. etc...

We propose in this paper a simple architecture for designing GDUI's. The objective is to provide high level tools, generic enough (i) to handle a large range of geographic applications (to adapt to a variety of GIS requirements) and (ii) to be independent of the database where the maps are stored. The starting point of our work was the interface model proposed in [Voi91, Voi92]. We decided to experiment its implementation, with emphasis on functionalities such as scrolling and zooming. Although we implemented simple zooming (magnifying data without resolution change) the ultimate goal we have in mind is the design and implementation of multiple scale representation. In a first step, only map display has been addressed (we left most of querying functionalities for a future work). Compared to the work in [Voi92], simplifying assumptions were made on several parameters related to the presentation of maps on the screen. In particular extremely simple legends and structures for the map display were chosen.

Our experiment [Rig93] presented below uses the GeO_2 [II92] data model. GeO_2 is a prototype of a simple geographic database implemented at IGN[1] with the object-oriented DBMS O_2 [BDK91]. In order to fulfill the requirement of database independence, a GDUI was not developed with the O_2 user interface tool O_2Look but in C++ [Sou86], X Windows [SG86] and OSF/Motif [OSF89] connected to O_2 through the O_2-C++ link [Tec92].

This paper is organized as follows. In Section 2, we describe a simple model and a four level architecture for implementing GDUI's. *Maps* are objects manipulated at the top level (in a vector form). They do not properly belong to

---

[1] *IGN* stands for the French "Institut Géographique National".

the interface but to the database. Then objects that belong to the interface world itself are considered. *Images* (in a vector form as well) represent the second level. Images are then translated into raster formats called *pixmaps* (third level) from which a subset included into a rectangle window is displayed on the screen to form a *presentation* (fourth level). Depending on the user query on objects displayed on the screen, either the presentation form or the pixmap form or the image form or the map itself are necessary for answering the query. As an example, both simple zooming and scrolling only require the pixmap form.

In Section 3, a detailed architecture is presented as it is currently implemented and some complex operations (e.g., display) are detailed. In Section 4, the problem of multiple scale display is studied. Section 5 presents our conclusions and future work.

2 Interface Architecture and Model

Maps are the basic objects stored in the database and to be manipulated by the end user. Maps have a geometric part and a descriptive part. Both should be available to the end user. What the user sees and interacts with is always included in windows as managed by window managers. As as matter of fact, what the user interacts with was defined in [Voi91] as a *mapget* (short cut for map widget). A mapget is the window hosting maps according to a given structure. As an example of structure suggested in [Voi91], the window can be split into several subwindows, each subwindow hosting more than one map (several layers overlay on the screen, or overlay of a map with a satellite image, etc.). In the sequel, for the sake of clarity, we assume the window is made of only one subwindow, and unless otherwise specified, there is only one map displayed in this window. A mapget is more than a window plus the presentation of the map displayed in it. A mapget is a specific widget since it allows operations specific to the geographic database end user. These of course depend on the end user requirements.

For an illustration of end user operations, see for example [Voi91]. We shall give a few of them that we arbitrarily split into 4 classes:

1. *Display*: Initially users want to display a map identified by a name (or an icon). Once it has been displayed (in a mapget), they might ask for the description of the displayed geographic objects or they might ask for a submap (or a set of objects) identified with the mouse: Select the lake including a point, select the road intersecting the rectangle, etc.

2. *Dynamic change of legend*: Maps are displayed according to a *legend*. This might be done once for ever for all users, or might be interactively changed during the session. Features such as map title, geometry naming, colors, texture, ..., belong to what is commonly called the map *legend*. In the sequel we restrict our attention to the graphic features of the map legend, namely colors and texture of zones and lines.

3. *Database querying*: The user might select maps or objects according to their descriptive features. Examples of such queries are: *"What are the cities with more than 2 Millions inhabitants"*, or *" what are the fields where corn is grown"*. The choice of geographic objects in maps can also be done according to their geometry (*"what are the fields of more than 1km2"*), their topology (*"what is the closest forest from Paris"*) or directly on the screen by using the mouse (see above). More sophisticated queries combine maps in order to create new maps (e.g. map overlay). Querying (except for selection by point, rectangle, ... of objects displayed on the screen) requires some knowledge of the logical structure of the maps in the database, such as the schema of a relational like implementation of the database.

4. *Miscellaneous*: There are many operations of primary importance that do not necessarily imply querying the database. As an example, users might try to modify the borderline of a country, without necessarily saving the result into the database. They might resize their map display (change its scale). They might zoom into it. The semantics of *zooming* is discussed in Section 4. *Scrolling* and *clipping* (keep the part of the map included in a subrectangle window) are other important operations.

As we shall see now the above classification is not so arbitrary but obeys an interface architecture that includes four levels from the map (first level) to the presentation (fourth level). As mentioned above, some end user operations (such as display a map identified by a name or give the name of the lake surrounding the point designed on the screen) involve the four levels. Other involve only one or two levels and therefore provide a faster answer. Besides, while in the top two levels data are vectorized, in the two low levels, data have a raster form.

2.1 Map and Legend

2.1.1 Maps

A geographic database includes maps. A map is a set of geographic objects (e.g. rivers or cities or countries). A geographic object has a type that depends not only on the application (GIS design according to user requirements) but unfortunately also on the DataBase Management System (DBMS) typing system: In particular, the structure of maps will depend on the technology used (relational vs object-oriented technology for example) and within a technology it will depend on the specific DBMS used. Anyhow, whatever DBMS and geographic data modelling are used, a map has two parts: One includes its description (represented by descriptive attributes: A river has a flow and a name), the other one (spatial attributes) describes the geometry and the topology of the map (e.g., a river is represented by a polyline). Note that, whatever the physical representation of basic (atomic) types such as points, lines, regions is chosen (raster, vector, spaghetti, etc.), in a given map, one knows which geographic(s) object(s) (river, road, town, county,...) is (are) associated with each point in the 2D space.

We shall not need here a precise description of the class of queries that can be performed on maps in the database. The interested reader is referred to [Abe89, Dav91, Ege91, SV92, SDMO87, Ha88, SPSW90, BDQV90, Güt88, ODD89, WH87].

Shortly speaking, a relational DBMS extended with spatial abstract data types (to answer spatial queries) or an object-oriented DBMS, has enough power for answering such queries. Usually the interface to such DBMS is the relational model (tables + relational algebra for queries on descriptive attributes) extended with spatial data types (a set of geometric and topologic types are specified together with the operations that can be performed on these types). There is no standard on these types that vary from one system to the other.

2.1.2 Legend

As said above we shall only consider in the *legend* graphic attributes (color, pattern, texture) associated with map attributes. As an example. consider a crops map in which corn should be painted in yellow, while wheat should be represented in brown and rye in black. Such legends can be simply represented by a (relational) table with the following columns (attributes): *theme, (map) attribute value, graphic attributes.*

One line (tuple) of this table could be: *crops, corn, yellow.* Legend tables are stored in a database that is not necessarily the map database. They are updated before the session (the legend is prepared for all displays of the application or for a specific map display) or might be changed during the session (interactive modification of a map display by the end user).

2.2 Image

This second level is the most abstract level of the interface. It corresponds to what an end user expects from an interface (at a logical level) independently of (i) the way maps are represented in the database and (ii) the details of presentation on the screen. The basic object at this level is called an *image*. An image is what the user sees from a map. It might be either its geometry only (with a specific legend) or its geometry together with a description displayed either closed to the geometry (e.g., road number) or separately.

An image is created from a map of the database. Note, however, that images might come from different databases. Although an image could come from a raster map (e.g., satellite image), we shall assume in the sequel that images are created from vectorized maps.

2.2.1 Image model

When designing interface tools, one should then specify an image model: Since what a user sees from a map is an image, querying the database is made via images. But operations on images are not only operations on maps: The superposition of several images in the same screen window does not correspond to an actual map overlay in the database (an actual creation of a new map that is the "intersection" of the source maps). Some applications require updating existing maps (e.g. the borderline of two adjacent counties) that shall not be necessarily integrated into the database.

For the sake of simplicity, in the sequel, we shall assume there is a bijection between the set of maps and the set of images, and that each image only includes (is projected on) the spatial attributes of its map. Then an image is a couple (*map, legend*).

With the above relational representation of legend, assuming the map is an (extended) relation as well, the creation of an image is a natural join between its map and its legend.

2.2.2 Separating the map level from the image level

One might argue that a specific image level is not necessary. Indeed, one might just project the map spatial attributes and join the result with the map legend before raster conversion and display. There are several reasons in favor of this intermediary level:

1. As seen above, the interface should be as independent as possible from a given DBMS. This was one of our design choices for the experimentation described in Section 3.

2. We have also seen above that operations on images include not only operations on maps but other specific operations.

3. *Performance:* Even though the interface were designed on top of only one given DBMS, and therefore dependent on its interface tools and on the DB structure, it is better to keep a separate representation of the spatial attributes of a map for two reasons: (i) the physical representation of images, i.e., the actual vector representation of spatial types (areas, polylines, etc.) is suited to fast displaying while the physical representation of spatial types in the database is chosen for efficient geometric and topologic operations on maps; (ii) some end user operations do not need to query the database, e.g., interactive change of legend. Then a "compiled" version of map m into a representation of spatial data types ready for fast display and denoted *convert(m)* saves a possibly costly operation of connection with the database, reading and converting the data from one structure to the other. In the sequel, *image (m,l)* denotes the image obtained from *convert(m)* and legend l.

2.3 Pixmap

The third level is not yet the final presentation on the screen. The choice of such an intermediate level was not clear at the first glance, but came out from technical considerations. More precisely, it is justified for two reasons:

1. It corresponds to the raster representation of the image obtained by low level interface tools, and thus to a ready-to-display structure. We shall call it *pixmap* (as it is in the X/Motif world). The image is converted into this raster format representation by using low level interface primitives applied to the basic spatial objects of the image (points, lines, areas, ...). In the following, $p = pixmap\ (i,\ s)$ denotes the raster conversion of image i into pixmap p of size s.

2. Operations at this level: From the pixmap, the final presentation is obtained by selecting a rectangle subwindow included in the pixmap, to be actually displayed on the screen. In particular, scrolling, simple zooming, and clipping applied with a rectangle window argument are efficiently performed at this level: They do not need to come back to the top levels (images and maps).

Several remarks are noteworthy:

1. With a given pixel of the pixmap, we cannot associate anymore (as at the image or map level) a geographic object.

2. Several images could be displayed on the same pixmap; [Voi91] proposes a stack structure for images sharing the same pixmap (allowing some to be hidden and with operations such as shift up, shift down, remove an image, add an image on top, (etc.). For simplicity reasons, we only consider in the following one image per pixmap.

3. Such an architecture exhibits a memory hierarchy for map representation (Figure 2): The database is the secondary persistent storage for maps. A first cache memory stores images which are converted into raster format in a second cache, namely the pixmap. Depending on the size of the caches and the strategies used, several scenarios may occur when the end user operation "gets out of the window". In particular, upon scrolling, one might get out of the pixmap. This "cache failure" necessitates then the loading of another image and its conversion into pixmap before resuming scrolling.

2.4 Presentation

The last level corresponds to the actual representation of the map on the screen. The extraction *visualize (p, win)* of pixmap p inside a rectangle window *win* is called a *presentation*. It corresponds exactly to the part of a pixmap a user sees on the screen (Figure 1).

2.5 End user operations and levels

In the introduction we gave a few end user operations we split into 4 classes. Let us look now which levels are concerned with each of these operations:

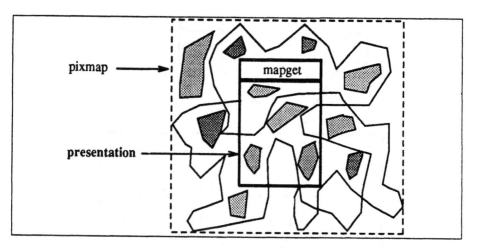

Figure 1:   A pixmap and a presentation

1. *display (m, l, win, s)* Display requires all four levels: Map m has to be fetched, loaded into an image $i = image(m,l)$ using map legend l; it is converted into raster format assigned to $p = pixmap(i, s)$, finally presentation $pr = visualize(p, win)$ is obtained.

2. Most of the queries to the database produce at the map level a target map that has then to be displayed. An example of query is "give the name of the road pointed out on the screen". *Clipping (m, win')* is an example of query on map m that does not need to produce a new map, but only involves the creation of a presentation on window win' from the already computed pixmap associated to the image of m: *clipping(m,win') = visualize (pixmap (image (m, l), s), win')*.

3. Interactive change of legend involves only the image level and not the map level: Indeed, only the legend has changed (and not the geometry). Therefore, we do not need the expensive conversion *convert (m)* which translates the geometry of map m as represented in the database into a structure ready for display. A new pixmap is created from *image (m, l')* where l' denotes the updated legend. From this pixmap p a new presentation *visualize (p, win)* is displayed.

4. Equivalently to clipping, scrolling and simple zooming on a presentation pr requires only the third level: We only have to consider pixmap p associated with map m and recompute a presentation: *scrolling* is equivalent to *visualize (p, win(dx,dy))* where $win(dx,dy)$ denotes the window obtained after scrolling. Simple zooming (magnifying window win into win' without changing the resolution) denoted *zooming (m, win')* is equivalent to *visualize (p, win')*.

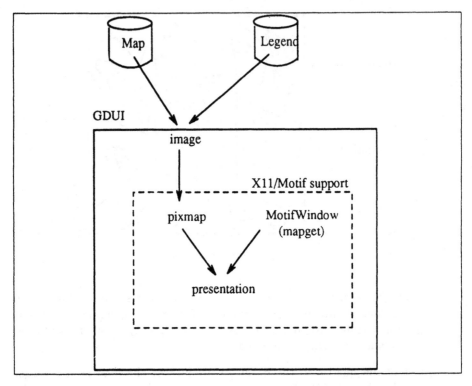

Figure 2: · A four level description of the GDUI

3 Prototype Implementation

This section is devoted to the description of a prototype kernel for displaying and querying maps stored in a database. We first describe the architecture as well as the tools we used at a support level. We then detail a few functionalities, such as the display of maps, scrolling and zooming.

3.1 Architecture

This prototype includes a support level (low level graphic tools facilities and DBMS functions) and a Graphical Database User Interface (GDUI) level, which represents the kernel of our implementation. At the support level, it uses on the one hand the O_2 DBMS for storing and querying the map database and on the other hand, X Windows (version X11 R4) [SG86, MIT87] and OSF/Motif [OSF89] for coding the interface. The functional architecture of the kernel is displayed in Figure 3.

Maps are stored in a vector mode as objects in the O_2 DBMS according to the GEO2 model ([II92], [DRS93a], [DRS93b], [Dem92]. The GEO2 database allows various functionalities implemented as methods of geographic objects, such as index facilities and selection by pointing.

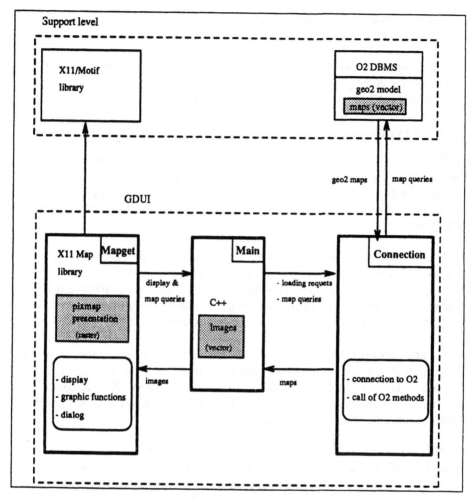

Figure 3: GDUI functional architecture

The main module, written in C++, deals with images. It communicates with (i) the X11 Map Library in which operations on pixmaps and presentations are defined and (ii) the connection module which is the interface with the database. These two modules are detailed below.

3.1.1 Connection O2/C++

Exchanges between O_2 and a C++ environment are provided by an utilitarian program called $O_2 Link$ [Tec92]. Two cases have to be considered:

1. The external application uses O_2 to handle data stored in the database and it sends messages to O_2 objects or even access directly to the values of these objects. The *export* function renders O_2 objects accessible to the external world.

2. The external application is connected to O_2 to render C++ objects persistent. The function used in this case is *import*.

The interesting case for us is obviously the former one, but we could use the latter one for storing our temporary C++ image objects in case they are too large to be kept in main memory during a whole session.

3.1.2 X11/Motif

X Window (or *X* to be short) [SG86] is a powerful windowing system that defines a network protocol for client-server applications and that provides a set of libraries for the development of such graphical interfaces. X applications can handle text as well as pictures and are well suited for the design of geographic interfaces.

X offers a programmer many tools to design a graphic user interface. The low level library is the *Xlib* [Nye92], which contains a wide set of C routines, opening full access to all the capabilities of the X protocol. Writing a whole application using Xlib would imply a long and difficult programming task, and would result in a difficult-to-read and difficult-to-maintain application.

Xt Intrinsics (or Xt) [NR90] is a Xlib-based toolbox that spares programmers most of the tedious programming tasks while offering them resources to customize the looks and the behavior of their applications. Xt implements an object oriented layer designed for the creation and the manipulation of high level interface components: The *widgets*.

The use of Xt implies its association with a widget set, i.e., a set of predefined interface components such as menus, buttons, popup windows, etc. All the widgets of a widget set have a consistent behavior that helps the end user to learn the environment and that offers a common presentation (*look and feel*), such as the well-known shadow border around all Motif widgets. Among all existing widget sets, the currently most used ones are Motif [OSF89] [Hel91] and Open-look. They can also be considered as toolboxes because they provide additional routines defining even higher level components than the ones offered by Xt Intrinsics. However, it is worthwhile noting that the use of a high-level toolbox such as Motif does not mean that Xlib routines do not have to be used: Indeed, the definition and the manipulation of drawing tools and of graphic attributes are only accessible via the Xlib library.

We chose to use Motif because of the large and attractive choice of widgets it offers. One of the characteristics of all widget sets is that there are extensible: Programmers can define their own widget class that implements a particular behavior. Our purpose is to define a *mapget* class as such a widget class and to add it to the Motif widget set.

3.1.3 Mapget Class

Mapget is implemented as a sub-class of MainWindow (Motif class). New attributes and new methods are then added in order to manage specific geographic interface operations. For instance one of the attributes is of type pixmap with some associated resources such as the size of the pixmap, and some methods manage the copy of subparts of the pixmap into the window.

Since Mapget is a subclass of MainWindow, an instance widget w of Mapget class is a window known by the window manager. Hence operations usually applied to such windows are inherited and can be applied on w as well, e. g. *move* or *iconify*. Other operations that modify the contents have to be redefined on a window of the Mapget class. This is the case for instance for *scroll, zoom, resize*. The scroll and zoom operations are studied in this paper. The resize operation is simpler and implies creation of a new presentation, according to the new size of the w window.

In the following sections, we will use the $w.a$ notation to refer to attribute a of mapget w. For instance, $w.window$ is the window attribute of mapget w.

3.2 Functionalities

End users are interested in querying and editing maps by means of operations such as the ones described in Section 2. These operations are implemented on a particular level among the four defined in the model (maps, images, pixmap, presentation). The level(s) on which operations will be performed is of course transparent to the end user but of primary importance for us. We describe below the detailed architecture through three examples of functions: Map display, scrolling and simple zooming.

We chose to prefix all function names by module names, i.e., by one of the following symbols:
DB: For database functions,
Connection: For Connection module functions,
Main: For Main module (C++) functions,
Mapget: For X11 Map editing library functions associated with a mapget,
GDUI: For general end user map editing/querying functions.

3.2.1 Display

The display of a map is implemented by function *GDUI: Display (m, l, w, s)* whose parameters are defined below. Display consists in the following steps:

1. Create widget w instance of class Mapget. This creation associates with w a pixmap p of size s, initialized to white pixels.

2. Read a map m in the database and create an image i according to legend l (function *Main:CreateImage(m, l)*).

3. Assign to pixmap p the raster translation of image i, according to size s (function
   Mapget:WriteInPixmap (i, s)).

4. Visualize the presentation in the window corresponding to widget w, by copying the corresponding part of pixmap p
   (function *Mapget:Visualize(w.p, w.window))*.

The corresponding functions are detailed below.

⋆ Creating an image

```
Function  Main:CreateImage (m: map, l: a-legend): image;
/* Creates image i corresponding to map m and to legend l.
This operation can be seen as a join between the database and the legend,
with a projection on the following attributes: (i) geometric attributes
that include standard geometric data (areas,lines, points),  denoted
i.geometry, and the bounding_box of the map (a rectangle), denoted i.B,
(ii) graphic attributes (color and texture) denoted i.graphic_attributes.
CreateImage corresponds to the function image(m,l) of Section 2. */
```

The *CreateImage* operation implies an access to the database to get a map whose geometry obeys a Spatial Data Model (SDM), and converts it to an interface specific image format. For a description of the latter, see [Rig93]. This implies the writing of DBMS and SDM dependent code.

The correspondence module currently implemented is composed of two parts:

1. A set of special methods added to database objects in order to read and convert image formatted geometric data (function *convert* of Section 2).

2. A C++ module using O_2Link features to extract data from the O_2 map database (using the previous methods) and to send them to the main module.

Note that only this set of functions has to be rewritten if another DBMS is used. We plan to study a more independent connection protocol and model conversion.

⋆ Creating a pixmap

```
Function Mapget:WriteInPixmap (i: image, s: rectangle): pixmap;
/* Translates image i into raster format assigned to a pixmap p of size s.
This function is implemented as a method of the Mapget class and is applied
to a pixmap associated with a widget of class Mapget. It corresponds to
function pixmap(i, s) of Section 2. The size s of pixmap p is customizable
and defined when the widget is instantiated. For the sake of simplicity, we
suppose below that attribute i.geometry contains only lines represented by
list of points.
Due to implementation choices that are explained below, a change of scale is
```

applied to the geometry of the image objects. The computation of the changing
factor uses two parameters: The bounding-box attribute B of image i (i.B) and
the size s of pixmap p. A change of scale is applied when the image data is
copied into X11 ready-to-display structures. */

```
/* Local declarations */
x_line: list(XPoint); /* XPoint: X structure for representing Point geometry */
i_line: list(IPoint); /* IPoint: C++ image structure for Point geometry */
ratio: integer;


/* Compute the change scale factor, using bounding-box i.B and pixmap size s */
ratio = s / i.B;


/* Compute the change of scale of each line in image i and put the result
   in the X-structure x_line */

  for each i_line in i.geometry
    x_line = Change-of-scale (i_line, ratio);

    /* Call the Xlib function XDrawLines to write x_line into pixmap p.
       Additional parameters are:
          - Display: An X variable representing the screen,
          - X_translate (i.graphic_attribute): Translation in X11 format
                                  of the graphic attributes of the image.
          - XtNumber(x_line): Number of points in the x_line list.
          - CoordModeOrigin: Specific X11 parameter. */
    XDrawLines (Display, p , X_translate (i.graphic_attributes),
                x_line, XtNumber(x_line), CoordModeOrigin);
    endfor;
end Mapget:WriteInPixmap;
```

A first implementation choice was to have image *i* fit exactly into pixmap *p*.
Therefore the ratio to be applied to the coordinates is computed using bounding
box *i.B* and the pixmap size *s*.
Alternate solutions are:

- To ask the user for the scale factor and create the pixmap accordingly, as
  long as there is enough memory for the pixmap,

- or to take into account both the user scale factor and the memory limitations, which renders operations such as scroll rather complex (see below).

⋆ Visualizing a presentation

```
Function Mapget:Visualize(w.pixmap: pixmap,w.window:rectangle):presentation;

/* This function copies on the screen, inside window w.window of mapget w,
the corresponding part of pixmap w.pixmap. Let window.height and window.width
be the parameters representing w.window size. Let cur-x and cur-y be the
```

relative address of the top-left corner of w.window inside w.pixmap. Using
these variables to extract a rectangle from w.pixmap and copy it onto
w.window, we obtain a presentation, as defined in Section 2. In the initial
display of a map, x and y are always null. Then the left-corner of w.window
is the left corner of the pixmap as well. */

/*The Xlib function XCopyArea extracts a rectangle sub-pixmap and copies it
on the window. The Display and graphic_attributes parameters are the ones
already used to create the pixmap from the image. */

```
XCopyArea(Display,                          /* Specific X11 parameter */
          w.pixmap, w.window,               /* Copy from pixmap to window */
          graphic_attributes,               /* Specific X11 parameter */
          cur-x, cur-y,                              /* the rectangle
          cur-x+window.width, cur-y+window.height,         coordinates */
          0, 0); /* Start the copy at the top left corner of the window */
end Mapget:Visualize;
```

Visualizing part of a pixmap inside a rectangle window is extremely fast and
automatically performed by X11. This operation at the pixmap level is somehow similar to a *clipping* operation but we chose another terminology because
"clipping" refers to a database operation on maps. Indeed, it consists in creating a new map that corresponds to the intersection between a source map and
a rectangle [SV89]. One could use this database clipping operation to restrict
a large map to a smaller submap instead of loading the whole map into an image.

The visualizing operation was implemented as a method of the Mapget class.
It is used each time it is necessary to refresh the screen. This is the case for
instance with Expose events: Suppose another window overlaps the Mapget window. Then the window manager sends an Expose message to the Mapget widget,
which has to redisplay its contents and then uses the Visualize method.
Visualize is also used in Scrolling, as explained in the next section.

⋆ General display function

Function GDUI:Display (m: map, l: a-legend, w: Mapget, s: rectangle)

/* This is the end-user function for displaying a map m in a given mapget
w according to a particular legend l; the pixmap associated with the mapget
is defined upon the mapget creation, with a size s that is a user-defined
parameter but that cannot be interactively changed in this version.
It instantiates widget w as a member of the Mapget class, using the Intrinsic
function XtVaCreateWidget. The pixmap w.pixmap is created during this operation
with size s. The variable application_window refers to the top window of the
interface. The PixmapSize resource is initialized with value s. */

```
w = XtVaCreateWidget ("mapget", MapgetClass, application_window,
                XtPixmapSize, s, NULL);
```

```
/* Create image i corresponding to map m and legend l */
i = Main:CreateImage(m, l);

/* Assign to pixmap w.pixmap the raster translation of image i */
w.pixmap = Mapget:WriteInPixmap(i, s);

/* Create a presentation by displaying the part of the pixmap associated
with mapget w. The initial presentation matches the top-left corner of the
pixmap with the top-left corner of w.window. */
presentation = Mapget:Visualize(w.pixmap, w.window);
end GDUI:display;
```

3.2.2 Scroll

As said in the above description of the *WriteInPixmap* function, we chose to apply a scale transformation on the geometric data contained in the image structure, such that the whole map extracted from the database fits into the pixmap.

This permits the full use of the efficient X11 scrolling. In this situation the programmer's task is limited to the association of a pixmap containing a raster picture with a Motif widget that belongs to the *ScrolledWindow* class. The X internal mechanism is then responsible for the management of scrolling. The Mapget class is a subclass of the ScrolledWindow. It then inherits the resources and functions for scrolling. This operation can also be seen as a dynamic raster clipping with an intersection value changing each time one scrolls inside a window. As a matter of fact, each time the user drags the scrollbars associated with the ScrolledWindow, the corresponding part of the pixmap is displayed. Limitations are defined such that scrolling stops when a pixmap border is reached.

Scrolling utilizes the *Visualize* operation, using the dx/dy intervals given by the end user when he/she moves the mouse up/down and left/right.

```
Function Mapget:Scroll(w: Mapget, dx: integer, dy:integer)
/* The two resources of the Mapget class cur_x and cur_y, already described
   in the Visualize function, are denoted as usual w.cur_x and w.cur_y. */
   w.cur-x = w.cur-x + dx;
   w.cur-y = w.cur-y + dy;
   presentation = Mapget:Visualize(w.pixmap, w.window)
end Mapget:Scroll
```

In the current implementation, the user is given a complete presentation of the geographic data in one pixmap and can then visualize the whole image using only X11 scrolling. Yet a pixmap must fit in main memory and therefore its size growth is limited by hardware characteristics. If the map is large and the pixmap size is small, i.e., if the ratio pixmap.size / image.bounding-box is smaller than a given threshold, one might get a pixmap whose visualization has a too small resolution to be acceptable by the user.

One might then decide to create the pixmap from a smaller part i' of the image obtained by clipping: $i' = clipping\ (i,\ r)$, using as a parameter rectangle $r = max\text{-}s\ /\ min\text{-}ratio$ where $max\text{-}s$ is the maximal size of a pixmap and $min\text{-}ratio$ is the smallest acceptable ratio (for ergonomic reasons). Scrolling out of the pixmap generates clipping on another part i'' of image i on a rectangle r', neighbor of r: $i'' = clipping(i,\ r')$.

3.2.3 Simple zoom

Zooming is an end user request that takes as parameters:

- A rectangle r drawn by the user within the presentation in order to define the region of map m that has to be zoomed.

- A scale factor f, defining the magnifying factor to be applied to the set of geometric objects selected in the rectangle.

The zoom operation consists then in the following steps:

1. *Clipping* of the map using the rectangle parameter. A new map, $m' = clipping(m,\ r)$, subpart of m, is then created. The rectangle parameter r becomes the bounding-box of the new map m'. As a matter of fact (see above the Visualize operation, Section 3.2.1), zooming is performed on the raster form of the image. There is no actual clipping in the database.

2. Magnification of the geometric part of all objects contained is the new map, according to the scale factor parameter f. The bounding-box r of the new map m' is represented by its top-left coordinates $xmin$ and $ymin$. Considering a point and its coordinates $xold$ and $yold$ in the initial map m, the coordinates $xnew$ and $ynew$ in the new map m' are:
$xnew = (xold - xmin) * f$
$ynew = (yold - ymin) * f$.

3. Display of the new map in a new window (see Figure 5).

X automatically provides zooming through function $xmag$. Therefore the X user does not see the logical decomposition into these steps. We shall see in the following section that for multiple scale representation, more complex zooming is eventually needed which implies a clipping operation that should be processed at the image/database level and therefore that cannot take advantage of the fast X raster zooming.

```
Function Mapget:Simple-Zoom (w: mapget);

/* First, ask for the zoom rectangle. This rectangle is drawn by the user on
   the presentation, i.e., the part of the pixmap displayed into the window of
   mapget w. */
        r = Mapget:GetRectangle(w);
        f = Mapget:GetFactor(w);      /* Gets the zoom factor */
```

```
/* Call to the X function xmag to magnify the part of the screen corresponding
   to the selected rectangle. */
      Xmag (r, f);
end Mapget:SimpleZoom;
```

4 Multiple Scale Map Display

4.1 Improvement of the current implementation

The previous section showed that at the interface level, 2 modes exist for representing spatial data: Data are vectorized at the image level while pixmaps are raster images. Zooming is an example of operation that, as discussed above, raised important issues relative to the advantages of a vector structure versus a raster structure. We first discuss the limitations of a raster-implemented zoom, then we propose an improvement which leads to a vector-based zoom with added functionalities.

4.1.1 Current limitations

The currently implemented zoom is a trivial raster zoom as directly provided by X. It allows to magnify a rectangle subset of the pixmap: The set of pixels of the pixmap is unchanged and each pixel is magnified. This operation does not fulfill the following goals:

- Improve the resolution, i.e., allow one to visualize smaller details. As implemented, the resolution remains the same whatever the transformation factor is.

- When increasing the scale, add information on the map already displayed at a smaller scale[2]. The current zoom just allows one to see the same information at a larger scale.

4.1.2 Multiple Scale Visualization

An operation that fulfills such requirements allows to implement multiple scale visualization [BA90, Lic79, LOL86]. It needs the access to geographic objects either in the database or at the image level. Whether it requires the image level or an access to the map database is discussed in Section 4.3.2. Let us say now that a fundamental interface requirement is efficiency.

The following sections discuss the implementation of multiple scale visualization and the choice of a structure providing efficiency.

---

[2]The terminology "small" or "large" for a scale applies to the ratio *displayed coordinates* / *real coordinates*.

4.2   Multiple Scale Visualization (MSV) specifications

MSV corresponds to a composition of two operations:

- Geographic information filtering.

- A transformation applied to the filtered information.

4.2.1   Filtering

The filtering criterion applied to a geographic object has two parts (i) a geometric criterion that takes into account a threshold from which the object to be displayed is too small to be seen, (ii) a semantic criterion, applied to an object even though it did not "pass" the previous threshold test, which is based on an object attribute called "display importance".

The geometric criterion is the most intuitive one: As long as we get closer from a geographic object, we distinguish its geometry in a more accurate way. This phenomenon is simulated by providing the end-user a more detailed representation of the map when the scale is growing.

"Display importance" is the other essential criterion. For instance, entities like monuments or crossings might be so small that their geometry cannot be represented at the current scales, but they are interesting enough to be represented on the map display anyway. Of course the representation below a given threshold (which is not necessarily the above geometric threshold), becomes then independent of these objects' geometry.

4.2.2   Geometric transformations

The transformation is the modification of an object's geometry according to the display scale. Assume a lake has to be displayed. Its representation can be:

1. A polygon with its exact boundary,

2. a polygon with a simplified boundary,

3. a point (with a label or/and an icon),

4. nothing: The lake disappears because its geometry is no more representable.

The second representation is related to generalization problems [Mul90] in cartography and is not our focus here because of its complexity. We concentrate on the transformations from 1 to 3 and from 3 to 4.

4.3 MSV implementation

4.3.1 MSV functionalities

From an end-user point of view, MSV is still the zooming function $Z(m, r, f)$ described in the previous sections. Yet it is now implemented as follows:

1. Zoom parameters are selected. As for raster zooming, they consist in a rectangle r drawn in the mapget associated with map m, and a zoom factor f.

2. A clipping query $C(m, r)$ returns an image $I0$ composed of the intersection of rectangle r with either map m or image $i = image(m, l)$, depending of the level of implementation chosen for the clipping operation (see Section 4.3.2).

3. The zoom factor f is applied to $I0$, giving a new image $I1 = Zf(I0, f)$ with magnified geographic objects. Note that we cannot use the *WriteInPixmap* function given in Section 3, because (i) the scale factor is now enforced by the user and no longer dependent on the pixmap size, and (ii) to apply filtering, we need to recognize geographic objects and thus use a vector structure (image level). Filtering consists of the four following steps:

   (a) A first filter compares the size of the $I1$ objects with the spatial threshold SpT. $I1$ is split into 2 new images: $\{I2, I3\} = Filter1(I1, SpT)$. $I2$ contains objects of $I1$ with size greater than SpT, $I3$ contains the objects whose size is smaller than SpT.

   (b) A second filter is applied to image $I3$. Objects of $I3$ whose display importance is greater than semantic threshold SeT are kept in image $I4 = SemanticFilter(I3, SeT)$. Other objects are definitively rejected. Recall we made the assumption that the geometry of the $I4$ objects is always a point.

   (c) A join with the legend gives to the $I4$ objects new graphic attributes that can be either a label or an icon. This new representation is consistent with the fact that the object's geometry is no longer displayable at the current scale.

   (d) The union $I5$ of images $I2$ and $I4$ is displayed in a new mapget.

4.3.2 Clipping implementation

Either clipping is efficiently implemented in the geographic database (by means of spatial indices associated with the map), or it has to be performed on the image itself. Besides the possibly heavy cost of (i) the connection to the database and (ii) the clipping operation itself, another drawback of a database clipping is to reprocess the join with the legend for the whole image imported, while if clipping is made on the image itself, image $I2$ does not have to be computed. However, clipping at the image level implies some improvements on the image structure in order to guarantee its efficiency.

A candidate solution would be to use a spatial index for a fast access to the image geometric data. This index must be built each time a map is loaded to create an image at the interface level. The choice of the index type must therefore take into account, not only the improvement it provides to rectangle intersection, but also the time complexity for index construction.

5 Conclusion and Future Work

We proposed in this paper a model and an architecture for designing Geographic Databases User Interfaces (GDUI). Four separate levels for representing geographic objects were exhibited: (i) from *maps* stored in the database as well as legends, (ii) *images* are created and loaded in the interface; (iii) *pixmaps* are raster representations of images (iv) from which a subset called a *presentation* is visualized on the screen.

This four levels decomposition demonstrates that better interface specifications are possible, which provide:

1. More genericity; they are a first step toward better toolboxes for GDUI design,

2. more independence and therefore more reusability of the interface toolbox: Independence between the toolbox from the one side and (a) the DBMS and (b) the low level interface software from the other side.

We then described our GDUI kernel implementation based on the proposed architecture. It uses C++ and X/Motif connected to the geographic database GeO$_2$ stored with the O$_2$ DBMS.

We plan first to extend the kernel functionalities to (i) handle a more complete set of user queries to the database, (ii) include user interactive legends choice and update, (iii) display presentations with a richer structure of mapgets such as the one suggested in [Voi91].

The proposed architecture exhibits a memory hierarchy for map display: A first cache for storing images and a second cache (pixmap) for a raster representation of the image. Operations such as scrolling on a large map might lead to cache failures: Only a subpart of the large map can be represented as a pixmap at a time. We plan to investigate and implement strategies for such cache failures.

Another important issue was raised concerning the independence of the GDUI design from the map database model and implementation. In particular GDUI's should accept maps (to be displayed and queried) coming from several distinct databases. In the reported kernel implementation, the internal representation of a map in the vectorized FEIV model of GeO2 [DRS93b] is converted into an image vectorized structure ready for fast displays. We plan to design a more independent connection protocol and model conversion.

Last but not least, a first proposal was made for multiple scale map display. We plan to investigate a more complete model for multiple scale map representation, querying and display.

Acknowledgments
Thanks go to B. David and L. Raynal from IGN for helpful discussions and for having provided the GeO$_2$ database.

References

[Abe89]   D. J. Abel. SIRO-DBMS: A database toolkit for geographical informations systems. *International Journal of Geographical Information Systems*, 3(2):103–116, 1989.

[BA90]   J. T. Bjorke and R. Aasgaard. Cartographic zoom. In *4th International Symposium on Spatial Data Handling*, 1990.

[BDK91]   F. Bancilhon, C. Delobel, and P. Kannelakis, editors. *The O$_2$ Book*. Morgan Kaufmann, 1991.

[BDQV90]   K. Bennis, B. David, I. Quilio, and Y. Viémont. Géotropics: Database support alternatives for geographic applications. In *Proceedings of the 4th International Symposium on Spatial Data Handling*, 1990.

[Dav91]   B. David. *Modélisation, représentation et gestion d'information géographique, une approche en relationnel étendu*. Ph.D. thesis, University of Paris VI, 1991.

[Dem92]   P. Demay. Representation Orientée-Objet des Cartes Topologiques. Technical report, CNAM (Conservatoire National des Arts et Métiers), Paris, 1992. In French.

[DRS93a]   B. David, L. Raynal, and G. Schorter. GeO2, version 4. IGN internal publication, 1993.

[DRS93b]   B. David, L. Raynal, and G. Schorter. GEO2: Why objects in a geographical DBMS? In *SSD'93*, 1993.

[EF88]   M. Egenhofer and A. Frank. Towards a spatial query language: User interface considerations. In *VLDB*, 1988.

[Ege91]   M. Egenhofer. *Spatial query languages*. Ph.D. thesis, 1991.

[Güt88]   R.H. Güting. Geo-relational algebra : A model and query lan-
            guage for geometric database systems. In *Conference on Extending
            Database Technology (EDBT '88)*, pages 506–527, 1988.

[Ha88]   J.R. Herring and al. Extensions to the SQL language to support
            spatial analysis in a topological database. In *GIS/LIS*, 1988.

[Hel91]   D. Heller. *Motif Programming Manual*, 1991.

[II92]   Cogit lab. IGN. GeO2: Structure de données. IGN internal publica-
            tion, 1992.

[Lic79]   W. Lichtner. Computer-assisted process of cartographic generaliza-
            tion in topographic maps. *Geoprocessing*, 1, 1979.

[LOL86]   F. Leberl, D. Olson, and W. Lichtner. ASTRA: A System for Au-
            tomated Scale Transition. In *Photogrammetric Engineering and Re-
            mote Sensing*, 1986.

[MIT87]   MIT. *X Toolkit Library - C language Interface, X Protocol Version
            11*, 1987.

[Mul90]   J.C. Muller. Rule based generalization: Potentials and impediments.
            In *4th international symposium on spatial data handling*, 1990.

[NR90]   A. Nye and T.O. Reilly. *X Toolkit Intrinsics Programming Manual*,
            1990. Second Edition for X11, Release 4.

[Nye92]   A. Nye. *Xlib Programming Manual*, 1992. Third Edition for X11,
            Release 4.

[ODD89]   B. C. Ooi, Ron Sack Davis, and K. J. Mc Donell. Extending a DBMS
            for geographic applications. In *5th Int. Conf. on Data Engineering*,
            1989.

[OSF89]   OSF. Motif 1.0 programmer's guide. OSF Journal, 1989.

[Rig93]   P. Rigaux. Etude et implementation d'une interface pour SGBD
            Geographique. Technical report, CNAM (Conservatoire National
            des Arts et Métiers), Paris, 1993. To appear.

[SDMO87]   R. Sack-Davis, K.J. McDonell, and B.C. Ooi. GEOQL: A query
            language for geographic information system. In *Australian and New
            Zeland Association for the Advancement of Science Congress*, 1987.

[SG86]   R.W. Scheifler and J. Gettys. The X Window System. *ACM Trans-
            actions on Graphics*, 5(2):79–109, 1986.

[Sou86]   B. Soustrup. *The C++ Programming Language*. Addison Wesley,
            1986.

[SPSW90] H. J. Schek, H. B. Paul, M. H. Scholl, and G. Weikum. The DASDBS project: Objectives, experiences, and future prospects. *IEEE Trans. on Knowledge and Data Engineering*, 2(1):pp. 25–43, 1990.

[SV89] M. Scholl and A. Voisard. Thematic map modeling. In *Design and Implementation of Large Spatial Databases (SSD'89)*, pages 167–192. Lecture Notes in Computer Science No. 409, Springer-Verlag, 1989.

[SV92] M. Scholl and A. Voisard. Object oriented database systems for geographic applications: An experiment with O_2, 1992. in *The O_2BOOK*, pages 585-618, Bancilhon, Delobel, Kanellakis (eds.), Morgan Kaufmann, San Mateo, California.

[Tec92] O2 Technology. O_2 Link, 1992. Chapter 11 of the O_2 Documentation.

[Voi91] A. Voisard. Towards a toolbox for geographic user interfaces. In *Advances in Spatial Databases (SSD'91)*. Lecture Notes in Computer Science No. 525, Springer-Verlag, 1991.

[Voi92] A. Voisard. Geographic Databases: From Data Models to User Interfaces. Ph.D. thesis, University of Paris at Orsay, 1992. In French.

[WH87] T. C. Waugh and R.G. Healey. The GEO-VIEW design, a relational database approach to geographical data handling. *Int. J. of Geographical Information Systems*, 1(2):101–118, 1987.

Appendix: Screen hard-copies

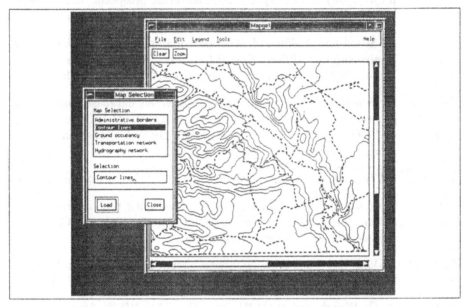

Figure 4: A Mapget.

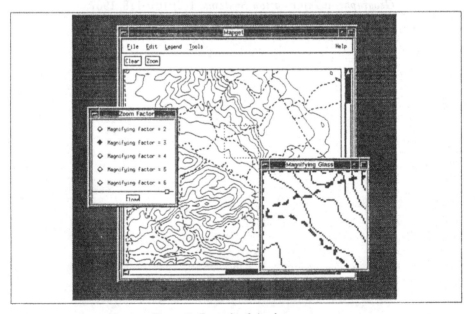

Figure 5: Example of simple zoom.

Metaphors Create Theories For Users[1]

Werner Kuhn

Department of Geoinformation, Technical University Vienna
Gusshausstrasse 27-29, A-1040 Vienna (Austria)
Email: kuhn@geoinfo.tuwien.ac.at

Abstract

The notion of a spatial information theory is often understood in the sense of a theory underlying the design and implementation of geographic information systems (GIS). This paper offers a different perspective on spatial information theories, taking the point of view of people trying to solve spatial problems by using a GIS. It discerns a need for user level theories about spatial information and describes requirements for them. These requirements are then compared with various views on metaphors held in computer science and cognitive linguistics. It is concluded that a cognitive linguistics perspective on metaphors best matches the requirements for user level theories. Therefore, the user's needs for theories of spatial information should be dealt with by explicitly crafting metaphors to handle spatial information by human beings. The paper discusses traditional and possible future metaphor sources for spatial information handling tasks.

1. Introduction

The past decade has been characterized by the quest for stronger theoretical bases to support the work of researchers and developers in the area of geographic information systems (GIS). A widespread recognition that the field of GIS was suffering from the lack of appropriate spatial theories [Dangermond, 1986; Frank, 1987] spurred major national and international research efforts [Abler, 1987; Shepherd, Masser, Blakemore, & Rhind, 1989] which have since produced significant advances in our understanding of spatial information, its modeling, management, use, and diffusion.

These efforts toward spatial information theories have been most successful where they dealt with the traditional issues of representing information about space and spatial relations in formal and thereby implementable ways [Egenhofer & Herring, 1991; Günther & Schek, 1991]. They have furthermore established interdisciplinary networks dealing with the more elusive issues of how to communicate spatial information [Frank & Mark, 1991; Mark & Frank, 1991; Mark, Frank, Kuhn, & Willauer, 1992], how to describe its quality [Beard & Buttenfield, 1991; Goodchild & Gopal, 1989], and how to assess its value [Onsrud & Masser, in press]. All these efforts have considerably expanded the theoretical basis for GIS work and continue to do so for novel issues such as modeling temporal aspects [Frank, Campari, & Formentini, 1992] or dealing with cultural and institutional contexts of GIS use [Campari, 1990; Onsrud & Rushton, 1992].

---

[1]Support from Intergraph Corporation for research on user interfaces is gratefully acknowledged.

Reviewing this body of work, one might identify an emphasis on certain aspects of theory and a tendency to overlook others. This paper is based on the observation that most spatial information theories are oriented toward *machines* rather than toward their *users*. They emphasize formalism (which is a must for their implementation on machines), but are less concerned about explanations. After contrasting this traditional orientation of spatial theories with a proposed user orientation of theories (section two), the paper outlines the key requirements for user level theories (section three), shows how various views of metaphors match these requirements (section four), and presents a catalogue of source domains for GIS tasks (section five).

While the need for a better theoretical understanding of GIS user interface issues has long been recognized, this paper contributes a view of spatial information theories going beyond standard formalization approaches. It is intended to contribute a novel way to understand user interface metaphors for GIS and to clarify their relation to overall system design.

2. Spatial Information Theories - For Whom?

Most spatial theories developed so far have been optimized with regard to consistency within themselves and completeness for some domain of interest. The notion of theory motivating these concerns is that of a formal theory, founded on mathematical logic [Mendelson, 1964]. The main goal of these theories is implementability, making them *theories for designers and implementors*.

For example, spatial data models and data structures have been and continue to be formalized extensively. There have been rigorous treatments of geometric and topological data models at multiple levels of sophistication, ranging from simple raster or point and line based models to complete topological models [Egenhofer & Herring, 1991]. The corresponding spatial theories include spatial statistics, Euclidean geometry, differential geometry, graph theory, and simplicial topology. While tremendous progress has been achieved along these lines, much work remains to be done to address issues like transformations between data models [Herring, Egenhofer, & Frank, 1990] or modeling spatial data at multiple resolutions [Oosterom & Bos, 1989], to name only two important directions.

Thus, formalizing domains with mathematical rigor is of foremost importance and value in dealing with spatial information and is in no way being questioned here. Formalization by itself, however, is not enough to arrive at useful theories. Concentrating on the mathematical aspects of a theory underlying an implementation can lead designers to neglect usability aspects of the resulting system and can produce undesirable effects at the user interface. Theoretical refinements sometimes burden the users with an additional load of concepts they have to master if they want to use a system effectively. Depending on the tasks and users, some concepts of a formal theory may be completely irrelevant or even unintelligible to users.

Topological data models represent, in fact, an example of a domain where highly sophisticated implementations have sometimes been garnished with rather puzzling user interfaces. To acquire geometric data by manual digitizing or to do spatial analyses involving multiple layers of data, users suddenly had to cope with concepts like topological primitives (nodes and edges) or artifacts (slivers and gaps) which had nothing to do at all with their tasks. Thus, satisfying the needs of the mathematical theory underlying a system (e.g., "building the topology") can impose heavy cognitive loads on users and disrupt their work flow [Owen, 1993]. For users, the internal consistency of a theory has little meaning, and completeness too often means an even larger or more abstract set of concepts to deal with.

Other cases in point are the manipulation of spatial data sets as files or the usual implementations of security and access mechanisms. Both require the user to cope with concepts at the operating system level which have nothing to do with the applications at hand and are notoriously difficult to learn. Since these concepts are also highly system dependent, there can be only limited knowledge transfer to other platforms.

What is needed, therefore, are *spatial information theories serving the users* of systems, in addition to those for designers and implementors. The following section proposes to understand the notion of a spatial information theory in the sense of a scientific theory, in order to find characteristics besides mathematical rigor which offer a stronger user orientation. In particular, it points to the explanatory role of scientific theories as a key factor in making systems easier to understand and use.

3. Theories That Explain

An essential function of a scientific theory is to explain, i.e., to account for certain phenomena being observed. With this explanatory power of theories projected into the future, theories are also able to predict occurrences and their properties. For example, Copernican astronomy was able to account for differences between the observed planetary positions and those predicted by the Ptolemaic system. Scientific theories are indeed born from this very need to reconcile observations with predictions [Kuhn, 1962].

Since users have a strong need for explanations and predictions of phenomena occurring at the user interface of a system, exploring this aspect of theories appears promising. The whole literature on user interface issues seems to tell us the same story over and over: that users have a tremendous and often unsatisfied need for reconciling observations with expectations in the process of using a system [Norman & Draper, 1986]. In fact, research about the mental models which users form of a system has recognized that users develop theories (or models) of their own to satisfy this need [Carroll, Mack, & Kellogg, 1988; Norman, 1988]. The situation reminds one strongly of the process identified by Thomas Kuhn in the history of science, where novel theories have normally emerged after clear failures in normal problem solving activity [Kuhn, 1962].

From this perspective, the state of the art in GIS user interfaces suggests a large *theory deficit*: GIS users struggle with the lack of explanations for the actual and potential behavior of their systems. They often do not and should not have to share with designers the knowledge of the spatial theories which went into the design of these systems. But in order to communicate with a system, they need to share some kind of a conceptual background with it.

Thus, the question to be answered is how one arrives at theories which help users understand and communicate about spatial information handling, reconciling what they see with what they expect. Clearly, these theories will not be of the kind of abstract and complex edifices underlying data models or data structures. Rather, they should offer to users coherent sets of intuitively understandable concepts, thereby making a system, to paraphrase Einstein, as simple to use as possible, but not simpler.

This paper claims that *metaphors* create precisely this kind of user level theories. They provide the desired explanatory power by mapping between implementation level theories and coherent sets of familiar concepts. In order to sustain this claim, a view of metaphors has to be postulated which is not (yet) commonly held in the context of information systems. The remainder of this paper therefore discusses various views of metaphors and their potential to create user level theories, i.e., theories that explain.

4. Views Of Interface Metaphors

The discussion of user interface metaphors in the literature has been sparked off primarily by the developments at Xerox PARC, in the seventies, of office automation technology. The highly inventive concepts grouped around a physical office or desktop metaphor [Smith, Irby, Kimball, Verplank, & Harslam, 1982] which nowadays dominate the user interfaces of most workstations and PC's drew the attention of designers and users to the significant effect which metaphors can have on the usability of computer systems.

The coincidence of these conceptual developments with the appearance of new technology has produced a technology oriented understanding of interface metaphors (discussed in 4.1.) while scientific and philosophical treatments have reinforced an understanding of metaphors as analogies (4.2.). Only during the past few years has a broader, cognitive perspective on interface metaphors begun to appear in the human-computer interaction literature (4.3.). It is this latter view of metaphors which has the constructive power needed to create appropriate user level theories.

4.1. Metaphors Are Visual Representations

Because the desktop metaphor appeared on the market hand in hand with the new graphics capabilities of raster displays (windows, icons, menus, and pointing devices; WIMP), this interaction technology is often seen as the essence of interface metaphors. Indeed, interface metaphors have been described as visual representations [Ziegler & Fähnrich, 1988]. Furthermore, the fact that icons may not be useful or may be used inappropriately to represent certain ideas has been advanced to claim that the "metaphor business has gone too far" [Nelson, 1990].

This conception of *metaphors as icons and associated graphic widgets* resembles the view, in traditional linguistics and literary theory, of metaphors as rhetoric or decorative ornaments of language. In both these views, metaphors are only a surface treatment, a matter of choosing certain symbols (words or icons) to achieve a more "colorful" dialogue.

Obviously, an interface metaphor has to be communicated to the user, often primarily by visual means. For example, the desktop metaphor is communicated by icons representing office objects like documents, folders or a trash can as well as by gestures and menu symbols representing operations like move, cut and paste, or throw away. These symbols of an interaction language play the same role as the words used in metaphorical speech or writing. The choice of these symbols is essential for the power of a metaphor, but it should not be confused with choosing the metaphor. It doesn't matter much how we design icons and menus if they represent inappropriate metaphors.

For example, when analyzing the success of the desktop metaphor, one should not ignore the basic structural change in interaction introduced by founding it on physical office concepts. Inverting the traditional command structure from verb-object to object-verb and designing a coherent set of (office) objects and associated incremental operations (verbs) have been the crucial innovations coming with that metaphor. WIMPs allowed for an attractive and intuitive visual representation at the interface, but the metaphor could (and did to some extent) exist without them.

The view that interface metaphors are essentially visual ornaments of an interaction language does not provide much assistance in establishing the kind of user level theories we are looking for. On the contrary, it either assumes a separate, unspecified design process determining the concepts to be presented to the user, or it neglects this need for a mapping from implementation level to user level concepts altogether. The state of the art in user interfaces of commercial systems (GIS and others) seems to reflect this belief in the power of WIMPs to boost system usability. The persisting usability problems [Davies & Medyckyj-Scott, 1993], however, suggest to look for alternative views of metaphors.

4.2. Metaphors Are Analogies

A predominant perspective on interface metaphors presupposes *similarity* between the source, a familiar domain of experience, and the target, a less familiar software domain. Designers, in this view, have to find familiar source concepts similar to some given abstract target concepts. The target domain of the desktop metaphor, for example, would consist of files, file directories, and actions such as organizing files in directories. The source would contain office documents matching files, and folders matching directories [Carroll, et al., 1988].

At a closer look, however, the supposed target domains often reveal themselves as a blend of metaphorical concepts. They provide evidence that thought, action and language around computers have always been structured metaphorically. Terms like file, directory, copy, move, memory, queue, stack, tree, read, write etc., have, of course, been borrowed from non-computational human experiences and activities. They have not only supplied the vocabulary, but also structured the functionality of systems, long before the advent of the desktop metaphor.

The physical office as a source domain for operating system metaphors may be similar to these previous (conscious or unconscious) choices of sources. The target domain, however, is not predetermined, having to be matched by a similar source domain. It is rather amorphous and only receives its structure by projection from a chosen source domain.

By the same token, there is no similarity between manual map overlay procedures and spatial analysis operations, establishing a map overlay metaphor.

Rather, by adopting that metaphor, the structure of map overlay operations is deliberately chosen as the pattern after which GIS analysis operations are being modeled. Thus, the issue is whether such a structure imposed on the target by the source is useful for dealing with the tasks at hand, rather than whether one can find a source domain which is similar to the target.

4.3. Metaphors Create Ontologies

The key concern in finding a "useful" target structure is that the operations should lead to effective ways of solving common application tasks. In other words, the source domain has to *generate an appropriate ontology*, i.e., a set of concepts appearing at the user level.

Office objects and operations, for example, seem to provide a useful ontology for operating systems of personal computers. An alternative source domain might involve Dictaphones and secretaries, based on an agent paradigm [Negroponte, 1989]. Clearly, there would be no similarity with files and directories, but the resulting ontology might be equally or more appropriate for some users and tasks than that created by the desktop metaphor.

In this view, the role of metaphors in human-computer interaction is to provide a coherent structure for a computational domain which may be abstract and amorphous or ill-structured before. The view derives directly from an understanding of metaphors in language and thought in general that has been advanced by cognitive linguists over the past decade [Johnson, 1987; Lakoff, 1987; Lakoff & Johnson, 1980]. Metaphors, in this perspective, are not an optional ornament of communication, but play a central role in our cognition, permeating not only language, but thought and action.

If we understand interface metaphors in this way, they become an essential ingredient of any user interface and the primary design consideration, long before any surface level issues of iconic representations are dealt with [Kuhn & Frank, 1991]. They take on a creative role, acting as "sense makers" by establishing a coherent and familiar set of concepts at the user level. Therefore, this perspective on metaphors allows us to see them as *creating theories for users*.

5. Sources For GIS Metaphors

Recognizing the creative nature of metaphors raises the question of how to find appropriate metaphors for a task domain. The last part of this paper therefore discusses possible metaphor sources for key spatial information handling tasks. The user interfaces of most GIS are based on some sort of a map metaphor. Its most prominent entailment is that of map overlays for spatial analysis tasks, a concept derived primarily from planning (5.1.). However, maps are a pertinent concept throughout all kinds of GIS related tasks, making a mapping perspective the dominating view of metaphors in GIS (5.2.). Alternative views on GIS metaphors include those based on engineering (5.3.), perception (5.4.), and cooperation (5.5.).

5.1. Planning Metaphors

Many spatial analysis tasks (e.g., queries like "which parcels lie in the flooding zone?") are dealt with in terms of overlaying map transparencies (e.g., a cadastral and a flooding zone map sheet) and intersecting polygon boundaries. This procedure is not determined by the tasks as such but imposed by a map metaphor. By choosing a source domain among candidates like maps, map libraries, map overlays and light tables, a GIS designer creates a map-related ontology in the previously amorphous "geographic software domain".

Adopting a map overlay metaphor for the analysis of geographic data implies that the users keep track of data sets (in terms of maps) and explicitly combine them to solve problems, rather than having the system deal with these issues. The resulting ontology includes maps, scales, regions or polygons, etc. Depending on the user community, these concepts may be appropriate or not [Gould & McGranaghan, 1990]. However, they are the result of consciously creating a user level theory and by no means predetermined by the application domain.

5.2. Mapping Metaphors

In the early days of GIS, the technology was primarily understood as an automated mapping tool. Beyond the specifics of the map overlay metaphor, maps have consequently dominated our understanding of most GIS tasks for a long time. Geographic information has been and is still widely understood as information on maps. Storing geographic data in a system means storing maps or map layers to most people. Entering data, asking queries, and producing output is strongly associated with map-based operations.

While the problems inherent in such a view in the context of storing spatial data have been recognized and largely removed (witness the move toward topological data structures), they are considerably harder to deal with at the human-computer interaction level. Maps are a very convenient and familiar means of communication for most people and most spatial data handling tasks, making it difficult to recognize the tremendous potential lying in non-map-based theories of interaction with spatial phenomena [Kuhn, 1991].

5.3. Engineering Metaphors

Engineering (particularly surveying) represents a tradition of handling spatial data which has always assigned high importance to a variety of other information repositories and communication forms than maps: sketches, registers, measurement records, legal and other regulations (constraints). Consequently, map based interface metaphors are unlikely to completely satisfy engineers in their interaction with GIS. While their output needs may not differ much from other professions, it is the input and database maintenance tasks that demand concepts going beyond digitizing coordinates and lines.

These requirements are closely related to those in computer aided design (CAD), though in many ways more stringent. Of primary importance are flexible means to support design tasks, such as sketching and dimensioning tools. For these purposes, metaphors derived from "back of the envelope sketches" have been proposed and implemented [Kuhn, 1990]. For the tasks of maintaining and editing measurement and constraint databases, appropriate metaphors have yet to be developed.

5.4. Perception Metaphors

Another alternative to map-based metaphors is to conceive of spatial tasks in terms of visual metaphors, leading to tools which allow the users to view a model of the territory and selectively visualize spatial entities and phenomena which are not necessarily visible in nature (e.g., land parcels and flooding zones). A detailed discussion of metaphors for these tasks can be found in [Kuhn, 1991].

Generalizing this approach leads to a whole new class of perception based metaphors, relying on our visual, auditory, and tactile - maybe even olfactory and gustatory - senses. These kinds of metaphors are currently becoming essential for handling human-computer interaction in the context of virtual reality technology [Aukstakalnis & Blatner, 1992].

Perceiving the world with our senses is a powerful way of directly communicating information and has obvious appeal for information systems like GIS which model the outside world. Virtual reality systems allow their users to expand the range of channels carrying information between them and a system. In addition to the currently dominating visual channel, audio signals and tactile input as well as output, all in multiple dimensions, are already becoming state of the art. Furthermore, reality can be superimposed with planned, hypothetical or otherwise not directly perceivable entities.

5.5. Cooperation Metaphors

Finally, another new technology demanding innovative metaphors is that of computer supported cooperative work, CSCW [Grudin, 1991]. The idea of cooperation among human beings supported by computers as media rather than as machines is quite different from previous conceptions of human-computer interaction. It requires a broad range of new metaphors which help users make sense of this new style of communication.

Shared information sources and work spaces, protocols for taking turns in discussions, and support for non-linguistic communication (e.g. facial expressions) are only some of these requirements. They need metaphorical grounding in order to become intelligible and usable to the growing community of people within and outside the GIS field who might benefit from this technology. Clearly, GIS tasks are by nature highly interdisciplinary and generally involve the cooperation of a number of specialists, making them ideal candidates for CSCW applications.

6. Conclusions

The main thesis of this paper has been that there is a need for spatial information theories serving the users. Most theories developed in the past have been oriented toward system design and implementation and have often fallen short of providing appropriate support at the user level. The notion of theory underlying them has been that of a formal mathematical system, aiming at consistency and completeness. While these are important goals at the user level as well, the broader notion of a scientific theory offers the additional characteristic of explaining observed phenomena. This explanatory function seems to be what GIS users need most from spatial information theories, in order to make sense of what is happening at the user interface.

Contrasting three popular, but very different views of metaphors, the paper has argued that only a cognitive linguistics perspective on metaphors is offering the necessary power to create coherent theories at the user level. The essence of this argument has been that the explanatory function of metaphors cannot be achieved through visual presentations or through analogy but through the conscious design of an ontology based on one or more carefully chosen source domains. By discussing a few important generic source domains for GIS metaphors, the paper has attempted to demonstrate the practical implications of its thesis.

Obviously, there remains a great deal of work to be done if one adopts the position advanced here. Future research could start with the following three steps: Primarily, careful analyses of the conceptual models of existing user interfaces should provide evidence for the need of more coherent user level theories in the form of specific case study materials. Secondarily, protocol analyses of practical GIS use situations are likely to reveal the kind of metaphorical explanations which users form by themselves when confronted with phenomena which are too abstract or difficult to understand for them. Thirdly, iterative redesigns of existing user interfaces, based on the observed user behavior and on a conscious choice of interface metaphors will allow for an experimental testing of the expected usability gains.

A deeply rooted belief that more powerful systems necessarily mean more difficulties in using them has long prevented the user community from requesting systems that are easier to learn and use. This situation is somewhat reminiscent of the growing complexity in scientific theories at times when they become less and less suitable to explain observed phenomena (cf. the complicated system of compound circular motions in Ptolemaic astronomy). Thus, a need for novel user level theories becomes apparent. The necessary breakthroughs in system usability are likely to result from theoretically sound conceptions of system use, either at the level of task specific metaphors or of generalized GIS use paradigms [Kuhn, 1992]. Fortunately, the adoption of novel metaphors and paradigms for GIS is being pushed by the rapid expansion and diversification of the potential user community.

References

Abler, R. (1987). The National Science Foundation National Center for Geographic Information and Analysis. *International Journal Of Geographical Information Systems, 1*(4), 303-326.

Aukstakalnis, S., & Blatner, D. (1992). *Silicon Mirage: The Art and Science of Virtual Reality.* Peachpit Press.

Beard, K., & Buttenfield, B. (1991). *Visualization of the Quality of Spatial Data* (Position Papers Initiative 7. National Center for Geographic Information and Analysis (NCGIA).

Campari, I. (1990). Accuracy vs. spatial statistical data: the Mediterranean region. In Proc. *EGIS'90*, (pp. 134-144). Amsterdam: EGIS Foundation.

Carroll, J. M., Mack, R. L., & Kellogg, W. A. (1988). Interface Metaphors and User Interface Design. In M. Helander (Eds.), *Handbook of Human-Computer Interaction* (pp. 67-85). North-Holland: Elsevier Science Publishers.

Dangermond, J. (1986). GIS Trends and Experiences. In D. F. Marble (Ed.), *Second International Symposium on Spatial Data Handling*, (pp. 1-4). Seattle, WA; July 5 - 10, 1986.

Davies, C., & Medyckyj-Scott, D. (1993). The USIS Project: Surveying User Opinion on GIS User Interfaces. In J. Harts, H. F. L. Ottens, & H. J. Scholten (Ed.), Proc. *EGIS'93*, 1 (pp. 474-483). Genoa, Italy: EGIS Foundation.

Egenhofer, M. J., & Herring, J. R. (1991). High-level spatial data structures for GIS. In D. Maguire, D. Rhind, & M. Goodchild (Eds.), *Geographic Information Systems: Principles and Applications* (pp. 227-237). Longman Publishing Co.

Frank, A., Campari, I., & Formentini, U. (Ed.). (1992). *Theories and Methods of Spatio-Temporal Reasoning in Geographic Space*. Springer Verlag, Lecture Notes in Computer Science vol. 639.

Frank, A. U. (1987). Towards a spatial theory. In R. T. Aangeenbrug & Y. M. Schiffman (Ed.), *International Geographic Information Systems (IGIS) Symposium: The Research Agenda*, II . Crystal City, VA, November 15-18: NASA.

Frank, A. U., & Mark, D. M. (1991). Language Issues for Geographical Information Systems. In D. Maguire, D. Rhind, & M. Goodchild (Eds.), *Geographic Information Systems: Principles and Applications* Longman Publishing Co.

Goodchild, M. F., & Gopal, S. (Ed.). (1989). *Accuracy of Spatial Databases*. Bristol: Taylor & Francis.

Gould, M. D., & McGranaghan, M. (1990). Metaphor in Geographic Information Systems. In K. Brassel & H. Kishimoto (Ed.), *Fourth International Symposium on Spatial Data Handling*, (pp. 433-442). Zurich, Switzerland; July 1990.

Grudin, J. (1991). CSCW: The convergence of two development contexts. In *CHI'91: Conference on Human Factors in Computing Systems*, (pp. 91-98). New Orleans, LA: ACM.

Günther, O., & Schek, H.-J. (Ed.). (1991). *Advances in Spatial Databases (Second Symposium, SSD '91, Zurich, Switzerland, August 1991)*. Springer Verlag.

Herring, J., Egenhofer, M. J., & Frank, A. U. (1990). Using Category Theory to Model GIS Applications. In K. Brassel & H. Kishimoto (Ed.), *4th International Symposium on Spatial Data Handling*, 2 (pp. 820-829). Zurich, Switzerland: IGU.

Johnson, M. (1987). *The Body in the Mind: The Bodily Basis of Meaning, Imagination, and Reason*. Chicago: The University of Chicago Press.

Kuhn, T. (1962). *The Structure of Scientific Revolutions*. Chicago: University of Chicago Press.

Kuhn, W. (1990). From Constructing Towards Editing Geometry. In *ACSM/ASPRS Annual Convention*, 1 (pp. 153-164). Denver, CO; March 18-23, 1990.

Kuhn, W. (1991). Are Displays Maps or Views? In D. Mark & D. White (Ed.), *Tenth International Symposium on Computer-Assisted Cartography (Auto Carto 10)*, (pp. 261-274). Baltimore, MD.

Kuhn, W. (1992). Paradigms of GIS Use. In Proc., *5th Int. Symp. on Spatial Data Handling*, 1 (pp. 91-103). Charleston, S.C.

Kuhn, W., & Frank, A. U. (1991). A Formalization of Metaphors and Image-Schemas in User Interfaces. In D. M. Mark & A. U. Frank (Eds.), *Cognitive and Linguistic Aspects of Geographic Space* (pp. 419-434). Dordrecht, The Netherlands: Kluwer Academic Publishers.

Lakoff, G. (1987). *Women, Fire, and Dangerous Things: What Categories Reveal About the Mind*. Chicago: The University of Chicago Press.

Lakoff, G., & Johnson, M. (1980). *Metaphors We Live By*. Chicago: The University of Chicago Press.

Mark, D. M., & Frank, A. U. (Ed.). (1991). *Cognitive and Linguistic Aspects of Geographic Space*. Dordrecht, The Netherlands: Kluwer Academic Publishers.

Mark, D. M., Frank, A. U., Kuhn, W., & Willauer, L. (1992). *NCGIA Research Initiative 13, Report on the Specialist Meeting: User Interfaces for Geographic Information Systems*. Technical Report 92-3. NCGIA Santa Barbara.

Mendelson, E. (1964). *Introduction to Mathematical Logic*. D. Van Nostrand.

Negroponte, N. (1989). An Iconoclastic View Beyond the Desktop Metaphor. *International Journal on Human-Computer Interaction, 1*(1), 109 - 113.

Nelson, T. H. (1990). The Right Way to Think About Software Design. In B. Laurel (Eds.), *The Art of Human-Computer Interface Design* (pp. 235-243). Addison-Wesley.

Norman, D. A. (1988). *The Design of Everyday Things*. Doubleday.

Norman, D. A., & Draper, S. W. (1986). *User Centered System Design - New Perspectives in Human-Machine Interaction*. Hillsdale N.J.: Lawrence Erlbaum.

Onsrud, H., & Masser, I. (Ed.). (in press). *Diffusion and Use of Geographic Information Technologies*. Kluwer Academic Publishers.

Onsrud, H., & Rushton, G. (1992). *Institutions Sharing Geographic Information: Report of the Initiative 9 Specialist Meeting* (Technical Paper No. 92-5). National Center for Geographic Information and Analysis.

Oosterom, P. v., & Bos, J. v. d. (1989). An Object-Oriented Approach to the Design of Geographic Information Systems. *Computers & Graphics, 13*(4), 409-418.

Owen, P. K. (1993). Dynamic Function Triggers in an on-line Topology Environment. In J. Harts, H. F. L. Ottens, & H. J. Scholten (Ed.), *EGIS'93*, 2 (pp. 1249-1256). Genoa, Italy: EGIS Foundation.

Shepherd, J. W., Masser, I., Blakemore, M., & Rhind, D. W. (1989). The ESRC´s regional research laboratories: An alternative approach to NCGIA? In E. Anderson (Ed.), *Auto-Carto 9*, (pp. 764-774). Baltimore, MA: ASPRS & ACSM.

Smith, D. C. S., Irby, C., Kimball, R., Verplank, B., & Harslam, E. (1982). Designing the Star User Interface. BYTE, p. 242-282. ·

Ziegler, J., & Fähnrich, K. P. (1988). Direct Manipulation. In M. Helander (Eds.), *Handbook of Human-Computer Interaction* (pp. 123-133). North-Holland: Elsevier Science Publishers.

Using a Landscape Metaphor to Represent a Corpus of Documents

Matthew Chalmers

Rank Xerox EuroPARC, 61 Regent St., Cambridge CB2 1AB, U.K.
chalmers@europarc.xerox.com

Abstract. In information retrieval, sets of documents are stored and categorised in order to allow for search and retrieval. The complexity of the basic information is high, with representations involving thousands of dimensions. Traditional interaction techniques therefore hide much of the complexity and structure of the modelled information, and offer access of the information by means of isolated queries and word searches. Bead is a system which takes a complementary approach, as it builds and displays an approximate model of the document corpus in the form of a map or landscape constructed from the patterns of similarity and dissimilarity of the documents making up the corpus. In this paper, emphasis is given to the influences on and principles behind the design of the landscape model and the abandonment of a 'point cloud' model used in an earlier version of the system, rather than the more mathematical aspects of model construction.

1 Introduction

Bead is a prototype system for the graphically-based exploration of information. The underlying notion of the system is one of our most familiar metaphors: spatial proximity to represent similarity in some more abstract interpretive framework [8]. We represent the relationships between articles in a bibliography by their relative spatial positions. We attempt to place similar articles close to one another and dissimilar ones further apart. The emergent structure is a model of the corpus: a landscape or map of the information within the document set. This 3D scene can be used to visualise patterns in the high–dimensional information space. The aim is to make interaction with a database of information more graphically–oriented, and to move away from interaction styles requiring knowledge of query languages and the database material itself. This allows people to move from cognitive problem–solving to more natural sensorimotor strategies, and to support more exploratory and cumulative modes of use.

The modelling techniques involved were the main focus of a description of an earlier version of Bead, published as [4]. That paper describes the kind of numerical techniques used and their combination with text analysis techniques to make 3D 'point clouds' of graphical data. An example point cloud is shown in Figure 1. This type of graphical structure is common in statistical graphics [5]. The raw material was a small

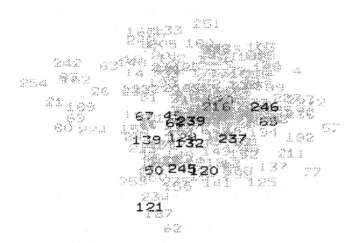

Figure 1. A point cloud constructed from articles in an human-computer interaction conference, CHI91. A search on the words 'information retrieval' leads to some of the document numbers being highlighted. Documents mostly lie in a central cluster, but documents matching the search are mixed throughout the corpus. This is more apparent in an animated display but occlusion and overall scene complexity still inhibit corpus comprehension.

bibliography of articles on HCI. Since then, the interface style has progressed to the use of a more accessible graphical structure, and the issues underlying this progress are the subject of this paper.

Low dimensionality is in accord with our everyday experience. We are used to a space of three physical dimensions wherein we perceive individual characteristics of objects and also their patterns and interrelationships. Given our lives on the surface of the earth, our experience is of a world with greater extent in the horizontal thean the vertical: one might even call our everyday world '2.1–dimensional'. Physical spaces of high dimensionality are unfamiliar to most of us, and it is generally more difficult to present, perceive and remember patterns and structures within them. If our activities depend on judgements based on both the individual characteristics of documents and their relative properties then we will gain by employing a representation which shows both in a space of familiar dimensionality.

In initial experimentation within EuroPARC this was found to be a problem of the 'point cloud' representation which, even with only three dimensions, did not easily afford overview of the entire set of documents. Occlusion of distant information, the lack of a fundamental ground plane and weakness with regard to other bases of everyday perception meant that it was difficult to become familiar with a significantly–sized corpus. The features that make graphical display most effective were not being brought to bear in making the information design effective and useful. Users found it difficult to orient themselves and navigate within the space, and consequently did not build up a useful mental model of the corpus. Instead they found occasional items of interesting data but had difficulty in assessing their relevance or significance in the wider context of the corpus. This seemed to echo a point forcefully put in [1]:

> Items of data do not supply the information necessary for decision–making. What must be seen are the relationships which emerge from consideration of the entire set of data. *In decision-making the useful information is drawn from the overall relationships of the entire set.* [Author's italics]

A goal of Bead is to represent a corpus of documents in a way which helps with tasks which rely on the relationships of the entire set of documents as much as the properties of individual members. A document must therefore play a dual role: it must act as an autonomous unit and it must also play a role as a component within a higher–level structure. We wish the corpus to have a layout which represents the more abstract patterns within it. The model of a corpus used in Bead emphasises patterns of thematic similarity as estimated by similarities in word usage. Individual documents take their places in these patterns by dint of the words used within them, but they also have important characteristics not used in constructing the overall layout. Labelling Bead as an information retrieval system is to take a narrow view. In a wider, more general sense, it is a system intended to aid in tasks which rely on consideration of the overall relationships of the themes and words active in a corpus, as well as the individual elements.

There is a point of view which holds that most information retrieval tasks are better performed if the overall relationships within a corpus are presented. As this view has spread, a more general model of the use of information systems than that traditionally used in Information Retrieval has arisen. The word 'retrieval' suggests an action by some agent to find and bring back information to a somehow detached or uninvolved user. The agent is given a specification of what is wanted, and the user waits for the results to be returned. However, it is becoming more accepted that people may not be able to express what they want to access in a corpus in a query language or even a natural language, for that matter, because they may not know exactly what they are looking for. Instead, they may be able to do what they want to do if they can begin by finding out roughly what is available in the corpus and, in an exploratory manner, refine and adapt their inquiries.

An individual document may initially appear 'relevant' but later be discarded if other documents better serve the interests of the user. There may be a great number of relevant documents, or there may be none: both are fair and on occasion appropriate results. There may be documents that are relevant in different ways, and this may lead to a continuation and adaptation of work as these different associations are assessed. Initially known documents may be dispersed among other unknown but potentially relevant ones. These examples stress the relative judgements that drive an adaptive and exploratory style of use that contrasts with and complements the less interactive 'retrieval' approach. This approach has been labelled information *access*, and to some extent reflects the increasing awareness of the importance of exploration and dynamism in perception and model construction [7, 14]. These ideas have also been influential upon the designers of other information interfaces e.g. the work of the User Interface Research and Intelligent Information Access groups at PARC [3, 10, 12, 13]. The following section goes slightly further into these and other areas of work that have been influential in attempts to make the design of Bead better suited to information access.

2 Influences and Comparisons

In this section, some of the issues, techniques and concepts behind the Bead system are presented and discussed. From the basic infrastructure of information retrieval techniques to influential examples and metaphors, a wide range of areas of study have had their effect on the choice of information displayed and the structure of the display: collectively, the *information design*.

2.1 Information Retrieval

As in most Information Retrieval systems[1], a document is essentially represented as a list of the words which occur within it, and some numeric measure of the relative frequency of each occurring word. The most simple measure is a one for a word occurring in a document, and a zero for non–occurrence. For reasons which include efficiency, not every word is actually used: common 'noise words' such as 'the' and 'or' are discarded.

This representation has a geometric interpretation. It is as though each document is a point in space. The large number of words means a large number of dimensions to this space: each unique word in a corpus of documents defines one dimension, and the frequency of occurrence for each word is a coordinate for that dimension. Word frequencies can be further weighted to take account of extra information. 'Noise words' are effectively given zero weight. In the case of the bibliographic data used in Bead, words in the 'keywords' and 'title' sections of a bibliography entry are given more weight than those from the abstract. Documents are close in this type of high-dimensional space if they have roughly the same words occurring with roughly the same weights. The assumption — generic and fundamental to all of Information Retrieval — is that if this is so then the documents are most likely to be similar in themes and topics i.e. spatial distance corresponds to some degree with thematic similarity.

Since the number of words in a corpus of documents can easily run into thousands, the number of dimensions is too large to display directly. We can get some information

---

1. [15] is one introduction to the basics of Information Retrieval techniques and systems.

about what documents are close to each other by the more traditional means of information retrieval: in effect we get a list of the documents near to one or more documents known *a priori*. The system can then find the nearest (most relevant) documents and return them to the user, usually in order of spatial distance (relevance ranking). A simple view of a query to an information retrieval system is that we make a fake, temporary document with just the keywords we give and with chosen (presumably high) weights, find where that document is placed in the high-dimensional space, and then return a list of nearby documents. Again, we get an idea about documents close to the item of information we placed into the space, but we obtain little global information. We find out about the locality near to the query, but we gain no idea about documents further away (involving different words) and how they relate to each other.

Note that we have to start off with some clear, *a priori* knowledge of what we want to find out. Unfortunately this is not so often the case, for example when we want to explore and browse our way towards whatever it is we come to decide is of interest or relevance. We can only do this by many repeated samples of localised regions, and in ourselves building up some sort of cognitive model of what there is in the corpus and how they fit together. Defining such samples means either having many well-understood documents in the corpus from which one can work outwards, or knowing how to put the words together in a query (i.e. how to use a query language) to sample in the region you want to know more about. It would be better for those who do such sampling less often (or who know less about the corpus or the query language) if some of the cognitive load could be taken on by the representation of the corpus.

Another approach using the same underlying representation is to partition the space into some number of regions which share roughly the same words and weights. We can then show representative members or the highly weighted words which typify each region. We therefore obtain a concise but more approximate representation which we can show in a list. We now get some idea of the overall range of the documents in the corpus, but our accuracy is limited because we can only write so much about each one on a screen or page. We could then choose one or two of these selected regions to look at more closely, perhaps by gathering in the members of the regions and then trying to spread them out again into another list of regions to select from, scatter out again and so on with ever more refined choices of documents. This is a rough description of the Scatter/Gather technique described in [6].

This approach is better for someone browsing the corpus because at each stage they get some idea of the overall contents of the corpus presented to them, and they need not know a query language. They choose one or more relevant members of a list, and in so doing move themselves closer to their goal. By reducing dependence on initial knowledge of keywords and query languages, Scatter/Gather is intended to favour information access more than information retrieval.

The model of the document corpus in Bead is similar to Scatter/Gather in its focus on the perception of the global structures and relationships of documents rather than the techniques associated with a hidden, high–dimensional representation. The model of the individual document is one which creates these larger–scale aggregate structures. In Bead, each individual document has as the dominant factor in its behaviour the resolution of the similarities and differences with all other documents. By making visible the setting of each document within the larger–scale structure of the corpus, browsing and exploration is aided. For this to work, the layout must impart to the user relationships and structures which 'make sense' and yet it must also let users orient themselves, navigate and examine the corpus. In other words, Bead should make the corpus 'imageable'.

2.2 Design of Complex Spatial Structures

This term is drawn from a work on the theory of a certain type of complex spatial structure of varied use and interpretation, namely the city. In *The Image of the City* [9], Lynch described properties of cities which led to people being able to orient themselves and navigate within the city so as to carry out tasks related to the spatial structure e.g. finding the way to some location or sketching out a rough city map. Lynch carried out surveys of people performing such tasks, and collected and analysed the results to come up with some ideas about the characteristics needed for imageability. These included: landmarks visible from most of the region, which allow for orientation; a delineated border (and perhaps borders of subregions) which serve as reference points; viewpoints, so that parts of the city region can be overviewed before travelling into the more local neighbourhoods; a skeleton of routes into and through the region which one can use to go to particular local areas and from which one's knowledge of local detail can grow and be fleshed out; and consistency of local texture (e.g. building styles) so that one can determine from a street's local detail information about the subregion one is in.

Another study of the interrelationship of the design of spatial structures with ease of use is published as [2]. This study looked at situations where the design of public buildings was such that signposting was necessary to overcome the difficulties users had in use and navigation. He suggests that "in many cases signposting is an admission of design failure. It reflects the fact that there are many situations in which the designer cannot rely upon the knowledge or experience of the user in finding the way" and in assessing various buildings in use, "those buildings with many short, winding corridors came out worst, whereas... the building which consists essentially of a large, open, partly differentiated space came out the best." In another situation, an airport terminal, Canter noted how the linear sequence of functions a traveller undergoes (check-in, customs, boarding, &c.) allowed a building design which was easily understood and used. He found that when the pattern of use of the building was simple in structure then the building could conform to this pattern, but for more varied or general use it was best for the structure to be largely open and visible, allowing users to gain an overview and plan their routes so as to suit their tasks.

Romedi Passini is one author who has noted how planning and wayfinding can be a dynamic activity, and many aspects of architectural design do influence our ability to find our way through buildings and landscapes [11]. While problems of this sort may often be broken down into subgoals, various types of information available may be used to progress towards a desired goal. In the simplest situations, sensory information is enough to guide us towards our goal e.g. what we seek is directly visible. In slightly more complex situations, some degree of memory use is needed in combination with sensory information in order to achieve a goal. In the most complex situations, the sensory and memory information need to be used and manipulated to create new information and achieve the goal. As memory and cognitive load increases, the task is made more difficult. All three levels may interact in complex and dynamic ways, but if we can design the framework wherein goals and activities take place so that more reliance is made on sensory than memory and manipulated information, then we will make the information tasks of users less difficult.

In many situations, however, we cannot determine our goal *a priori*, and we must rely on our skills in the dynamic interpretation of information. We continually orient ourselves, comparing expected and visible information, and so reassess our plans and expectations as we move towards a final goal. Another significant point is the danger of information overload, where information is presented in such volume and with such low discrimination of importance that one is swamped, and cannot find one's way through to a final goal. Even though the pertinent information may be available, it is lost amongst the clutter of other useless or distracting information.

2.3 Information Design and Perception

This latter point of visual clutter is one of the main issues put forward by Edward Tufte in [16]. Too many information designs are counter-productive in that attempts to enrich the information presented to a user lead to distracting or confusing results. In the case of the point cloud used in the first version of Bead, it was often the case that the complexity of the 3D patterns meant that the data cluttered itself, and even small amounts of detail presented at each data point led to a great morass of visual clutter. To improve the information design, such clutter should be reduced or tailored so that one is more selective in the way that one displays detail as one examines different areas of an information display.

In the studies of perception, there is a significant body of work that explores the dynamism that is an essential and everyday characteristic of our behaviour [7]. Gibson argues that direct perception, memory and interpretation act together as we explore our environment, build up our mental models, and plan uses and actions within that environment. We resolve visual ambiguities, direct our attentional focus, add areas of the environment to our memory, and perceive shapes and uses all in relation to our body and its movement.

When we look at one area close up, the detail there has as its context the neighbouring regions which extend on out to the boundary or horizon. Continuity of movement over a landscape, coupled with perspective viewing, allows one to incrementally refine one's attentional focus down to more local areas, while smoothly adding more information to the context or periphery of view. We select areas to examine in detail by moving closer, while distant areas are seen in a less detailed (or more abstract) way. There is continuity between the close and detailed, and the far and abstract. Reference is continually made to the ground plane of the landscape, as perceived by the variations in size and texture of objects as well as larger-scale features such as the horizon and directionality of light. Although areas of the environment become hidden as we move, we are experienced in maintaining a mental model which lets us return to or otherwise use such areas.

Note that in the case of point clouds and other 'strongly 3D' structure, the environment is more complexly structured with many occlusions and obstructions of view. Without references such as a horizon and a consistent ground plane, information gained by overview and exploration is more difficult to come by. Our skills in perception mental model–making, as honed on our everyday '2.1D' world, become more difficult to employ.

In this section the fundamental data to be accessed has been described, along with the basic notion of using a landscape metaphor for data representation, and some issues relevant to the design and perception of such landscapes have been presented. The next section describes some ways in which these different threads have been, or are being, woven together to improve the information design.

3 Improving the Information Design

In coming to advance the design of Bead's information display, the collective effect of the issues pointed out in the preceding section suggested a move away from strongly 3D structures and towards map-like (2D) or landscape-like ('2.1D') structures. The decision was made to sacrifice the greater exactitude of relative distances that can be gained in full 3D, and the modelling process of Bead was changed accordingly. An attempt is under way to use the greater accessibility and familiarity of the landscape metaphor to display a corpus of documents. A grayscale snapshot from the colour display is shown in Figure 2, below.

In the type of visualisation system common to 3D graphics, viewpoints are easily available since one can place one's eye arbitrarily in space so as to get an oveview of any region of interest. Similarly, one need not really have 'routes' in or through the corpus, as one is free to move above the landscape. In trying to introduce other useful characteristics, though, the structure and appearance of the modelled data must serve to provide such features as landmarks and borders, and the positions of documents must be done so that it is apparent that there is local consistency of themes and topics. Individ-

385

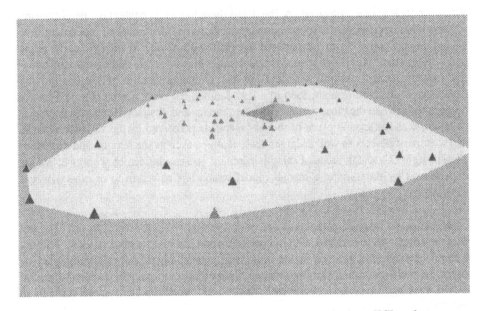

Figure 2. An overview of a landscape constructed from articles in an HCI conference, CHI91. Documents matching the word 'interface' are marked in light green, while the remainder are black. Matching documents mostly lie in a central cluster, befitting the frequency of use and significance of this theme in the corpus. Documents less central in this regard lie more on the periphery of the 'island' of documents.

ual documents are shown as coloured markers placed within the setting of the landscape and they consequently produce collective patterns of density and locality.

The open landscape affords an overview of the patterns and structures of the corpus. Perspective viewing heightens the visual effect, emphasising nearer documents but also fitting in with our everyday framework of vision and perception. The lack of more literal (and constraining) 'ways in' such as routes or other surface features, however, suggests that more abstract and flexible means should be provided. This also means that one should be aware of some of the variety of means by which people initially approach such bodies of information..

A small and informal pilot study was performed at EuroPARC as part of an initial assessment of the information design. Two years' articles of CHI were laid out using three different sets of layout rules, making six layouts in all. For the purposes of the study, layouts were printed out on large sheets of paper, with titles and keywords shown for all articles. The polygons and shading of the animated window-based interface — the standard tool for interaction in Bead — were left out of the printed 'maps'.

The first layout rule used only the keywords of the articles, and the consequent patterns produced separated, dense clusters of sometimes naively-grouped documents e.g. the various senses of a relatively common word such as 'design' would cause all related documents to be close together. The second rule used abstracts of articles as well as keywords, and so disambiguated some word uses but led to relatively uniformly dispersed layout patterns with few clusters or gaps. The last layout rule was close to random positioning, although a few pairwise distances were set exactly, so as to offer some local instances of 'sense'. These layouts were presented in turn to each of seven EuroPARC researchers who were familiar with the journal involved. They were asked to describe the qualities (or lack of them) in the layouts, and to rank the layouts in order of overall preference.

In the absence of a genuine task or 'way into' the information, general features such as physical features such as dense clusters, exceptionally close pairs and the overall centre of the layout were initially examined. Other initial activities included the location of familiar or relevant articles and scanning for consistency of documents in randomly-chosen subareas of the layout.

Overall, the keyword-based layouts were most favoured, while abstract-based and near-random layouts were less favoured to a roughly equal degree. Keyword-based layouts were said to be best because the clumps and gaps served as reference points to access the layout, to return to areas previously examined, and to assess distances between clumps of related documents. While the random layouts had some of this useful 'texture', the overall sense of document positioning was weak. Overall sense of positioning and disambiguation of word usage was better in the abstract-based layouts, but the lack of reference features (such as clumps and gaps) meant that the subjects found it difficult to put together an overall model of the corpus.

Apart from physical features such as patterns of local density, word searching provides perhaps the most significant 'way in' to the landscape of documents, affording access to its basic patterns and themes, and providing the initial experiences which support later browsing and exploration. When a search for a keyword is made in the animated Bead interface, matching documents have their colour changed (see Figure 3, below). The resulting patterns of colour show the distribution of matching documents in the corpus of documents. The patterns show how discriminating the search is and how matching documents are distributed throughout the corpus. The latter point tends to show the areas where there are different uses or aspects of the words searched for.

The landscape model means that one can immediately obtain an overview of the entire set which is the basis for such judgements. More information such as the title and keyword list are shown when an individual document (matching or non-matching) is selected with the mouse.

From such initial searches and selections, one gains basic knowledge which one can use to browse and explore other nearby documents. These might be close to those previously inspected but may not have matched one's initial search. One can find documents which are relevant but do not contain familar keywords, and one can move beyond one's initial search by using interesting words found as one explores. One can

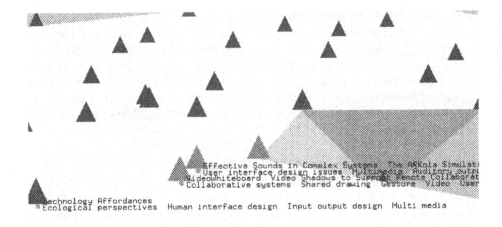

Figure 3. After a search for the word 'collaborative', some documents near to the 'valley' region at the centre of the corpus have been highlighted. Browsing with the mouse reveals the title of a nonmatching document (Technology Affordances) that is near to one of the matching papers (Effective Sounds in Complex Systems) and may be of interest as the two papers have more general issues in common as well as an author, Bill Gaver.

shift smoothly from browsing to searching and back again, all the while maintaining continuity and context by reference to the static landscape framework.

The Bead 'Viewer' program displays a scene showing a corpus landscape, and commands for movement (e.g. zoom in) and searching can be invoked either by typing in commands or by using a set of buttons arrayed above the display. The same Viewer commands can also be invoked via messages sent from other concurrently executing programs. (System architecture issues are discussed in more detail in [4]). Movement, viewing and mouse selection, along with simple searches for word occurrences, make up the means of access to the model. This small set seems straightforward to use; the limitations on use are pushed over more towards the model itself. Given a good corpus model, it should not be necessary to be an expert in query languages or on the database itself — the ideal is that this load should be taken on by the system.

The modelling process of Bead sometimes leads to peaks and valleys, and there is always a surrounding contour or 'shore'. These serve as natural reference points (or, more literally, landmarks) and are important in orientation and navigation. Slopes are shaded in accordance with their gradient, so that steep slopes are darker then gentler slopes and flat areas. The peaks and valleys also show areas where the 'fit' of the documents is rough. This may mean that the system could not find a good layout in 2D, although such a layout may exist. Alternatively, it could be that there is an inherent conflict in the set of desired mutual distances of documents, and therefore no planar layout can be found. In either case, the roughness of fit is an important property which should be conveyed to the user. This type of metalevel information is not often shown in an information display, although it does convey an extra dimension of the modelled information which could be influential in judgements based on proximity and implied similarity of documents.

The shore delimits the corpus and is usually made up of documents which are less strongly associated with any central theme (or themes) of the corpus. In the modelling process, such documents are pushed out to the physical periphery, which adds to the consonance of the model. Lastly, areas of density show clusters of strongly related documents, serve as landmarks and reference points, and offer bases for initial exploration of the corpus model.

Although more realistic shading and texturing could make the landscape more 'naturalistic', such colours and textures might be better put towards informational content rather than the 'framework' of the landscaping. Indeed, there is a danger that adding such detail might produce distracting visual clutter, obscuring information content. In all information design, one must choose which dimensions of the display should be used for information content (e.g. the position and colour of individual documents), which for a framework which conveys the basic style of interpretation of the information (e.g the shading of the slopes of the hills and valleys) and which should be unused so as to avoid clutter. Adding more information to the landscape design in this way is one of the topics of current work discussed in the next section.

4 Conclusion

The system described attempts to give access to a body of complex, high–dimensional information by using a model or metaphor of a landscape. The display design is directed towards a more exploratory and dynamic style of use than that of most traditional information retrieval systems, and it tries to take advantage of our natural sensorimotor skills by presenting a corpus as a mostly open landscape. This model supports overview and browsing, as well as searches for keywords represented by the colouring of individual documents. It is intended that the dynamics of movement, selection and searching can be combined in a way which favours information access rather than information retrieval.

Current work focuses on the modelling algorithms which construct the landscape, and how low-level issues such as word weighting affect the accessibility and quality of the resulting model. Apart from pairwise properties such as relative distance of documents, the emergent properties of the layout such as patterns of density and overall 'texture' are significant features in the perception and use of the modelled corpus. A related topic is the extension of the construction techniques to more robustly and efficently handle larger–sized corpora. Other lines of interest are directed towards the enrichment of the information content of the display while avoiding excessive and counter–productive visual clutter, making the 'information space' able to be shared by people working on different workstations, and improving the tools for movement and navigation within the space.

Ideally, the modelled information space should reflect the documents and document–related activities of the people using the system. It should help to guide and should be guided by their work with other systems and in other media. In this way, Bead might offer new possibilities for computer–based support of both individual and collaborative work in a semantically rich virtual environment, and might take a useful place amongst the more general environment of everyday work.

5 Acknowledgements

Thanks are due to my fellow researchers at EuroPARC and PARC for discussion and comment, in particular Bob Anderson, Victoria Bellotti, Bill Gaver, Gifford Louie, Diane McKerlie, Tom Moran, Abi Sellen and Andreas Weigend.

This is an expanded (by ~50%) version of a paper titled 'Visualisation of Complex Information' submitted for presentation at the East-West Human-Computer Interaction Conference, to be held in Moscow in August 1993.

6 References

1. J. Bertin: Graphics and Graphic Information Processing, Walter de Gruyter, Berlin, 1981.
2. D. Canter: Wayfinding and Signposting: Penance or Prosthesis? In: Proc. NATO Conf. on Visual Presentation of Information. Published as: Information Design, Easterby & Zwaga (eds.), Wiley, 1984, pp. 245–264.
3. S.K. Card, G.G. Robertson & J.D. MacKinlay: The Information Visualizer, an Information Workspace. Proc. CHI'91 (New Orleans, Louisiana, 28 April - 2 May, 1991), ACM, New York, pp. 181–188.
4. M. Chalmers & P. Chitson, Bead: Explorations in Information Visualisation. In: Proc. SIGIR'92, published as a special issue of SIGIR Forum, ACM Press, pp. 330–337, June 1992.
5. W.S. Cleveland & M.E. McGill (eds.): Dynamic Graphics for Statistics, Wadsworth & Brooks/Cole statistics/probability series, Belmont, CA, 1988.
6. D.R. Cutting, J.O. Pedersen, D. Karger & J.W. Tukey: Scatter/Gather: A Cluster–Based Approach to Browsing Large Document Collections. Proc. SIGIR'91, published as a special issue of SIGIR Forum, ACM Press, pp. 318–329.
7. J.J. Gibson: The Ecological Approach to Visual Perception, Lawrence Erlbaum, 1979.
8. G. Lakoff & M. Johnson: Metaphors We Live By, University of Chicago Press, 1980.
9. K. Lynch: The Image of the City, MIT Press, 1960.
10. J.D. MacKinlay, G.G. Robertson & S.K. Card: The Perspective Wall: Detail and Context Smoothly Integrated. Proc. CHI'91 (New Orleans, Louisiana, 28 April - 2 May, 1991), ACM, New York, pp. 173–180.

11. R. Passini: Wayfinding in Architecture, Van Nostrand Reinhold, New York, 1992.
12. R. Rao, S.K. Card, H.D. Jellinek, J.D. MacKinlay & G. Robertson: The Information Grid: A Framework for Information Retrieval and Retrieval-Centred Applications. Proc. UIST'92 (Monterey, California, November 1992), ACM, New York, pp. 23–32.
13. G.G. Robertson, J.D. MacKinlay & S.K. Card: Cone Trees: Animated 3D Visualizations of Hierarchical Information. Proc. CHI'91 (New Orleans, Louisiana, 28 April – 2 May, 1991), ACM, New York, pp. 189–194.
14. D.M. Russell, M.J. Stefik, P. Pirolli & S.K. Card: The Cost Structure of Sensemaking. Proc. InterCHI'93 (Amsterdam, April 93), ACM, New York, pp. 269–276.
15. G. Salton: Automatic Text Processing, Addison-Wesley 1989.
16. E. Tufte: Envisioning Information, Graphics Press, 1990.

From Interface to Interplace:
The Spatial Environment as a Medium for Interaction

Thomas Erickson

Human Interface, Advanced Technology Group,
Apple Computer, Inc., Cupertino, CA 95014

Abstract. Today's human-computer interfaces are cumbersome, sterile, and uninviting; they stand in stark contrast to the richness and depth of the everyday world. The thesis of this paper is that spatial environments have great potential as interface metaphors, particularly as computers begin to serve as a medium through which human-human interaction occurs. One section of the paper focuses on ways in which MUDs—text-based, multi-user dialog systems—use spatial metaphors to support social interaction. Then the paper examines how real spaces structure and enrich human interaction, drawing on observations from the literature on urban design, landscape architecture, and related disciplines. Ultimately, we hope that a better understanding of these issues can lead to the development of spatially-based interfaces which support human-human interaction.

1. Introduction

I believe people are beginning to use computers in a radically new way. As computing and communications merge, and as networks increase in their bandwidth and pervasiveness, we will witness the transformation of computers from simple work processors into a networked medium which supports communication, cooperative work, and social interaction. By simultaneously providing a framework for structuring social interaction, and a collection of props for facilitating and enriching interaction between individuals, computing as a social medium will have a far reaching effect on our daily lives.

The thesis of this paper is that spatial information theory can not only provide the underpinnings for effective Geographic Information Systems (GIS), but can also provide the foundation for this new use of computing technology. Just as spatially based interfaces are natural candidates for supporting the retrieval, manipulation, and understanding of spatial information, so are they likely candidates as interfaces for computing as a social medium. However, such an application requires that we focus on different properties of space. In particular, we need to understand the properties of space which are entwined with human interaction. Although this investigation initially leads us a away from the theoretical terrain typically explored by designers and implementors of GIS, ultimately we will end up in the same place. For users do not retrieve and manipulate geographic information in a vacuum—ultimately their purpose is share it with others so that common goals might be achieved.

This paper makes a start at investigating the properties of spaces that enable them to serve as a frameworks for communication, cooperative work, and social interaction. To avoid getting lost in abstraction, I begin by giving an example of how a physical space functioned as an effective interface for coordinating the activity of a large number of people engaged in a single task. Next I turn to the realm of human computer interaction, and discuss an example in which space is used—in a purely metaphorical way—as an interface. Having established that space—both real and metaphorical—can play a positive role in supporting and structuring human interaction, we examine ways in which real spaces structure and enrich interaction. This is done by drawing upon observations from urban design, landscape architecture, and the sociology and anthropology of situated behavior. I conclude by suggesting that designers of interfaces for GIS may want to keep in mind that the interfaces they are designing for spatial information might ultimately evolve into interfaces of a much more general nature.

1.1   Space as an Interface

In December of 1989 I had an striking experience which, for me, came to foreshadow the way in which a user interface of the future might function. Oddly enough, no computer was involved.

I was participating in a workshop aimed at producing a book. About 30 authors had gathered to mutually critique papers for a period of three days. At the end of the workshop the organizers decided to try something unusual: all 30 authors would meet and jointly create an organization for the book from scratch—the authors would decide on what the book sections should be, and how the chapters should be ordered within them.

We gathered in a room, each author with a copy of his or her chapter. To start the process, someone had spread some pieces of paper on the floor, with (possible) book section names written on them, and authors were asked to put their chapters near an appropriate section. After this, the procedure was simple: anyone could pick up any chapter and move it elsewhere; anyone could change the name of a book section; anyone could propose a new section by writing a name on a new piece of paper.

Although it seemed like a recipe for chaos, and was in fact characterized by a lot of milling about and simultaneous conversations, the process was exceptionally effective. In about 30 minutes, 30 people had come up with an mutually agreed upon organization for a book of 30 chapters, with everyone participating in the discussion. It seemed to me to be a human analog of the long predicted agent-based computing systems, in which distributed agents, each possessing incomplete knowledge, cooperatively interact to accomplish a task.

The key to the success of this process lay in its spatial nature. Several phenomena seemed to be important:

First, it seemed important that the process was carried out in a bounded space, dedicated to the task. We could see activity as happened: it was clear when disagreements arose; it was evident when someone was making a major change; and we could tell when the organizational process settled down because the general level of conversation and movement settled down.

Second, the participants had assigned meanings to parts of the space. This was done by writing the names of proposed book sections on paper, and by positioning chapters relative to the proposed sections. This method of assigning meaning also allowed participants to convey ambiguity and novelty. Sometimes participants would position a chapter halfway in between two sections, indicating that they weren't sure to which it belonged. Sometimes participants would put chapters in an area of the room without a section name, suggesting that those chapters belonged together but fit no existing sections.

A consequence of having a shared, meaningful space was that actions in that space were meaningful, and often triggered discussions and explanations. For example, when someone went to move a chapter to another area of the room (i.e. assign it to another book section), there would be one or more people around (the audience) with whom the mover would discuss the rationale for the move. The consequence of this discussion was that either:

a) mover and audience would agree on the move,
b) mover and audience would decide to change the name and definition of the category so that the chapter better fit where it was, or
c) the audience would convince the mover that the chapter was indeed in the right place. In all of these cases, a result was that there was a greater shared understanding of the section names and definitions, and of the gist of each chapter.

Finally, physical constraints shaped the way in which people could participate in the organizational process. The fact that the chapters and section names were spread all over the floor had an important impact: it meant that no one could dominate the organization of the book. Those who had strong opinions about where their chapters belonged, tended to hover near their chapters, ready to 'defend' them from would-be reorganizers; those who had ideas about the arrangement of the book as a whole had to flit about from section to section, thus giving up any strong control over where their chapter (or any single chapter) was positioned.

There were a variety of other elements of the situation which influenced the success of the interaction, but these are sufficient to establish the point that properties of space can usefully structure and facilitate interactions.

Now, this is very nice, but the skeptical reader may wonder to what extent this applies to computer interfaces. Is this another paper talking about how wonderful virtual reality will be? No. While virtual reality may be an important interface medium of the future, I want to focus on interfaces in general. I claim that a better understanding of how properties of space support human interaction has direct applications to the more mundane, two-dimensional interfaces with which we interact today. To make this point, we turn to an example of a spatial interface which is entirely metaphorical.

2   MUDs: Virtual Places as Frameworks for Interaction

In this section we examine MUDs. MUDs are examples of systems which use space as a way of organizing information and coordinating activity. They are particularly interesting because they do this entirely through metaphor: no spatial representations of space (either 3-D or 2-D) are used; the entire interface is textual, and yet MUDs manage to draw upon some properties of space of facilitating interaction.

2.1   What are MUDs?

MUD is an acronym which stands for Multiple User Dungeon, though the D is mapped to many words, including Dimension, and Dialog. In technology-centered terms, a MUD is a multi-user database typically used as an infrastructure for gaming or chatting; MUDs may also be thought of as text-based virtual realities, or multi-user interactive fiction programs. But these descriptions all fall a bit short of the mark. The key attribute of MUDs is that they are environments which support social interaction: people can converse, chance meetings can occur, 'rooms' can be constructed, 'objects' can be created and interacted with, and events of all sorts can be organized and conducted. MUDs act as social media, providing frameworks for social interaction, and collections of objects which catalyze, facilitate and enrich interactions. Although MUDs have primitive interfaces, and allow interaction only in the most cumbersome way, I believe they are the precursors of a new way of using computers.

2.2   The Origin of MUDs

MUDs grew out a genre of text-based adventure games, popular in the 1970s on time-shared minicomputers. These were text-only games in which the player explored a dungeon, hunting for weapons and treasure and battling monsters, all by typing in simple command strings. Success at these tasks increased the player's score, which translated into increased rank and power. A typical dialog between the user and the adventure game might look like this:

```
User: Look
Game: You are in a large room; there is an arched
      doorway to the north, and a sword lying on the
      floor.
User: Pick up sword
Game: You pick up the sword. You feel a tingling in your
      arm. The sword appears to be enchanted.
```

```
User: Go north
Game: You are in a round room. There are doors to your
      left and right.
```

And so on.

MUDs evolved from this type of game, the first being implemented in 1979. In such adventuring MUDs, in addition to obtaining weapons, treasure, and killing monsters, one could encounter other players—potentially logged on from other machines on the Internet—and converse with them, go on adventures with them, or kill them, as inclination dictated. Often, it was the latter. In 1989 James Aspnes of Carnegie Mellon University constructed a MUD without monsters or weapons or scoring, a place where people could gather to talk and extend the MUDs structure by using a built in programming language (see [3] for a more extensive history).

2.3   MUDs Today

Recently, MUDs have captured the interest of researchers in computer science, media, communications, and human computer interaction. In particular, much interest has been stimulated by the creation of MUDs at Xerox PARC [4] and the MIT Media Lab [2]. Although there are a variety of types of MUDs in existence, I will focus on conversation-oriented MUDs, rather than those oriented towards adventuring; most of my examples will be drawn from MediaMOO, the MUD at MIT.

MUDs have become quite popular, in spite of the fact that their interfaces are cumbersome. For example, Curtis [4] estimates that about 750 players connect to the PARC MUD every week, with 12 to 35 being connected simultaneously at any time of day. In the first year and a half, over 3500 different players connected from over a dozen countries. More generally, there are 200 to 250 MUDs active on the Internet; Curtis calculates that with a conservative estimate of 100 players per MUD per week, there are at least 20,000 people using MUDs every week.

For a more complete description of MUDs and related programs, the reader should see [4] and [3]; Bruckman and Resnick [2] provide a description of MediaMOO. For our purposes, it will be sufficient to describe some basic attributes of MUDs, and show what a typical interaction would look like.

Characteristics of MUDs. MUDs oriented towards supporting conversation have three characteristics:

MUDs have Geographies: MUDs use geographical metaphors. LamdaMOO, the mud at Xerox PARC, is described as a rambling old house with many rooms. Other MUDs may be based on the metaphor of a medieval town, a dungeon, or of a fictional world if that MUD is oriented towards role-playing. MediaMOO, the MUD at MIT, is unusual in that its geography is partially based on the real world. Its core is a model of the Media Lab building, and—through connections over a virtual Internet—other areas are based on the layout of other Internet sites, Apple's R&D building being one example.

MUDs are Extensible by their Participants: Most conversational MUDs allow their participants to add to the MUD's structure. In these MUDs, players can build their own rooms, create objects that other players could interact with, and even program autonomous (ELIZA-like) agents which roam around the MUD, accosting or otherwise interacting with other players. Note, again, that this is all done via text. 'Building' a room or an object or an agent simply means writing a program that gives appropriate textual responses to textual actions: 'looking' at a room, agent, or object displays its description; doing some other sort of action produces some textual description of what happens.

MUDs Support Complex Social Interaction: MUDs have social hierarchies, which are usually based on knowledge: more experienced participants know the structure of the mud, and can help newer participants find places, program things, and so on. Participants often demonstrate their prowess by creating neat spaces and objects. A variety of complex social behaviors have been reported on MUDs. These range from the formation of friendships and romances, to cases in which participants pretend to have different genders, appearances, and personalities. These, and other interesting social behaviors, are described in Bruckman [2] and Curtis [4].

An Example of an Interaction on a MUD. The text below is a simulated interaction between a user, 'Tom' and the MediaMOO MUD program. The text resembles the script of a play, with actions and comments by Tom preceded by "You", and actions by other participants preceded by their names; the rest of the text (unlabeled) is generated by the MUD program: it usually consists of descriptions of spaces or objects in the MUD. Text in curly brackets {} are commands typed by Tom, and are not seen by other players.

```
Amy is here.
Amy says, "now you can go to com, and then to apple."
Amy says, "there's also a router that'll jump you
    around."
Amy says, "type 'help internet' for more info."

{Say Thanks much! See you later.}
You say, "Thanks much! See you later."

Amy says, "see you around..."
Amy steps into The Root Lounge.
```

In the above script, Amy, another participant, has just appeared. Knowing that Tom is new, she's telling Tom how to travel from MIT to Apple, by using the MUD's virtual Internet. Tom uses the "Say" command to reply. (This script would look different on Amy's display: instead of "Amy says", she would see "You say", and so on.) In addition, anyone in the current room or space, can 'hear' the conversation.

The next fragment of script shows text generated by the MUD that describes what the participant, Tom, sees, and what he may do. Then it shows Tom specifying an action ("out"), and the MUD responding with a new description of what is happening to Tom (what another participant, if present, would see), and what Tom sees:

```
You see the back of a computer screen here.
Obvious exits: down to com and out to Apple Computer
   R&D Atrium

{out}
Your packets gather in a glob, and then flow into the
   screen! You feel yourself rematerializing.
Apple Computer R&D Atrium
You are in a glass atrium, four stories tall.
Offices look out from the walls.
Obvious exits: jeremy to Jeremias' Office
```

In this fragment of script Tom has traveled through the Internet, and rematerialized in the Apple Computer R&D Atrium (this description matches the appearance of the real Apple R&D Atrium). After describing what the participant sees, the MUD describes possible actions (above, typing "jeremy" will go to Jeremias' Office), as well as describing objects which may be interacted with, as shown in the next bit of script:

```
You see an ancient Apple I board hooked up to a
   monochrome monitor here. You see a sign here.

{look sign}
If you want to connect your office to the atrium,
   contact jeremias (Jeremy Bornstein).
[You finish reading]
```

Here Tom has read the sign, which contains text that another participant (Jeremias) has previously entered. Note that just as the R&D Atrium in the MUD has a real world counterpart, so 'jeremias corresponds to Jeremy Bornstein, the person who has control over what can be added to this part of the MUD).

This is enough of a description to get across the most basic things that happen in a MUD: the user can converse with other participants; the user can move through the MUD, seeing new parts of it; and the user can interact with objects in the MUD.

2.4   Ways in Which MUDs Encourage Interaction

MUDs use several interesting techniques to encourage interaction. One is to sponsor social events. For example, MediaMOO was officially opened with an inaugural costume ball on January 20, 1993, timed to coincide with the Clinton inaugural ball in Washington D. C.. Over 60 people from 5 countries logged onto MediaMOO during the ball, and participated in the festivities.

In this case, a second method for encouraging social interaction was used: spaces tailored to particular kinds of interactions were constructed (that is, the participants understood the meanings of the spaces). For example, a ballroom was constructed, with a dance floor and orchestra. When on the dance floor, participants could issue a "dance" command, and the MUD would generate a description of them dancing (often a description that was amusing or unusual). Other spaces were a bar, a lounge, a dressing room, all of which suggest particular types of action or interaction.

Another interaction generator is the participant's need for information. MUDs work in spite of their interfaces, not because of them. Although there is a help system, the easiest way to find out how to do something is to ask someone else. For example, it was possible for participants to teach the orchestra to play a particular song; the easiest way to learn this was by asking others who already knew how.

I hope that this conveys a bit of the flavor of what a MUD is like. There are two things which are difficult to convey through this sort description: First, although having a textual conversation with someone is cumbersome, it feels very realistic—much more so than an email conversation. Perhaps this is because it is occurring in real time. A second factor is the overlap between MediaMOO and the real world: entering an area of the MUD where the text descriptions resonate with your knowledge of the real place is surprisingly engaging.

Regardless of your impression of what the experience of participating in a MUD is like, the key point is that even in this purely textual interface, properties of spaces have been used to structure, facilitate, and generate social interactions. It is quite easy to imagine other types of interfaces—using two or three dimensional representations of spatial environments—which could also benefit from a greater understanding of ways in which spaces can support interactions.

3 Properties of Space Which Affect Interaction

The goal of this section is to examine ways in which real spaces function to structure and enrich human interaction. While there is little consensus on design principles for real (or virtual) spaces, my hope is that this examination will alert us to some of the untapped possibilities of spaces, and to suggest some productive directions for design explorations.

3.1 Objects can Generate and Catalyze Interactions

One of the ways in which spaces can be enriched is by what I call evocative objects. An evocative object is one that encourages people to interact with it. My favorite example is a sculpture in Seattle titled Waiting for the Interurban. The piece consists of five life size figures representing ordinary individuals (e.g., a workman with a lunch pail, a mother and child, a businessman, etc.). The figures are in various attitudes of waiting, and are, in fact, located beside the tracks at a train station. What is remarkable is the way in which people interact with the sculpture:

...the public response has been hearty enthusiasm. People have climbed on the statues, placed flowers in the woman's hand, kissed the workman's cheek (as revealed by a smear of lipstick) and strapped a leash to the dog as required by city law. In cold weather, the sculptures receive scarves and warm hats; on holidays like Halloween and New Year's they are decked out in masks and party outfits; and when Mt. St. Hellens erupted, they were given dust masks. [6]

The ability of an object to capture people's attention, and to encourage interaction with it is remarkable. It would be fascinating to see a study of pieces of public art, and an analysis of why some seem to evoke activity, and others seem to be sterile. Understanding more about this power of objects could have multiple applications. Museums could use it to design better exhibits; schools could create better demonstrations; retailers could design better displays; conference exhibitors could design more effective booths.

Notice also that, in this case, people are not just interacting with the object. They are also interacting indirectly with one another through the medium of the object: they are making comments on their shared experience, telling jokes, editorializing. As one Seattle resident says about Waiting for the Interurban: 'Whenever I go by the sculpture I smile: I go by the sculpture to see what's happening.' [6]

Objects can do more than just encourage interaction with themselves. Objects can also catalyze direct interactions between people. Another example from the domain of public art is a piece called "Dancers' Series: Steps." This is a series of eight bronze panels etched with diagrams of dance steps embedded in the sidewalk along a street. Each panel uses arrows and numbered outlines of men's and women's shoes to illustrate the steps for a particular dance.

It is not difficult to guess the reaction this sculpture elicits:

Each series of dancer's steps traces the process of human motion, [eliciting] from some viewers an irresistible urge to pause and mentally reenact the dance combinations, and from others a spontaneous spurt of dance fever! Those congregating around the steps or actually dancing attract other passersby, creating a chain reaction of participation and shared amusement. [6]

William Whyte, a sociologist who has studied how people interact in spaces, uses the term "triangulation" for this tendency of certain objects to catalyze interactions between people [18]. It is interesting to reflect upon the vast range of objects, spaces, and events in our culture whose basic purpose is to promote interaction between people. Everything from ice-breaking party games, to dinner parties, to scientific conferences are directed towards that end. And as demonstrated by MUDs, most of them seem perfectly appropriate for implementation in a virtual environment.

3.2   Spatial Constraints can Generate Activity

In an article on urban design, Christopher Alexander writes of the corner of Hearst and Euclid, in Berkeley, California [1]. Hearst and Euclid is an ordinary corner with sidewalks, a stoplight, a drugstore, and a newspaper machine in the store's entrance. As pedestrians wait for the light to change, they study the headlines and perhaps decide to buy a paper. When the traffic is moving, people wait and tend to buy papers; when the light turns red and traffic stops, pedestrians hurry across the street, and are less likely to buy papers. In a real sense, the traffic light is helping to sell papers by making people pause.

What makes the news rack-stoplight system function effectively is a variety of constraints. Obviously, physical constraints are operating: pedestrians don't want to get run over crossing against the light, and most would rather not have to run across the street. In addition, there are social constraints. While flagrant violations of traffic lights by pedestrians are not uncommon, neither is it uncommon to watch a pedestrian wait patiently for a light when no cars are in sight. Such social constraints are the reason, although it is a subtler effect, that we may say that the news rack helps people obey the traffic light: if there is something of interest, people are less likely to violate the relatively weak social constraints on obeying traffic lights. More generally, if an evocative or catalytic object is present, the stoplight increases the likelihood that the object will be noticed and will have its effect. For example, a particularly outrageous newspaper headline may cause one pedestrian to exclaim in disgust, prompting a bystander to agree.

Note that these constraints may have non-local effects. Traffic lights, because they cause pedestrians to bunch up, create a rhythm in the flow of people which may have an impact at some distance away. For example, Whyte [18] has observed that best predictor of whether people will stop to look in a store window is if other people are looking in it; thus, a traffic light acting to bunch up a sparse stream of pedestrians may act to increase the incidence of window shopping in the next block.

Spatial constraints can be powerful ways of generating activity. Designers of virtual environments might very well want to build in constraints. Although such behavior runs against the grain (someone who has just spent considerable effort making certain actions possible is not typically inclined to turn around and impose artificial limits), it may nevertheless be an important way of making a virtual space a more productive environment. Urban designers recognize that putting a high speed freeway through the center of town may transform it into a barren, impoverished place; interface designers may need to come to a similar recognition.

3.3   Spatial Elements may be used to Structure Activity

Even minor features of the physical environment can structure behavior in subtle ways. In a study of automated teller station (ATM) use, Marine [10] observed that people waiting to use an automated teller station typically left a gap between the head of the line and the person using the machine. This in itself isn't surprising: entering a secret code to withdraw cash is regarded as private activity. What is surprising is that the lines of users usually formed behind a crack in the pavement, which happened to be at a reasonable distance from the ATM. An obviously accidental environmental feature served to structure the behavior of ATM users.

This type of phenomenon is apparently well known to building contractors. Don Norman, a cognitive scientist who has done extensive work on human interface design, reports that he recently had a section of his driveway re-poured, and that the contractor doing the work suggested putting in a colored border between the old and new sections. The contractor explained that it would act as a natural boundary that people who used the driveway to turn around in would not venture beyond. Norman lives on a dead-end, popular beach street, and gets about ten turnarounds a day: he reports that it works. On telling this story to his neighbor across the street, his neighbor reported being told the same thing by his contractor, and pointed to the cement entryway to his brick driveway [12].

In addition to these apparently unconscious uses of environmental features, people quite commonly use features of the environment to their own ends. For example, in a study of ways in which children play in cities, Dargan & Zeitlin [5] cite many examples of how children have re-purposed spaces and objects: "There were dozens of games going on at any given time. Whatever was handy would be a base—a dead pigeon, a car, or your little brother could be second base."

The moral here is clear. People use features of spaces in unexpected ways. It is the profusion of objects, the apparent haphazardness and disorder of spaces, that makes them adaptable and usable.

3.4 Place: Space with Meaning

An important attribute of space is that people understand a lot about particular types of space—they see meaning in space. I like to use the word "place" to refer to space plus meaning. At the most basic level, people understand that particular types of places have very generic functions. Closets, drawers, and cabinets are for storing things; libraries are places for storing, exhibiting, and browsing information and artifacts; living rooms, ballrooms, and atriums are public spaces; bedrooms and offices are typically private spaces for individuals. MUDs use this type of knowledge of space to help participants know what to expect.

Other types of place meaning have not been exploited. Places often reflect their history. Urban designer Kevin Lynch describes a multitude of ways in which cities reflect their history, from deliberately constructed monuments, to traces of vanished city walls reflected in the city's grid of streets [9]. Similarly, objects, buildings, cities and landscapes can reflect their physical context: the forces which shape them, the local materials out of which they are constructed, the situated activities to which they are responses—this is called vernacular architecture and landscape. Such places feel as if they have grown out of their surroundings, and convey a unique sense of place (see [8]). Entwined with all this are meanings which the inhabitants project onto a place. A field where a famous battle was fought, the birthplace of a celebrity, or simply the first house built in a town become invested with meaning, and serve as concrete, shared symbols to those who live there.

3.5 Ritual and Place: Places can Suggest Types of Interaction

Closely associated with the fact that places have meanings, is that places often have activities associated with them. One way of capturing this is through the concept of ritual. Ritual is useful because it connects three important elements of human interaction: participants, repeated sets of actions, and artifacts or spaces. Various theories of ritual and its forms and functions have been developed by anthropologists studying religious behavior (e.g. [17]). More recently, the concept has been applied to every day behavior such as consumption [15], tourism [16], and leisure activities [11].

A good example of the importance of place and ritual can be found in a description of the redevelopment of the town of Manteo, South Carolina [8]. The designers approached the project by interviewing community members, and asking them to describe places at which they frequently spent time. They developed a map of the 'sacred structures' of the community that should be preserved. These sacred structures were not actually sacred (usually), but were places like the corner store, the local post office, even a parking lot. What made these places special were the interactions that took place: chatting with neighbors at the post office, negotiating business in the drug store. These were rituals in the sense that they were sets of particular actions tied to particular places, and carried out at particular times. Because of the rituals tied to these places, the community had come to identify itself with them.

As in the material world, ritual can support interaction in a spatial interface by providing frameworks which encourage strangers to chat, act in particular ways, and share information with one another.

4 Conclusion

Today's computer interfaces, whether they be to spreadsheets, spatial data, or MUDs, are primitive, cumbersome, sterile, and uninviting. They stand in stark contrast to the richness and depth of our everyday world. While there has been a little work on applying some of these ideas to human computer interaction (see [7] and [19], in addition to the work on MUDs already cited), we have barely scratched the surface. Understanding more about space, and the properties which enable it to generate, catalyze, and structure human interaction, is a promising direction for further work.

I'd like to conclude with one last perspective from urban design. Stuart Perez has written about what he calls incidental architecture, about "the fences, flowerpots and checkered table cloths; window curtains, mailboxes and trash; cars and vendors and the sounds of each," all the things which make up the unnoticed fabric of our everyday existence. He suggests that perhaps architects and urban designers ought to change the way they think about what they do.

> What we plan and design—the facades, the bulk, the surfaces and detail— is the armature on which the sensory life of the city is built. Like an armature, our architecture succeeds or fails on its ability to support this stuff of life, not by cleaning it up and putting it away but by knowing what it is, understanding how it works and setting a place that makes room for it to happen. [14]

As those of us in computer science, human computer interaction, and related disciplines work with spatial information theory to realize our near term goals, it is important to recognize that the embodiment of spatial information in computing systems may ultimately serve as an armature that will support rich and engaging forms of human interaction.

5 Acknowledgments

Thanks to Amy Bruckman and her colleagues at the MIT Media Lab for creating MediaMOO and opening it to outside researchers. Thanks to members of the Apple Human Interface community, particularly the erstwhile urban design reading group, for many discussions of architecture, urban and landscape design, and its application to interface design. Thanks to Amy Bruckman and Jonathan Cohen for comments on a draft of this paper.

References

1. Alexander, C. A City is Not a Tree. Design After Modernism (ed. J. Thackara). New York: Thames and Hudson, Inc., 1988.

2. Bruckman, A & Resnick, M. Virtual Professional Community: Results from the MediaMOO Project. To be presented at the Third International Conference on Cyberspace, May 1993. (Available via anonymous FTP from media-lab.mit.edu: /pub/MediaMOO/Papers/MediaMOO-3cyber-conf.)

3. Bruckman, A. Identity Workshop: Emergent Social and Psychological Phenomena in Text-Based Virtual Reality. Unpublished manuscript, 1992. (Available via anonymous FTP from media-lab.mit.edu, /pub/MediaMOO/Papers/ identity-workshop.)

4. Curtis, P. Mudding: Social Phenomena in Text-Based Virtual Realities. Proceedings of DIAC '92. Available via anonymous FTP from parcftp.xerox.com, pub/MOO/papers /DIAC92.

5. Dargan, A. & Zeitlin, S. City Play. New Brunswick and London: Rutgers University Press, 1990.

6. Fleming, R. L., Von Tscharner, R. PlaceMakers: Creating Public Art that Tells You Where You Are. Boston: Harcourt, Brace, Jovanovich, 1987

7. Hill, W. C., Hollan, J. D., Wroblewski, D. A., & McCandless, T. P. "Edit Wear and Read Wear." In Proceedings of the ACM Human Computer Interaction Conference. Austin, TX: ACM Press, pp. 3-9, 1991.

8. Hough, M. Out of Place. New Haven: Yale University Press, 1990.

9. Lynch, K., What Time is this Place? Cambridge: The MIT Press, 1972.

10. Marine, L., Study of the Opportunistic Use of Environmental Structure to Organize Behavior, Unpublished manuscript, 1990.

11. Moore, Alexander. Walt Disney World: Bounded Ritual Space and the Playful Pilgrimage Center. Anthropological Quarterly, Vol. 53, #4, 207-218, 1980.

12. Norman, D. A., Personal communication, 1991.

13. O'Guinn, T. C. & Belk, R. W. Heaven on Earth: Consumption at Heritage Village, USA. Journal of Consumer Research, Vol. 16, September 1989.

14.   Perez, Stuart. Incidental Architecture. Places, Vol. 7, #1, Fall 1990.

15.   Rook, D. W. Ritual Behavior and Consumer Symbolism. Advances in Consumer Research, Vol 11 (ed. Thomas C. Kinnear). Provo, UT: Association for Consumer Research, 279-284, 1984.

16.   Theilmann, J. M. Medieval Pilgrims and the Origins of Tourism. Journal of Popular Culture, Vol. 20, #4, 1987.

17.   Turner, V. The Ritual Process: Structure and Anti-Structure. Chicago: Aldine, 1969.

18.   Whyte, W. H., City: Return to the Center. New York: Doubleday, 1988.

19.   Wroblewski, D. A., McCandless, T. P. & Hill, W. C. Advertisements, Proxies, and Wear: Three Methods for Feedback in Interactive Systems. Natural Dialogue and Interactive Student Modeling[tentative title]. Heidelberg: Springer-Verlag, in preparation.

A Keystroke Level Analysis of Manual Map Digitizing[1]

Peter Haunold
Werner Kuhn

Department of Geoinformation
Technical University Vienna
Gusshausstrasse 27-29
A-1040 Vienna (Austria)
Fax: (+43 1) 504 3535
Email: haunold@geoinfo.tuwien.ac.at
kuhn@geoinfo.tuwien.ac.at

Abstract

The acquisition of digital spatial data is a key economical factor in GIS projects. Transforming analog graphic data by manual digitizing is slow and therefore extremely expensive. The work reported here investigates the possibility of applying the Keystroke-Level Model to the modeling and optimization of manual digitizing tasks. This model predicts the time it takes an experienced user to perform routine tasks on a given system. It has been applied successfully for text editing tasks. Here, its suitability for manual digitizing is being tested and additional unit tasks are determined.

The suitability of the Keystroke-Level Model for manual digitizing tasks is analyzed in the context of a major national map digitizing effort. The agency responsible for topographic and cadastral mapping in Austria (Bundesamt für Eich- und Vermessungswesen, BEV) is currently developing a digital cadastral map (Digitale Katastral-Mappe, DKM). Scanned cadastral maps are being manually digitized on screen in order to transform analog graphic data into digital form. The article describes the principles of the model, the design of an experiment, and encouraging first results.

1. Introduction

Collecting GIS data is a difficult and time consuming process. Usually, existing analog map data have to be transformed into digital form by complex conversion processes. In many application areas manual digitizing is the primary method of acquiring map data for a GIS [Marble, Lauzon, & McGranaghan, 1984]. Automatic digitizing methods, e.g., scanning and vectorizing, are being developed, but so far they can only be used in prototype implementations or special cases [Crosilla, & Piccinini, 1991]. The resulting data are often of limited value because of data quality problems [Connealy, 1992].

---

[1] The contribution from the *"Bundesamt für Eich- und Vermessungswesen"* and the support from Intergraph Corporation for user interface research from Intergraph Corporation is gratefully acknowledged.

The slow and expensive nature of manual digitizing processes necessitates the investigation of the user interface characteristics of these operations [Kuhn, 1990a; Kuhn, 1990b; Frank, in Press]. Research on human-computer interaction has produced many methods and models for analyzing and developing user interfaces. The hypothesis underlying the work reported here is that one of these models, the Keystroke-Level Model [Card, Moran, & Newell, 1980a; Card, Moran, & Newell, 1980b] is applicable to manual digitizing tasks. If true, this would offer great potential for economizing on these very expensive operations [Haunold, & Kuhn, 1993].

The Keystroke-Level Model provides a simple method to predict the time for standard functions performed by expert users on computer systems. The time required for these functions is modeled by counting keystrokes and other low-level operations which have known average performance times. The prediction determines the time spent on performing each task. Individual tasks can then be optimized by comparing different execution strategies. The model has been used successfully for text editing tasks [Card, Moran, & Newell, 1983].

A study investigating the Keystroke-Level Model as a tool to optimize manual digitizing is run by the authors in cooperation with the Austrian agency responsible for topographic and cadastral mapping (Bundesamt für Eich- und Vermessungswesen, BEV). The suitability of the model is being tested in an experiment. This experiment compares predicted times from the Keystroke-Level Model with actual performance times. In the overall scheme of planning and optimizing the map data acquisition process, this project analyzes only elementary actions and their performance times. These actions are at the level of pressing a key, pointing to a target or homing the hand between cursor and keyboard. Some of them only take fractions of a second, but they are repeated very often.

The paper describes the Keystroke-Level Model (section two), an experiment for comparing predicted and observed performance times (section three) and early results on the suitability of the model (section four).

2. The Keystroke-Level Model

The Keystroke-Level Model provides a simple method to predict the time an expert user needs to perform a given task. The central idea behind the model is that the time for an expert to do a task on an interactive system is determined by the time it takes to do the keystrokes. Therefore, one can write down the execution method used for the task, count the number of keystrokes required, and multiply by the time a single keystroke takes in order to get the total time. Obviously other elementary operations must be added to the model. These operations are pointing, homing, mental preparing and a system response time [Card, Moran, & Newell, 1983].

A requirement for applying the Keystroke-Level Model is the dissection of functions into error-free repetitive tasks at the keystroke level. Only error-free tasks are predicted, because the model cannot predict where and how often errors occur. Previous experience shows that expert users rarely make errors in performing well-known tasks, therefore errors do not use much time, although this restricts the usability of the model. A second requirement is the subdivision of tasks into standardized sequences of performance steps, called unit tasks.

2.1. The Time Prediction Problem

The major goal of the Keystroke-Level Model is to predict performance times for sequences of unit tasks. The prediction problem is as follows:

Given: A task; the command language of a system; the motor skill parameters of the user; the response time parameters of the system; the methods used for the task.

Predict: The time an expert user will take to execute the task, provided he uses the method without error.

Given a large task, a user will break it into a sequence of small, quasi-independent unit tasks. The time to do a unit task consists of two parts, the *acquisition* of the task and the *execution* of the task acquired. The total time is the sum of the time for these two parts:

$$T_{task} = T_{acquire} + T_{execute} \qquad (2.1)$$

The acquisition for a unit task is a very complex process and cannot be modeled with the Keystroke-Level Model [Card, Moran, & Newell, 1983]. The model can only be used for tasks performed by experienced users who do not spend time thinking about the steps to perform, i.e. acquisition-time is assumed to be zero.

2.2. Operators[2] of the Keystroke-Level Model

The model uses simple keystroke level operators. The execution of one task can be described with four physical-motor operators, one mental operator and a system response operator. These operators are listed in Figure 1. The total execution time is the sum of the times for each of the operators:

$$T_{execute} = T_K + T_P + T_H + T_D + T_M + T_R \qquad (2.2)$$

The operators **K, P, H** and **D** are assumed to remain constant for each occurrence.

The Operator K. The most frequently used operator is **K**, which represents a keystroke or a button pushed on a mouse or on a cursor. It refers strictly to keys and not to characters. Therefore, pressing the "shift" or "control" key counts as a separate **K** [Card, Moran, & Newell, 1980b]. Time varies with the typing skill of the user; the range of typical values is shown in Figure 1. The total time for all keystrokes T_K used to do one task is:

$$T_K = n_K t_K \qquad (2.3)$$

Where n_K is the number of keystrokes and t_K is the time per keystroke.

---

[2] "Operator" is being used here in a mathematical sense, following Card *et al*. The person operating a system is called "user".

Operator	Description and Remarks	Time (sec)
K	Keystroke or button press.	
	Best typist (135 wpm)	0.08
	Good typist (90 wpm)	0.12
	Average skilled typist (55 wpm)	0.20
	Average non-secretary typist (40 wpm)	0.28
	Typing random letters	0.50
	Typing complex codes	0.75
	Worst typist (unfamiliar with keyboard)	1.20
P	Pointing to a target on a display with a mouse.	1.10
H	Homing the hand(s) on the keyboard or other device.	0.40
D(n_D,l_D)	Drawing (manually) n_D straight-line segments having a total length of l_D cm.	$.9n_D + .16l_D{}^e$
M	Mentally preparing for executing physical actions.	1.35
R(t)	Response of t sec by the system.	t

Fig. 1. The Operators of the Keystroke-Level Model, from [Card, Moran, & Newell, 1980b]

The Operator P. The operator **P** represents pointing to a target on a display with a pointing device. Pointing time t_P varies as a function of the distance to the target, d, and the size of the target, s, according to *Fitts's Law* [Card, English, & Burr, 1978]:

$$t_P = 0.8 + 0.1 \log_2 (d/s + 0.5) \ sec \qquad (2.4)$$

The time ranges from 0.8 to 1.5 sec, with 1.1 as an average time. This operator does not include the button press that often follows pointing.

The Operator H. Using different physical devices, the user will move his hands between them. This hand-movement is represented by the operator **H** ("homing"). A constant time t_H of 0.4 sec for movements between any two devices is assumed [Card, English, & Burr, 1978; Card, Moran, & Newell, 1980a].

The Operator D. The operator **D** represents the manual drawing of a set of straight lines using a device. **D** takes two parameters, the number of lines, n_D, and their total length l_D. An average value is given with:

$$t_D = 0.9n_D + 0.16l_D{}^e \qquad (2.5)$$

This is a very restricted operator. It assumes that drawing is done with the mouse on a system that constrains all lines to fall on a square 0.56 cm grid. Users also vary in their drawing skills.

The Operator M. Users spend some time "mentally preparing" to execute many of the physical operators just described. This mental preparation is represented by the operator **M**, which is estimated to take 1.35 sec on the average. The Keystroke-Level Model provides a set of rules regarding where to place this operator in action sequences [Card, Moran, & Newell, 1980b].

The Operator R. Finally, the Keystroke-Level Model represents system response time by the operator **R**. Response times are different for different systems, different commands within a system, and different contexts of a given command. The Keystroke-Level Model cannot contain estimates for response times t. Therefore, they must be measured for actual tasks. These times must be input into the model. The response time counts only if it causes the user to wait.

A Text Editing Example. An example taken from text editing illustrates the usage of the operators and the time prediction. Replacing a 5-letter word with another 5-letter word, where this replacement takes place one line below the previous modification is done as follows with a line oriented editor called *Poet* [Card, Moran, & Newell, 1980b]:

Jump to next line	**MK**[LINEFEED]	
Call Substitute Command	**MK**[S]	
Specify new 5-digit word	**5K**[word]	
Terminate argument	**MK**[RETURN]	(2.6)
Specify old 5-digit word	**5K**[word]	
Terminate argument	**MK**[RETURN]	
Terminate command	**K**[RETURN]	

Using the operator times from Figure 1 and $t_K = 0.2$ sec for an average skilled typist the time it will take to execute this task is predicted:

$$T_{execute} = 4t_M + 15t_K = 8.4 \; sec \qquad (2.7)$$

2.3. Special Operators for Manual Digitizing

To model digitizing tasks, we introduce two additional operators, K_C and P_S. The operator K_C is needed for a keystroke on a cursor with 16 buttons. It will be checked during the experiment whether the value t_K can be used or t_{K_C} must be defined differently.

A further pointing operator P_S is needed for a second, but shorter target-seeking when aiming at an already digitized point, using a function to snap to the point. The operator P_S is expected to take less time than the operator P and the time parameter t_{P_S} will be estimated during the experiment. The hypothesis that the operator P_S is equal for both actions will be tested in the experiment.

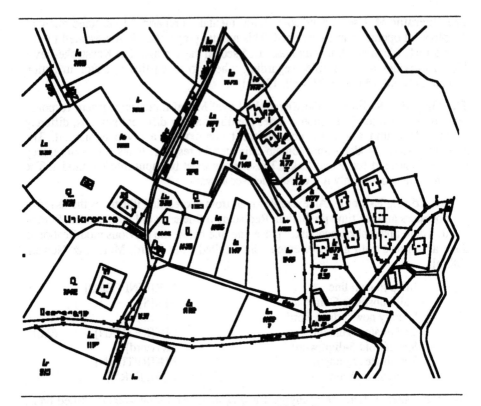

Fig. 2. A Digitized Cadastral Map Sheet

3. The Experiment

In the experiment currently underway, the Keystroke-Level Model is being tested for suitability for optimizing manual digitizing tasks. The predicted and the measured times for given unit tasks are compared. Also, the time parameters t_{K_C} and t_{P_S} are determined. This chapter describes preliminary tests as well as the planning, the execution and the evaluation of the experiment.

3.1. Preliminary Tests in the BEV

During some preliminary tests in the mapping agency, the method of manual digitizing on screen [BEV, 1991] was analyzed in order to find unit tasks. Manual digitizing on a screen means that a scanned picture of a map sheet (Fig. 2) is being digitized directly on screen. In this method, a total of 80 tasks was identified for digitizing boundaries (parcels, houses and land use), parcel-ID's and symbols. The tasks for changing and moving the views on screen and all copying and correcting tasks were also included. The most important and frequently used tasks were selected and defined as unit tasks for manual digitizing. This resulted in a total of 38 unit tasks including digitizing of boundaries, symbols, parcel-ID's and some important view

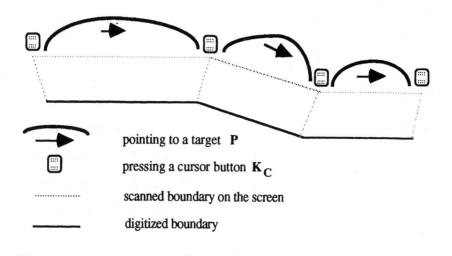

pointing to a target **P**

pressing a cursor button **K**$_C$

scanned boundary on the screen

digitized boundary

Fig. 3. Digitizing Vertices in a Boundary

manipulation functions. These 38 unit tasks represent some 50 % of all unit tasks, but they enable the user to perform 90 to 95 % of all tasks.

During the same preliminary tests, the sequences of keystroke level actions for all tasks were observed. These actions are keystrokes on a typewriter keyboard **K**, button presses on the cursor **K**$_C$, pointing to a target **P**, mental preparing **M**, the response time **R** and homing the hands between keyboard and cursor **H**.

An example is the task *"Digitizing a vertex in a boundary"* (Fig. 3). For this task the following sequence was detected:

Pointing to a vertex	**P**	(3.1)
Digitizing the vertex	**K**$_C$[Cursor button 0]	

The operators from Figure 1 (t_{K_C} = 0.28 sec) predict the time for this task with:

$$T_{execute} = t_P + t_{K_C} = 1.38 \ sec \tag{3.2}$$

3.2. Planning the Observations and the Experiment

The measurements of performance times were done entirely under production conditions at the agency. Expert users were observed at their regular work places. In consequence, the planning of the observations and the experiment was constrained by the available hard- and software configuration as well as by the need for precise time measurements.

For digitizing, the agency uses a program based on AUTOCAD Version 11.0, installed on an IBM PS/2-80 computer. The digitizing pad is a HOUSTON HIPAD

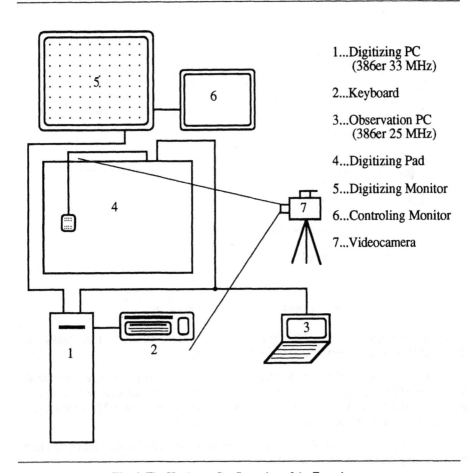

1...Digitizing PC
(386er 33 MHz)

2...Keyboard

3...Observation PC
(386er 25 MHz)

4...Digitizing Pad

5...Digitizing Monitor

6...Controling Monitor

7...Videocamera

Fig. 4. The Hardware Configuration of the Experiment

PLUS with a 16 button cursor. Digitizing itself is done on a 19" monitor (Sony GDM 1950). The hardware-configuration of the experiment is shown in Figure 4.

The agency requested that the experiment should not interrupt the production process. This represented a difference to the experiments done by Card *et al.* under laboratory conditions [Card, Moran, & Newell, 1980b].

A requirement of the experiment was to get time stamps at a resolution better than 1/10 seconds for each action on the cursor or on the keyboard. For each time stamp, a well-defined link to the cursor or keyboard action must be given. The coordinates of the cursor must also be recorded if a button press is done.

The idea of only videotaping the actions was dismissed, because the time spent on each action is too short to recognize its start and end accurately. Also the number of performed actions is too high for an efficient video analysis. Therefore a program to record the data was written. It had to be set up so that it did not interfere with AUTOCAD and the overall digitizing process.

The program was written in C++ and installed on a separate laptop computer. This computer was connected to the cable between digitizing pad and computer (see figure 4). Thereby, each signal sent by a button press on the cursor was recorded with its time stamp. Keyboard actions could not be recorded directly. As the necessary tool to split a keyboard signal was too expensive for this project (cost is around $ 3500), videotaping the actions on the keyboard **K** and homing **H** was necessary.

3.3. Executing the Observations

After two days of tests in December 1992, the experiment was executed in the beginning of January 1993. During the tests, some recorded actions were analyzed to see how the program and the hardware-configuration worked. Afterwards the program was modified slightly, to get better results for a faster and easier analysis.

The observations lasted two to four hours for each of the seven expert users. On this occasion the users digitized boundaries, buildings, parcel-ID's and symbols (see Fig. 2). A total of about 30,000 actions were recorded. These actions correspond to about 15,000 performed unit tasks. The users were videotaped for the whole observation time. For the first four users, the video camera focused on the digitizing pad and the keyboard during the whole observation time. While observing the last three users, the camera was focused on the screen during line digitizing. Thereby, the time for the operator **R** was specified and the times for **P** and **M** were examined.

3.4. Analysis of Observations

The analysis was done entirely in Microsoft Excel, running on a Macintosh Quadra 700. The action sequences were assembled from the unit tasks and the performance time for each task was determined.

At the moment, other analyses, like a statistical analysis of the task time distributions, are in progress. The distribution function is being estimated to get the quality of the measured time stamps. Then the performance times for the unit tasks can be determined. This has already been done for one user. Further analysis will compare measured and predicted times for the other users.

4. First Results

Our early analyses show a very good correspondence between measured and predicted times. This section presents the estimated time parameters of the new operators K_C and P_S, the comparison between measured and predicted time for two tasks, as well as the prediction results for other tasks, determined for one user.

4.1. The Operators K_C and K

Recent studies and applications of the Keystroke-Level Model have offered new operators as well as modified time parameters for existing operators [John, & Newell, 1989; Olson, 1987; Olson, & Nilsen, 1988; Walker, Smelcer, & Nilsen, 1991]. For actions on a 16 button cursor, we introduced the operator K_C. The time parameter t_{K_C} for one user, taken from the videotapes and the observation records, is not very different from the value given in Figure 1. The value t_{KC} for this particular user was estimated from about 2000 performed tasks to be 0.30 ± 0.03 sec.

In accordance with the research done on text editing tasks each user varies in his speed for pressing a key [Card, Moran, & Newell, 1980b]. We have estimated the observed user's time parameter t_K for pressing a key on the keyboard to be 0.40 ± 0.08 sec within about 70 tasks.

4.2. The Operator P$_S$

For modeling certain tasks in digitizing lines and houses, an operator P$_S$ has been introduced. An analysis of the videotapes showed the operator to be the same for two actions.

The first action is that of digitizing a vertex which is difficult to detect in the scanned map. The user extends the line on the screen beyond the vertex, in order to detect the accurate breaking point of the line. Then, a second pointing action to digitize the detected vertex follows. This second pointing is faster, because the distance to aim is shorter than the one assumed with the operator **P**.

The second action is that of aiming at an already digitized point. This action is used if a new boundary line starts or ends on a digitized point of another boundary or if a house is coincident with an existing digitized boundary. The user then performs a function to catch the existing digitized point. The distance to aim at this point is similar as in operator **P**, but when using the function to catch a digitized point, the size of the target gets larger. Therefore the user needs less aiming time and the whole pointing time decreases. Thus, operator P$_S$ requires less time than operator **P**.

Analyzing the videotapes, about 100 tasks were detected which included these two actions. The time parameter t_{P_S} was estimated to be 0.85 ± 0.18 sec on average for both actions.

4.3. Predicting Performance Times

To show how well the Keystroke-Level Model predicts performance times, the task from Figure 3, "*Digitizing a vertex in a boundary line*," is used here again.

The operators from the Keystroke-Level Model are:

Pointing to a vertex	**P**	(4.1)
Digitizing the vertex	**K**$_C$[Cursor button 0]	

Using the operators from Figure 1 and $t_{KC} = 0.30$ sec, the total performance time for this task is predicted to be:

$$T_{execute} = t_P + t_{KC} = 1.40 \; sec \qquad (4.2)$$

Analyzing this task, the videotape and the observation records show that a screen manipulation function done in between is changing the performance time. Thus the task "*Digitizing a vertex in a boundary line*" defined in the preliminary test, is split into two new tasks. The first one, called T-1, is performed like the one given in

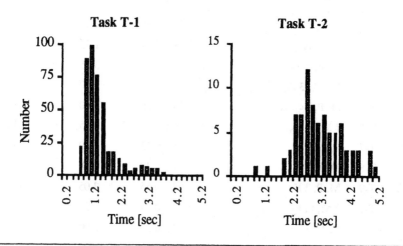

Fig. 5. Distribution Function of Task T-1 and Task T-2

Formula 4.1. The second, called T-2, differs in that the user has to reorient himself on the screen after a screen manipulation function. Therefore, an operator **M** for mental preparing must be added.

Based on this modification, the task T-1 corresponds to the task "*Digitizing a vertex in a boundary line*" with the operators given in Formula 4.2. An approximately normal distribution between 0.0 and 2.4 sec for this task can be seen in Figure 5. Therefore a mean value of 1.25 ± 0.36 sec is calculated to perform this task. The predicted time from Formula 4.2 is 1.40 sec, resulting in a difference of 11% between predicted and measured time. Based on our observations, we assume that the values larger than 2.4 sec belong to another task where the user needs a second pointing action to digitize the vertex. This assumption will be verified by doing statistical tests after all users have been analyzed.

The task T-2 is "*Digitizing a vertex in a boundary line after a screen changing function*" plus a mental operator **M** for orientation after a screen function (ZOOM and PAN). For this task, a modified sequence of actions must be given.

Therefore, the relevant operators from the Keystroke-Level Model are:

Orientation on the screen	**M**	
Pointing to a vertex	**P**	(4.3)
Digitizing the vertex	**K**$_C$[Cursor button 0]	

Using the operators from Figure 1 and t_{K_C} = 0.30 sec, the total performance time for this task is predicted to be:

$$T_{execute} = t_M + t_P + t_{K_C} = 2.75 \; sec \qquad (4.4)$$

Name of Task		Operators	$T_{meas.}$ [sec]	$T_{pred.}$ [sec]	Diff. [%]
NDGG	T-1	P, K_C	1.25 ± 0.36	1.40	- 11
NDGG	T-2	M, P, K_C	2.98 ± 0.52	2.75	8
NDGG	T-3	M, P, 2 K_C	3.28 ± 1.13	3.05	8
NDGG	T-4	M, P_S, 3 K_C	2.92 ± 0.63	3.10	- 6
NDGG	T-5	M, P, P_S, K_C	3.48 ± 0.65	3.60	- 3
NDGG	T-6	2 M, P, P_S, K_C	4.76 ± 0.58	4.95	- 4
NDGG	T-7	P, 2 K_C	1.65 ± 0.43	1.70	- 3
NDGG	T-8	M, P, P_S, 2 K_C	3.61 ± 0.49	3.90	- 7
NDGG	T-9	M, P, 2 K_C	3.52 ± 0.55	3.05	15
NDGG	T-10	2 M, P, P_S, 2 K_C	5.79 ± 0.24	5.25	10
NDGG	T-11	P, 3 K_C	2.12 ± 0.39	2.00	6
NDGG	T-12	P_S, 4 K_C	2.25 ± 0.26	2.05	10
NDGG	T-13	M, P_S, 4 K_C	3.36 ± 0.27	3.40	- 1
NDGG	T-14	M, P, 4 K_C	3.52 ± 0.32	3.65	- 4
NDGG	T-15	M, P, P_S, 4 K_C	4.57 ± 0.22	4.50	2
NDHG	T-16	P, K_C	1.29 ± 0.17	1.40	- 8
NDHG	T-17	P, 2 K_C	1.87 ± 0.25	1.70	10
NDHG	T-18	P_S, 2 K_C	1.36 ± 0.27	1.45	- 6
NDHG	T-19	P_S, 2 K_C, R	3.37 ± 0.30	3.20	5
NDSG	T-20	P_S, 4 K_C	2.12 ± 0.16	2.05	3
SYMBOL	T-21	P, 2 K_C	1.83 ± 0.47	1.70	8
NDGN	T-22	M, P, 2 H, 6 K, 2 K_C	6.72 ± 0.68	6.25	8
NDGN	T-23	M, P, 2 H, 5 K, 2 K_C	5.92 ± 0.83	5.85	1
NSGN	T-24	M, P, 2 H, 4 K, 2 K_C	5.40 ± 0.78	5.45	- 1
NRGN	T-25	2 M, 2 P, 2 H, 4 K, 3 K_C	8.59 ± 0.84	8.20	5
PAN	T-26	M, P, 2 K_C, R	3.40 ± 0.63	3.30	3
ZOOM	T-27	2 M, 2 P, 2 K_C, R	6.38 ± 1.13	6.05	5

Fig. 6. Comparison between Measured and Predicted Performance Times

The distribution function for task T-2 is also shown in Figure 5. The mean value to perform this task is calculated to 2.98 ± 0.52 sec, referring to an approximate normal distribution between 2.0 and 4.0 sec. Thus the difference between predicted and measured performance time is 8%. The range limitation from 2.0 to 4.0 sec is based on the same assumption as for task T-1.

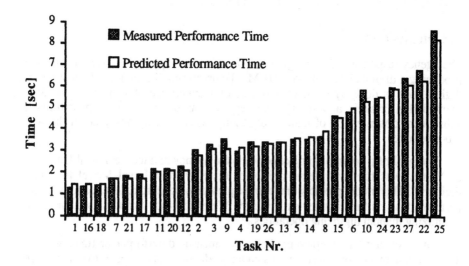

Fig. 7. Comparison between Measured and Predicted Performance Times (with tasks ordered according to their performance times)

4.4. Time Predictions for Other Tasks

The entire observation series has already been analyzed for one user. The observations covered two hours and a total of 1,500 tasks (3,427 actions) was recorded. After deleting non unit tasks (3%), incorrectly performed tasks (12%) and tasks repeated less than five times (8%), a total of 1,154 unit tasks (77%) remained. Twenty-seven unit tasks, containing the two tasks given here as examples, were analyzed. These tasks, their operators, and the differences between predicted and measured performance times for each of them are summarized in Figure 6 and shown graphically in Figure 7.

Tasks T-1 to T-15 (NDGG) and task T-20 (NDSG) are line digitizing tasks for acquiring boundaries. Tasks T-16 to T-19 (NDHG) are for digitizing buildings. Task T-21 (SYMBOL) is done for digitizing land use symbols. Tasks T-22 to T-25 (NDGN, NSGN and NRGN) are for digitizing parcel-ID's. The last two tasks, T-26 and T-27 (PAN and ZOOM) represent screen functions. The names of the tasks are taken from the function names used by the agency.

Comparisons between all measured and predicted performance times show a very high correlation. The differences for tasks T-1 and T-2, given as eleven and eight per cent respectively, are similar to the percentages for predicting tasks in text editing [Card, Moran, & Newell, 1983]. The other 25 analyzed unit tasks show differences ranging from one to 15 per cent between measured and predicted performance time. These results also compare well with the percentages given for text editing. We therefore conclude from the experiment that the Keystroke-Level Model accurately predicts the performance times for the 27 tasks observed.

5. Conclusions

The agency responsible for topographic and cadastral mapping in Austria is currently producing a digital cadastral map (DKM). Transforming the analog information into digital form is done by manual digitizing of scanned cadastral map sheets. Since manual digitizing is a time- and money-consuming process, the possibility of subsequent optimization of manual digitizing tasks is being analyzed in the work reported here.

The goal of a common project between the agency and the Technical University of Vienna was to show the suitability of the Keystroke-Level Model to manual digitizing. This model predicts times for sequences of keystroke level actions used to perform standard functions by expert users. We expect that it can be used for further optimizing manual digitizing operations.

An experiment to compare predicted and measured performance times was run. The experiment supplied observations under production conditions. Previous research on the Keystroke-Level Model was only done under laboratory conditions [Arend, 1990; Card, Moran, & Newell, 1983].

First results led to the addition of the operator K_C for a button press on a 16 button cursor and of the operator P_S for two special pointing actions. Their time parameters are estimated as $t_{K_C} = 0.30$ sec and $t_{P_S} = 0.85$ sec. The operators M, P and H had been taken from research on text editing and the time parameter t_K for the operator K was estimated to be 0.40 sec for one user.

The experiment evaluation is still going on. Preliminary analyses and results of 27 unit tasks show a highly accurate prediction of performance times for manual digitizing tasks. The average difference of seven per cent agrees with the differences found for text editing tasks analyzed by Card, Moran and Newell.

The analyses of the observations for the other six users are in progress. Varying time parameters for the operators K and K_C are expected for each user. In the completed analysis, the operators M, P and H, taken from text editing tasks, provided good results. Therefore, no separate estimation of these operators for the prediction of manual digitizing tasks was done. A detailed analysis of these operators may still result in differing time parameters, but the preliminary results provided no evidence for this.

Acknowledgments

The authors would like to thank the Austrian *Bundesamt für Eich- und Vermessungswesen* (BEV) for its help during the tests and the experiment. The cooperation of its personnel, especially of the digitizing staff, contributed much to the successful operation of the experiment. Furthermore, much encouragement from, and discussion with, Andrew Frank contributed to this article.

References

Arend, U. (1990). *Wissenserwerb und Problemlösen bei der Mensch-Computer-Interaktion: empirische Untersuchungen zur handlungsorientierten Gestaltung der Benutzeroberflächen an einem Datenbankprototypen.* Techn. Hochschule Darmstadt, Regensburg: S. Roderer Verlag.

BEV (1991). Die Anlegung der digitalen Katastralmappe (DKM), Erlaß des Bundesamtes für Eich- und Vermessungswesen vom 27. März 1991, GZ K 5108 / 90-21.603. 2.Auflage. Wien.

Card, S. K., English, W. K., & Burr, B. J. (1978). Evaluation of mouse, rate-controlled isometric joystick, step keys, and text keys for text selection on a CRT. *Ergonomics, 21,* 601-613.

Card, S. K., Moran, T. P., & Newell, A. (1980a). Computer text-editing: An information-processing analysis of a routine cognitive skill. *Cognitive Psychology,* 12, 32-74.

Card, S. K., Moran, T. P., & Newell, A. (1980b). The Keystroke-Level Model for User Performance Time with Interactive Systems. *ACM Communications, 23* (7), 396-410.

Card, S. K., Moran, T. P., & Newell, A. (1983). *The Psychology of Human-Computer Interaction.* Hillsdale, New Jersey: Lawrence Erlbaum Associates.

Connealy, T. (1992). A Complete and Innovative Raster to Vector Conversion Algorithm. *EGIS'92,* 1102-1111. Munich: EGIS Foundation.

Crosilla, F., & Piccinini, L. C. (1991). A new approach for pattern recognition in cadastral cartography. *Geo-Informations-Systeme, 4* (4), 14-20.

Frank, A. (in Press). The user interface is the GIS. In Medickij-Scott, D., & Hearnshaw, H. (Eds.), *Human Computer Interaction and Geographic Information Systems.*

Haunold, P., & Kuhn, W. (1993). Die Analyse von manuellen Digitalisierungsabläufen. *EVM., 70,* 37-40.

John, B. E., & Newell, A. (1989). Cumulating the Science of HCI: From S-R Compatibility to Transcription Typing. *Proceedings of the CHI'89 Conference on Human Factors in Computing Systems,* 109-114. New York: ACM.

Kuhn, W. (1990a). Editing Spatial Relations. In Brassel, K., & Kishimoto, H. (Eds.), *Fourth International Symposium on Spatial Data Handling,* 423-432. Zurich, Switzerland; July 1990.

Kuhn, W. (1990b). From Constructing Towards Editing Geometry. *ACSM/ASPRS Annual Convention, 1,* 153-164. Denver, CO; March 18-23, 1990.

Marble, D. F., Lauzon, J. P., & McGranaghan, M. (1984). Development of a Conceptual Model of the Manual Digitizing Process. *International Symposium on Spatial Data Handing, 2,* 146-171. Zurich, Switzerland: Geographisches Institut, Abteilung Kartographie/EDV.

Olson, J. R. (1987). Cognitive analysis of people's use of software. In Carroll, J. M. (Eds.), *Interfacing Thought - Cognitive Aspects of Human-Computer Interaction.* Cambridge, MA: Bradley Books/MIT Press.

Olson, J. R., & Nilsen, E. (1988). Analysis of the cognition involved in spreadsheet software interaction. *Human-Computer Interaction, 3,* 309-350.

Walker, N., Smelcer, J. B., & Nilsen, E. (1991). Optimizing Speed and Accuracy of Menu Selection: A Comparison of Walking and Pull-Down Menus. *International Journal of Man-Machine Studies, 35* (6), 871-890.

Critical Issues in the Evaluation of Spatial Autocorrelation

Yue-Hong Chou

Department of Earth Sciences
University of California
Riverside, CA 92521-0423, USA
hong@ucrac1.ucr.edu

Abstract. Spatial autocorrelation measures the degree to which a spatial phenomenon is correlated with itself in space. As such, it can be used as an indicator of the fundamental topological structure of the spatial relationship among geographic entities displayed on a map. Statistics of spatial autocorrelation are especially useful for characterizing the spatial pattern in the distribution of any phenomenon in question. This paper addresses three important issues pertaining to the evaluation of spatial autocorrelation: measurement of the study variable, definition of geographic units, and specification of spatial weighting functions. Theoretically, spatial autocorrelation is most adequately evaluated when the study variable is measured in either an interval or a ratio scale and geographic units are better delineated by polygons of homogeneous surfaces based on variables that are significant to the distribution of the study phenomenon. In evaluating spatial autocorrelation, weighting functions such as area, boundary length, distance, and their combinations must be examined carefully and specified whenever necessary.

1 Introduction

Spatial autocorrelation measures the degree to which a spatial phenomenon is correlated to itself in space. A positive spatial autocorrelation refers to a spatial pattern where geographic entities of similar values tend to be clustered on a map. A negative spatial autocorrelation implies a spatial pattern in which features of similar values scatter throughout the map. When no statistically significant spatial autocorrelation exists, the spatial pattern is considered random.

Statistics of spatial autocorrelation are useful for characterizing the spatial pattern of any phenomenon displayed on a map. In general, a spatial pattern is defined as how geographic entities are arranged in space and how they are related one to another. A spatial pattern describes the distribution of the entire body of all entities of the study phenomenon and may or may not be related to the geometric properties of individual entities.

The capability of presenting, describing, and analyzing spatial patterns is a prerequisite to the understanding of the complicated process governing the spatial distribution of a phenomenon. In a complex ecosystem, for instance, the distribution of a specific species tends to be affected by a number of interrelated factors. Any attempt to modeling the distribution in order to explain the underlying process requires the capability of characterizing spatial patterns in a scientific manner.

As a measure of spatial pattern, spatial autocorrelation can be considered as a basic element in the structure of spatial information. Three important issues pertaining to the evaluation of spatial autocorrelation are addressed in this paper: measurement of the study variable, definition of geographic units, and specification of spatial weighting functions.

2 Measurement of the Study Variable

Well defined statistics of spatial autocorrelation, including joint-count statistics, Moran's I coefficient, and Geary's c coefficient [7, 13, 14] and their procedures for significance testing are thoroughly discussed in [5]. The G statistics were more recently developed by Getis [9] and Getis and Ord [10]. Among these statistics, the I coefficient is employed in this paper because this statistic is most widely used in the literature. Nevertheless, the statements presented in this paper are not limited to the I coefficient and the issues addressed here are applicable to the evaluation of spatial autocorrelation using any of the above statistics.

Formally, the I coefficient is defined as:

$$I = \frac{n}{2a} \frac{\Sigma_i \Sigma_j w_{ij}(x_i - \bar{x})(x_j - \bar{x})}{\Sigma_i (x_i - \bar{x})^2} \tag{1}$$

where n is the number of geographic units; w_{ij} denotes the spatial relationship between the ith and jth geographic units, which equals 1 for adjacent units and 0 otherwise; $2a = \Sigma_i \Sigma_j w_{ij}$ is the total number of adjacent pairs. Procedures for testing the significance of spatial autocorrelation under both assumptions of normality and randomization are available in [5].

The value of the I coefficient is generally between -1 and 1 while theoretical extreme values of the coefficient can be identified [12]. A greater value close to 1 indicates a positive spatial autocorrelation and implies a clustered spatial pattern. A smaller value close to -1 indicates a negative spatial autocorrelation and implies a scattered pattern. When the value is close to 0, there is no significant spatial autocorrelation and the spatial pattern is random.

Figure 1 shows three general spatial patterns depicted in a 20-by-20 grid [2]. For simplicity, these patterns are evaluated on the basis of a binary nominal variable, i.e., a shaded cell represents the presence of a hypothetical phenomenon while a blank cell denotes the case of absence. Figure 1-A shows a clustered pattern in which all the shaded cells are grouped together to form a single cluster. In this case the calculated I coefficient is 0.947. Figure 1-B shows a scattered

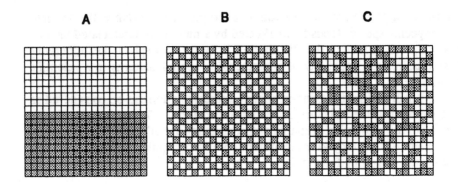

Figure 1: Three hypothetical spatial patterns: clustered (A), scattered (B), and random (C).

pattern in which every shaded cell is surrounded by blank cells and every blank cell surrounded by shaded cells. The I coefficient for this perfectly scattered pattern is equal to -1.000. In Figure 1-C, there is no significant pattern in the arrangement of shaded and blank cells. The pattern is relatively random with the calculated I coefficient equal to 0.168.

Before evaluating spatial autocorrelation, the first question one must carefully consider is how to measure the study variable x in equation (1). The examples in Figure 1 are all based on the simplest nominal dichotomy. Statistics of spatial autocorrelation could be biased due to the inadequate measurement of the study variable. For instance, if the variable represents the presence or absence of a specific species, then the variation within the class of presence could be much larger than that between the presence/absence classes. Some cells may have hundreds of observations yet they are classified in the same category as those with only one observation. In this case, environmental conditions of geographic units with very few observations may be much more similar to those with no observation than to those with numerous observations. For this reason, the study variable can be more adequately measured in either an interval scale or a ratio scale.

Figure 2 shows the distribution of wildland fires in the Idyllwild 7.5-minute quadrangle defined in the topographic map series of the U.S. Geological Survey. Shaded areas on this map represent surfaces that were burned during the period between 1911 and 1984. As geographic units are defined by a 20-by-20 grid, a cell which was burned during the study period is assigned a burn code equal to 1 while an unburned cell assigned a burn code equal to 0 if the simple binomial classification is employed. The calculated I coefficient is equal to 0.667, implying a significant positive spatial autocorrelation.

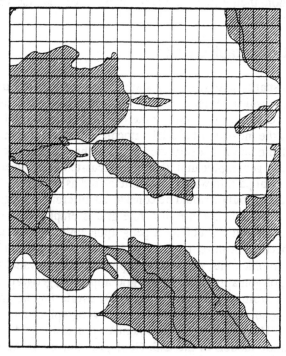

Figure 2. Fire activity in the Idyllwild quadrangle

Although different classification schemes are available for assigning binomial values, such as those developed for processing digital data of satellite imagery, the problem with dichotomous measurement remains unsolved no matter which method is used. For instance, in one method the burn code of a cell is determined by whether its centroid location was burned or not. Apparently, in this case a cell which was completely burned is treated as the same as another cell in which only a small proportion was burned.

In dealing with complicated spatial phenomena constrained by dichotomous classification, the study variable can be more adequately measured by modifying the binomial assignment with a suitable weight in order to convert the measurement into one of either interval or ratio scale. For instance, the binary code of fire activity may be adjusted by the proportion of burned area in each cell. Although the range of possible values of the study variable remains between 0 and 1, the study variable is no longer a simple binomial case. With this modification, the lower right cell in Figure 2 which was burned partially would be assigned a more reasonable value closer to an unburned cell than to a burned cell. In this case, the calculated I coefficient becomes 0.777.

In general, one should try to measure the study variable in either an interval or a ratio scale whenever possible. If the study variable is already coded in a binomial classification, then efforts should be made to convert the variable into a more reasonable measure.

3 Definition of Geographic Units

For characterizing spatial patterns using statistics of spatial autocorrelation, a second critical issue which requires careful consideration is the definition of geographical units. One must be aware of the fact that how geographic units are defined determines the measure of spatial autocorrelation. In a simple example where the study phenomenon is distributed in a number of separate clusters, if each cluster defines a geographic unit, then the calculated I coefficient would be negative and the spatial pattern would be characterized as scattered. The spatial pattern of the same distribution would be considered different if geographic units are defined differently. For instance, if a fine grid is employed to define geographic units, i.e., each grid cell defines a geographic unit, then the calculated I coefficient would become positive, implying a clustered spatial pattern.

In defining geographic units, two general approaches are available: the grid approach and the polygon approach. In the grid approach, geographic units are defined by a grid of arbitrary size and position. By overlaying the grid onto the map of the study phenomenon, one obtains a set of geographic units corresponding to the grid cells. This approach is convenient for processing gridded data such as the digital elevation models of the U.S. Geological Survey or the digital data processed from satellite imagery. Furthermore, programming of spatial autocorrelation statistics for the grid approach is much easier than for the alternative due to the regularity in the size of geographic units and the space between adjacent units, i.e., any topological relationship among geographic units can be systematically traced through row and column counts. Another advantage of this approach associated with the regularity in size and spacing is that spatial weighting functions based on area or distance are unnecessary.

In spite of the above advantages, some theoretical problems with the grid approach should not be neglected. First, the measure of spatial autocorrelation is affected by the size and resolution of the grid. In evaluating the spatial autocorrelation in the distribution of wildland fires, Chou systematically compared the performance of grids of varying size and identified a log linear relationship between map resolution and spatial autocorrelation [1]. The findings indicate that the I coefficient increases in a log-linear manner with the increased grid resolution. Second, as the spatial distribution of natural phenomena tends to be nonstationary, statistics of spatial autocorrelation may be altered by placing the grid in a different way. Since the position of the grid is arbitrarily determined, one has no way to know if a calculated coefficient is more valid than another derived from a grid placed differently.

Alternatively, the polygon approach defines geographic units according to some criteria that are independent from the study phenomenon. Geographic units defined in this approach tend to illustrate polygons of irregular shape and size. Also, the distance between adjacent units is a variable rather than a constant. This approach has been widely employed in studies that do not rely on gridded data. For example, in the classical study of measles epidemics in Cornwall [11], geographic units are defined by local authority districts. Another example is the study of international conflicts in which geographic units are defined by political boundaries [15].

For ecological and environmental applications, the polygon approach is generally more appropriate than the grid approach because polygons of homogeneous surfaces tend to reflect natural geographic units. The polygon approach requires one or more variables to be selected for defining geographic units. To do so, five important questions must be considered.

First, the polygons delineated from the selected variables must exhaustively cover the entire study area. If there are patches of unclassified surfaces, then spatial relationships among polygon features would be distorted.

Second, the delineated polygons should not illustrate a clear spatial pattern by themselves. Variables that generate a systematic distribution are not appropriate for defining geographic units. For instance, temperature zones delineated by isotherms that demonstrate a clear north-south pattern should be avoided. In many cases, a variable measured in a nominal scale is better than one measured in an interval or a ratio scale in defining geographic units.

Third, the polygons delineated from the selected variables must define geographic units that are mutually exclusive. For instance, in Figure 2 polygons delineated by wildland fires may overlap because fires may occur more than once in the same area. In this case, geographic units should not be defined by polygons representing wildland fires.

Fourth, the selection of variables for defining geographic units must also take into consideration the relationship between the selected variables and the study variable. In general, a variable which is significant to the distribution of the study phenomenon is more desirable than an irrelevant variable. In studying the distribution of wildland fires, both vegetation and topography are reasonable candidates for defining geographic units because both are believed to be related to fire behavior, i.e., fires behave differently on slopes of different gradient or aspect and on surfaces of different vegetation coverage. For such a study, census tracts are not a reasonable choice because these geographic divisions have no effect on wildland fires. However, census tracts may be an ideal choice for defining geographic units when a socio-economic phenomenon is to be analyzed using census data.

Fifth, in order to avoid the distortion of spatial relationships due to the varying size of units or the varying distance among geographic units, variables that delineate polygons of relatively homogeneous shape and size are preferred. Vegetation in a desert environment is not appropriate for defining geographic units since its distribution tends to illustrate a spatial pattern where a number of extremely little islands of different vegetation types are surrounded by a vast unvegetated area.

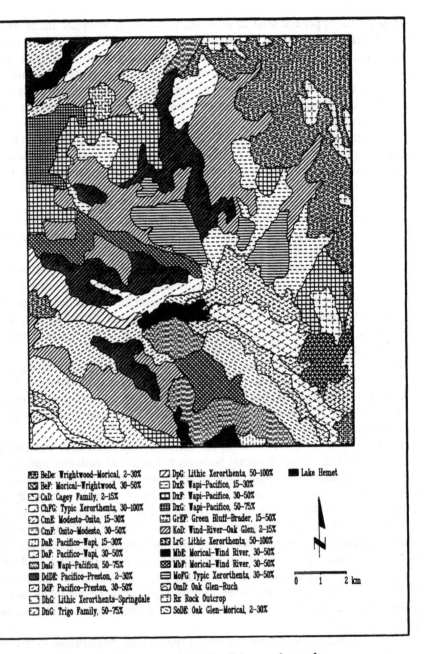

Figure 3. Soil map of Idyllwild quadrangle

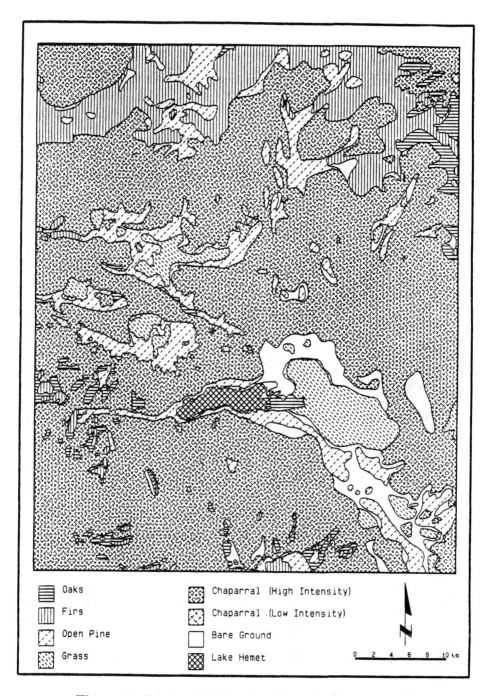

Figure 4. Vegetation of Idyllwild quadrangle

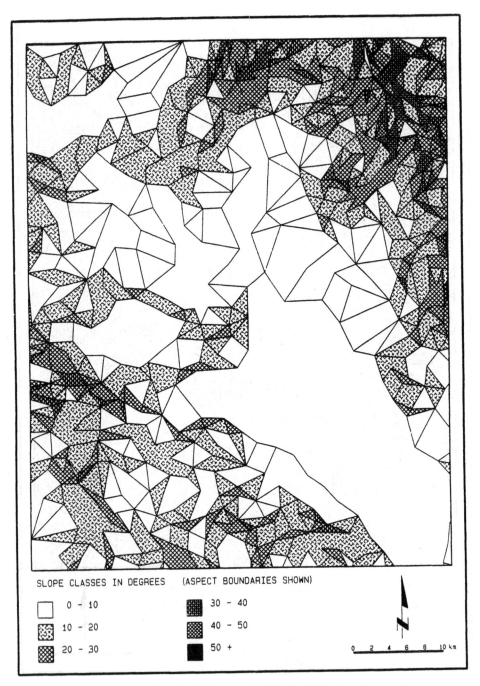

SLOPE CLASSES IN DEGREES   (ASPECT BOUNDARIES SHOWN)

☐ 0 - 10   ▨ 30 - 40

▨ 10 - 20   ▨ 40 - 50

▨ 20 - 30   ■ 50 +

0   2   4   6   8   10 km

Figure 5. Topography of Idyllwild quadrangle

Figures 3 and 4 are the soil and vegetation maps of the Idyllwild quadrangle, respectively. In terms of wildland fires, vegetation is shown to be significant in the distribution of fires [4]. On the other hand, polygons of homogeneous vegetation types vary much in size. Compared to vegetation, the soil map has the advantage that soil polygons are much more similar in size, yet soils are not as closely related to fire behavior. Figure 5 shows polygons of homogeneous topographic characteristics. Compared to the vegetation map, the topographic polygons are relatively similar in size. In addition, topography has been shown to be closely related to fire behavior. With these concerns, the topography map is most appropriate for defining geographic units in the study of wildland fires, at lease it is preferred over the grid map, the vegetation map, and the soil map (Figures 2-4).

4 Specification of Spatial Weighting Functions

Another important concern is whether the measure of spatial autocorrelation is affected by other geometric and topological variations. One must be aware of the problem of topological invariance, i.e., once the contiguity matrix has been specified, the size and shape of geographic units, and the relative strength of their relationships are completely ignored [6]. A feasible solution is to specify meaningful elements as weighting functions and incorporate them into the measure of spatial autocorrelation. Although generalized coefficients with spatial weighting functions and statistical testing are developed [5], how to specify such weighting functions remains to be studied [3, 8].

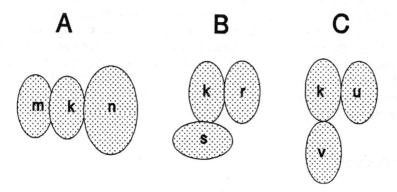

Figure 6. Possible spatial weighting functions: area (A), boundary-length (B), and distance (C)

Without a spatial weighting function, the evaluation of spatial autocorrelation is based on contiguity alone, the simplest spatial relationship. In this case, the spatial relationship between two geographic units (w_{ij}) in the I coefficient is coded as either 1 when they share a common border or 0 otherwise. This simplified spatial relationship fails to take into consideration effects due to other significant variations. Figure 6 shows three hypothetical examples of meaningful spatial relationships that require modifications on the contiguity weight.

Figure 6-A shows three units of different sizes. Assuming that the distance between K and N is equal to that between K and M, and that the length of the common border between K and N is equal to that between K and M, one can reasonably expect a more prominent relationship between K and N than between K an M because N is larger than M in size. For instance, spatial interaction is expected to be greater between K and N than between K and M if population density is homogeneous throughout the entire region. If the spatial relationship is represented by contiguity alone, the relationship between K and N would be identical to that between K and M. To take this variation into consideration, one must adjust the contiguity weight by area.

In Figure 6-B, one may expect a greater spatial relationship between K and R than between K and S due to the longer common border between K and R. In this case, contiguity must be adjusted by the length of common border. Likewise, Figure 6-C illustrates an example where the spatial relationship can be expected to be greater between K and U than between K and V because, other things being equal, the Euclidean distance between the centroid locations is shorter for the K-U pair.

In the study of wildland fires [3], six weighting functions were defined and compared: contiguity, area, boundary-length, distance, distance-area, and distance-boundary. The results indicate that area and boundary-length do not provide additional information about the distribution of wildland fires. Correlograms constructed for higher spatial orders suggest that the distance weight causes a scaling and smoothing effect on I coefficients. In other words, the distance weight exaggerates the measure of spatial autocorrelation and is useful for the identification of less evident spatial patterns. The roles of different spatial weighting functions vary over spatial phenomena. Therefore, it is important that meaningful spatial weighting functions be examined and specified whenever necessary.

5 Conclusions

Spatial autocorrelation, equivalent to Tobler's "first law of geography" [16] that "everything is related to everything else but near things are more related than distant things", is the most fundamental element in spatial information theories because it represents the basic relationship among features on a map. Indeed, spatial autocorrelation is the implicit assumption in general models of spatial interaction, such as the gravity model, that closer units affect each other more than units that are farther apart. Clearly, evaluation of spatial autocorrelation is a prerequisite to the understanding of the topological structure in the distribution of a spatial phenomenon.

The issues addressed in this paper are critical to the use of spatial autocorrelation for characterizing spatial patterns. The study variables must be measured adequately and geographic units must be defined appropriately. Also, weighting functions must be examined and specified whenever necessary. Furthermore, the effects of spatial weighting functions on statistics of spatial autocorrelation can be identified from correlograms that show higher-order spatial relationships. These tasks were difficult to accomplish in the past due to the difficulty in handling complex topological relationships. The recent development in geographic information systems (GIS) has made it possible to analyze complicated spatial relationships effectively [1]. As spatial autocorrelation can be effectively evaluated and spatial weighting functions carefully specified, further research in this area will make important contributions to the development of spatial information theory.

Acknowledgments

I am grateful to R.A. Chase of Forest Service, U.S. Department of Agriculture for supporting this study under grant PSW910004CA. The vegetation map was interpreted from aerial photographs by R.A. Minnich of University of California, Riverside. I also thank the editors and anonymous reviewers for thier helpful comments on an earlier draft.

References

1. Y.H. Chou: Map resolution and spatial autocorrelation. *Geographical Analysis*, 23, 228-246 (1991).

2. Y.H. Chou: Management of wildfires with a geographical information system. *International Journal of Geographic Information Systems*, 6, 123-140 (1992).

3. Y.H. Chou: Spatial autocorrelation and weighting functions in the distribution of wildland fires. *International Journal of Wildland Fire*, 2, 169-176 (1992).

4. Y.H. Chou, R.A. Minnich, and R.A. Chase: Mapping probability of fire occurrence in the San Jacinto Mountains, California. *Environmental Management*, 17, 129-140 (1993).

5. A.D. Cliff and J.K. Ord: *Spatial Processes: Models and Applications*, London: Pion (1981).

6. M.F. Dacey: A review of measures of contiguity for two and K-color maps, in B.J.L. Barry and D.F. Marble (editors), *Spatial Analyses: A Reader in Statistical Geography*, Englewood Cliffs, New Jersey: Prentice-Hall (1965).

7. R.C. Geary: The contiguity ratio and statistical mapping, *The Incorporated Statistician*, 5, 115-145 (1954).

8. A. Getis: A spatial association model approach to the identification of spatial dependence, *Geographical Analysis*, 16, 17-24 (1989).

9. A. Getis: Spatial interaction and spatial autocorrelation: a cross-product approach, *Environment and Planning A*, 23, 1269-1277 (1991).

10. A. Getis and J.K. Ord: The analysis of spatial association by use of distance statistics, *Geographical Analysis*, 24, 189-206 (1992).

11. P. Haggett: Hybridizing alternative models of an epidemic diffusion process, *Economic Geography*, 52, 136-146 (1976).

12. P. de Jong, C. Sprenger, and F. van Veen: On extreme values of Moran's *I* and Geary's *C*, Geographical Analysis, 16, 17-24 (1984).

13. P.A.P. Moran: The interpretation of statistical maps, *Journal of the Royal Statistical Society*, Series B, 37, 243-51 (1948).

14. P.A.P. Moran: Notes on continuous stochastic phenomena, *Biometrika*, 37, 17-23 (1950)

15. J. O'Loughlin: Spatial models of international conflicts: extending current theories of war behavior, *Annals of the Association of American Geographers*, 76, 63-80 (1986).

16. W.R. Tobler: A computer movie simulating urban growth in the Detroit region, *Economic Geography*, 46, 234-240 (1970).

A Directional Path Distance Model
for Raster Distance Mapping

Cixiang Zhan, Sudhakar Menon and Peng Gao

Environmental Systems Research Institute
380 New York Street
Redlands, CA 92374 , USA

Abstract. This paper proposes a directional model for weighted path distance mapping on raster data structures. The approach that is used to map the shortest path distance from a set of source cells to any background cell incorporates both non-directional and directional weights on the links between adjacent cells. Non-directional weights allow users to model impedance in terms of the traditional cell based friction costs used in raster distance mapping. Directional weights allow users to model non-symmetric travel costs incurred in traveling through value-gradients in continuous fields such as elevation, temperature or density, or incurred when traveling through a prevailing flow field. The model has been implemented as a cell-based raster distance mapping tool with user selectable directional factor functions, using both Dijkstra's and the Bellman-Ford's shortest distance algorithms. Two experimental results are included.

1 Introduction

Distance is a fundamental variable in spatial operations involving movement or regionalization based on spatial proximity. In a broad sense, distance can be viewed as a relation defined on a set of objects in a space, and can be represented by a relation matrix, in which the diagonal entries are zero, and others, if valid, are positive values [Gatrell, 1983]. In order for its value to be unique at every entry in the matrix distance must be a one-to-one mapping from an *ordered* pair of objects to a non-negative value. Although this notion of distance does not provide the mapping function to calculate distances, it allows great flexibility in defining a variety of distances.

The Euclidean distance between two objects fits this distance concept. The Euclidean distance between two objects requires no path for its definition, and is based solely on the locations of the two objects. The Euclidean distance between two objects is symmetric, i.e. $D(x,y) = D(y,x)$, and satisfies triangularity, i.e. $D(x,z) \leq D(x,y) + D(y,z)$.

The path distance from an object to another requires paths between the two objects. Since paths from an object to another may not be unique and distance along those paths generally are different, path distance does not satisfy the unique distance concept. The shortest path distance from one object to another, however, is unique, and satisfies this distance concept. Smith [1989] shows that the shortest path distance also satisfies triangularity. We will refer to the shortest path distance as the path distance in cases where there is no confusion.

The path distance between two objects may not just be the physical distance along the path. To each segment of a path, a cost per-unit of physical distance or length, such as gasoline consumption or time, may be attached. The weight attached to a segment of path may be directional or non-directional. By non-directional weights we mean that the cost per-unit length is independent of travel directions along the segment. By directional weights we mean that the cost per-unit length is dependent on travel direction along the segment. Without weights or with non-directional weights, the path distance is symmetric. When directional weights are imposed, the path distance becomes non-symmetric.

The task of distance mapping on a cell-based raster data structure, given a set of *source* objects (represented as sets of cells sharing a unique cell value against a background value) consists of computing, for each cell in the cell layer, the shortest distance from the set of objects to the cell. Note that in the presence of directional weights the value computed for a cell is not necessarily the shortest distance from the cell to the set of objects. Various distance mapping algorithms have been developed for the Euclidean and non-directional path distance mapping problems (Danielsson [1980], Tomlin [1980] [1990], Borgefors [1986], Eastman [1989], Menon et al [1991]). However, the problem of path distance mapping with directional weights on a raster layer has received much less attention.

Directional weights exist widely in actual situations, and in many cases can not be simplified as non-directional weights. The cost to travel up a sloping route is certainly different from that to travel down on the same route, and heat flows from locations of high temperature to location of low temperature. Driving a vehicle against the wind direction consumes more fuel than driving along the wind direction, and particles disperse most easily along the flow direction. In this paper, we describe a general model for weighted path-distance mapping on a cell-based data structure, which considers both directionally uniform weights such as friction cost, and directionally heterogeneous weights that can be modeled as resulting from value gradients in fields such as density, elevation and temperature, and/or prevailing flow directions.

2 Raster Directional Path Distance Model

In order to compute the shortest path distance in the cell-based raster data structure, we need to define objects, valid paths (links) between objects and distance formulae. The objects in a cell layer can be generally defined as all single cells with the location of a cell represented simply by its center. However, a set of cells with the same value, either connected or non-connected can also be viewed as a single object, and the distance to the object from a cell is the shortest distance from the center of the cell to the center of the nearest cell in the object. The valid paths from any cell to other cells must pass through the links from the center of the cell to the centers of its eight neighboring cells. The basic unit of a path is the link from a cell to its neighboring cells. Orthogonal and diagonal units have different physical lengths.

Having defined objects and links between objects, the total cost to pass along the basic units of a path must be defined. The path distance from a cell to another is simply the sum of the costs of all basic units along the path. The cost to pass along a basic unit of a path, or the distance along a basic unit is affected by many factors, such as friction (resistance), the surface slope between the two cells, the gradient corresponding to differences in some variables between two cells, and the directions of flow fields at both cells. We only consider the directions in a 2-D flow field to reduce the complexity. The general idea is to consider the effects of friction, surface slope, value gradient and horizontal direction as multiplicative factors to the flat (planimetric) physical distance along a basic unit. Starting with the flat physical distance, we gradually incorporate those factors into the formula of the path distance.

Assume the cost from cell F to one of its orthogonal neighboring cells, labeled T, is to be computed, as in Fig. 1. We further divide the basic unit between cell F and cell T into two halves, one from the center of cell F to its border point B that lies in the middle of the two centers, and the other from point B to the center of cell T. We will derive the distance from the center of cell F to point B.

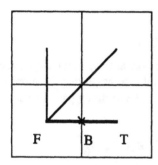

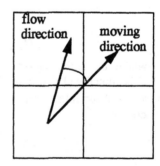

Fig. 1 Cell F and T, and their Middle Point B

Fig. 2 Relation Between Flow and Moving Directions

With assumption of square cells, the flat physical distance D_0 between cell F and point B is simply an half of the cell size. The surface distance factor S_{ft} is used to incorporate the effect of the difference in the surface elevation at the centers of F and T, and is computed as $1/(\cos a_z)$, where a_z is the zenith angle from the elevation of cell F to the elevation of cell T. The surface difference factor S is a non-directional weight because $\cos a_z = \cos(-a_z)$. The surface physical distance between cell F and point B is given by

$$D_{fb} = 0.5 \times S_{ft} \times D_0$$

Friction is assumed uniform within any cell, and is a non-directional weight, too. The friction term can be used to model any non-direction travel costs such as costs in traveling through dense vegetation. Denoting the friction of cell F as F_f, the cost distance between cell F and point B is

$$D_{fb} = 0.5 \times F_f \times S_{ft} \times D_0$$

The value gradient between two cells in a field, such as elevation and concentration, may lead to the cost of travel from cell F to cell T being different than the cost of travel from T to F. The value gradient factor V is a directional weight that is a function of the moving direction. After accounting for V_{ft}, the value gradient from cell F to cell T, the distance from cell F to point B becomes

$$D_{fb} = 0.5 \times V_{ft} \times F_f \times S_{ft} \times D_0$$

The horizontal direction factor H is used to account for the effect of a prevailing flow field on the path distance. The horizontal direction factor applied to the segment of the path in a cell is a function of the difference between the flow direction at the cell and the moving direction in the cell along the path segment, as shown in Fig. 2. The horizontal direction factor at cell F in the moving direction m can be determined by

$$H_{fm} = H(a_m - a_f),$$

where a_m is the angle of the moving direction, and a_f is the flow direction at cell F. The functional dependence of H on the relative angle may be represented either in terms of formulae or in using lookup tables. The resulting path distance from F to B becomes

$$D_{fb} = 0.5 \times H_{fm} \times V_{ft} \times F_f \times S_{ft} \times D_0$$

The path distance from cell F to cell T is the sum of the path distance from F to B and the path distance from B to T and is given by

$$D_{ft} = 0.5 \times (H_{fm} \times F_f + H_{tm} \times F_t) \times V_{ft} \times S_{ft} \times D_0$$

For diagonal neighbors, the only difference is that D_{ft} is $\sqrt{2}$ times larger. Note that H and F are functions of a particular cell, and V and S are functions of the differences in values or elevations of both cells. Note also that $S_t = S_{ft}$, but $V_{tf} \neq V_{ft}$. Fig. 3 shows some value-gradient functions and Fig. 4 show some horizontal direction functions.

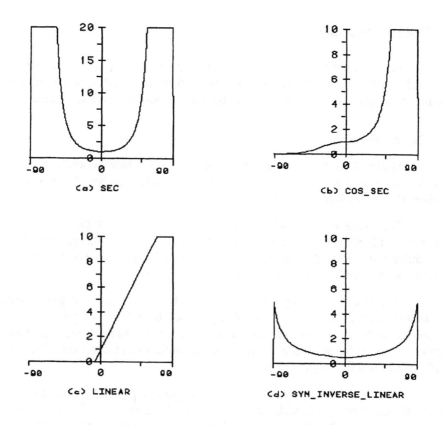

Fig. 3 Four Value-Gradient Functions

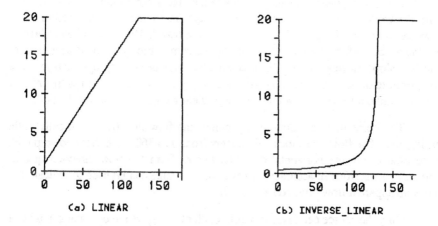

Fig. 4 Two Horizontal Direction Functions

3 Distance Mapping

To perform distance mapping using the above-defined path distance model, we need to define the necessary input cell layers, functions to calculate the different factors, the output cell layers and the algorithms to compute the shortest distance.

3.1 Input Cell Layers

The input cell layers to the path distance program include the source layer, the surface elevation layer, the friction cost layer, the value gradient layer and the horizontal direction layer.

The source layer provides the source objects, from which the shortest path distances to all cells in the working area are computed. The cells of source objects may have different codes to indicate different objects, and each object may have a set of cells that can be either connected or non-connected.

A surface elevation layer which is commonly available and storage-efficient, is used to calculate at each cell the zenith angles from the cell to its neighboring cells. To speed calculation, the program builds a lookup table with zenith angles as inputs and surface difference factors as outputs. The range of zenith angles are between 0^0 to 90^0 with a predefined fine angle interval.

The friction layer provides friction values per unit physical distance for each cell within the working area, with a unique value to indicate an impassable cell that acts as a barrier to any path. All non-directional friction cost variables derived from layers such as soils, vegetation cover and land use can be combined into a single friction layer.

The value gradient layer provides values for a continuous field at all cells within the working area. A non directional weight is computed as a function of the difference in values along a link in the direction of travel. The value gradient layer is the same as the surface layer when modeling non symmetric travel costs as a function of surface slope. In general the value gradient layer may be any appropriate layer, such as temperature or density, The value gradient for each basic unit is used as the input to the value-gradient function to get the value-gradient weight or factor for the unit.

The horizontal direction layer provides the flow direction at all cells in the working area. The flow directions are integers from 0 to 360, such that a lookup table can be used to get the horizontal direction factors from the flow direction quickly without much sacrifice in accuracy. A negative cell value can be used to indicate that there is no apparent flow direction at the cell.

Only the source cell layer is required, all other layers are optional and may be combined as needed to model a specific distance problem.

3.2 Value Gradient Functions and Horizontal Direction Functions

The value gradient function is used to derive the value gradient weight factor from the value gradient. The horizontal direction function is used to derive the horizontal direction factor from the relative horizontal angle. They are the key to producing directional weights, and need to be convenient and flexible for applications. In both cases a set of keyword accessible built-in mappings is provided for convenience. In addition, various function modifiers provide variations of the same mapping. Table inputs allow arbitrary user defined mapping functions.

A value-gradient function takes input gradient angles, which are the "zenith" angles from a cell to its neighboring cells in the value gradient layer, to produce value-gradient factors. The gradient angles ranges from -90^0 to 90^0, with angle 0^0 indicates no value difference between two cells. The built-in value-gradient functions include Binary, Linear, Inverse-Linear, Symmetric-Linear, Symmetric-Inverse-Linear, Cos, Sec, Cos_Sec and Sec_Cos, some of which are shown in Fig. 3. Function Binary outputs a finite value-gradient factor for an angle range starting from 0^0 and infinity for the rest of angles. Function Linear indicates that the output factor is a linear function of input angles with a specified slope and Y-intercept. Function Inverse-Linear is just the reciprocal of function Linear. Functions Symmetric-Linear and Symmetric-Inverse-Linear are similar to functions Linear and Inverse-Linear, respectively, but their output factors are symmetric to angle 0^0. Functions Cos and Sec indicate that the output factor is the cos and sec functions of input zenith angles, respectively. Function Cos_Sec and Cos_Sec use either cos or sec for their negative or positive zenith angles.

These functions may be modified by appropriate function modifiers. For instance, modifiers for the cut-angle set all output factors to be infinite beyond certain angles. This modifier can be used, for instance, to indicate that no road can be built beyond a slope. Modifier zero-angle-value sets the base output factor at gradient angle 0^0, and may be used to set the Y-intercept for the set of linear-type functions. Modifiers cos_power and sec_power apply a power to the trigonometric-type functions to make these function rise or drop sharply with respect to angle changes. A table of two columns, one as the input angle and the other for the value-gradient factor, can also be used as the value-gradient function.

The horizontal direction functions are quite similar to the value-gradient functions. They produce a horizontal direction factor from the relative angle between the horizontal flow direction and the moving direction along a basic unit or link. The horizontal direction functions are made symmetric to the flow direction as in practical application. Relative angles range from 0^0 to 180^0, with 181^0 to 360^0 converted to the angle range by subtracting 180^0.

The built-in horizontal direction functions include Binary, Forward, Linear and Inverse-Linear, some of which are shown in Fig. 4. Function Forward gives a value within (0^0 , 45^0), another value (normally higher) within (45^0, 90^0) and infinite beyond 90^0. Other functions and modifier are similar to that in the value-gradient functions.

3.3 Shortest Path Distance Algorithms

Due to the large number of nodes (cells) and links in the cell-based raster data the shortest path distance mapping requires extensive computation and a large memory to hold the entire data set in memory. The Dijkstra's algorithm [1959] sequentially processes each node whose path distance to the source is the smallest, until all nodes are processed. It needs less distance computation, but requires random access to the data set. When the entire data set can be held in memory, the Dijkstra's algorithm is quite efficient. Otherwise random disk input and output take much more computer time than distance computation and lead to an I/O bound implementation. The Bellman-Ford's algorithm [1958] iteratively propagates the path distance from each node to all nodes along its links and updates the distances of those nodes. It is parallel in nature and allows multiple sequential passes over the data set until the distances to all nodes does not change and convergence is reached. When the entire data set can not be held in memory, a block iterative approach can be taken: it is possible to stabilize the distance propagation in one small block of data at a time, and then to handle the distance propagation between blocks. Although it requires overhead in distance computation, it demands much less data input and output and is preferred for large data sets.

We combine these two algorithms in the implementation of the shortest path distance program using Dijkstra's algorithm to handle relatively small data sets, and the Bellman-Ford's algorithm to handle large data sets. Both the Dijkstra's algorithm and the Bellman-Ford's algorithm are designed for the single source shortest-path problem. In our distance mapping, the cells of outer boundaries of source objects form multiple sources, and the shortest path distance at a cell is simply the smallest of all the shortest path distances from all source cells to the cell.

3.4 Output Cell Layers

In addition to the distance layer, in which each cell stores the shortest path distance at the cell, two more cell layers, called back-link and allocation can be produced simultaneously without much overhead in computation. In the back-link layer each cell stores the link code which points to its predecessor along the shortest path from the source cell to itself. Due to the optimal substructure of a shortest path, the link code at each cell is unique, and a single layer with link codes 1 to 8 for all eight neighbors can be saved with the back links at source cells coded as 0. The back-link layer make it quite efficient to subsequently retrieve the shortest path to any cell, by following the back links starting from the specified cell until a source cell (code 0) is reached.

In the allocation layer each cell stores the code of the source object to which the cell is closest (the shortest path from the identified source object to the cell is smaller than the shortest paths from all other source objects to the cell). The allocation layer shows the how the study area is divided based on proximity into regions that belong to certain source objects, and is useful in regionalization based on distance.

4 Experimental Results

Two experiments were performed to show some applications of the directional path distance mapping. In the first experiment wind direction is used as the horizontal direction layer, the reciprocal of wind speed as the friction layer, and elevation as the surface layer, as shown in Fig 5. The horizontal direction function of type Linear with lower weights in the prevailing wind directions was used. Fig. 5 shows the resulting path distance layer with a source cell marked in white near the upper-right corner. Note that the isolines are extended from the source in the prevailing wind directions. The little tail of the dark low-distance region coincides well with wind directions, low friction and the valley in this area. The dispersion-like plume suggests that directional path distance mapping may be used to model some physical processes in dispersion.

In the second experiment an elevation layer around the San Jacinto mountain was used as both the surface layer and the value gradient layer in order to find the best routes from the peak of the mountain to various locations. The peak of the mountain was used as the source. A table with low weights for relatively flat and small down-slopes and high weights for deep slopes is used as the value-gradient function. Using the distance layer and back-link layer resulting from the distance mapping, the shortest paths from the peak to different locations were retrieved and were superimposed on the shaded relief layer, as shown in Fig. 6. Different value-gradient functions can be used for hikers with different capabilities. Using the origin of any travel (for instance, a base camp) as the source, the program can produce the recommended routes from the base camp to various locations in surrounding mountains. Land cover types (vegetation, snow cover and etc.) can be used as the friction layer to further improve performance.

5 Conclusion

This study has developed a directional path distance model that incorporates some frequently encountered types of directional weights. In the model, all weights that participate in the path distance calculation are incorporated as multiplicative factors,. This makes it possible to define and apply the factors involved in a complex precess separately and in a simple way. Directional factors can be entered into the program by using value-gradient functions and horizontal direction functions and function modifiers, which allow many physical phenomena to be modeled in proximity analysis. The input data set to the distance mapping is generally available or easy to produce, and can be combined flexibly for specific distance problems. The combination of the Dijkstra and the Bellman-Ford shortest distance algorithms makes the program work efficiently for both small and large data sets.

References

1. Bellman, R., "On a toutin problem," Quarterly of Applied Mathematics, 16(1), pp.87-90, 1958.

443

2. Borgefors, G., "Distance transformations in digital images," Computer vision, Graphics and Image Processing, Vol. 34, pp. 344-371, 1986.

3. Danielsson, P., "Euclidean distance mapping", Computer Graphics and Image Processing, Vol. 14, pp. 227-248, 1980.

4. Dijkstra, E. W., "A note on two problems in connect with graphs," Numerishe Mathematik, Vol. 1, pp. 269-271, 1959.

5. Eastman, J. R., "Pushbroom algorithm for calculating distances in raster grids," AUTO-CARTO 9, pp. 288-297, 1989.

6. Gatrell, A, Distance and space: A geographical perspective, Clearendon Press, 1983.

7. Menon, S., P. Gao, C. Zhan, "Grid: a data model and functional map algebra for raster geo-processing," Proceedings, GIS/LIS'91, Vol. 2, pp. 551-561, Atlanda Gorgia, 1991.

8. Smith, T. E., "Shortest-path distances: An axiomatic approach," Geographical Analysis, Vol. 21, No. 1, 1989.

9. Tomlin, D., The Map Analysis Package (MAP Manual), part of doctoral dissertation, 1980.

10. Tomlin, D., Geographic Information Systems and Cartographic Modeling, Prentice-Hall Inc., Englewood Cliffs, NJ 07632.

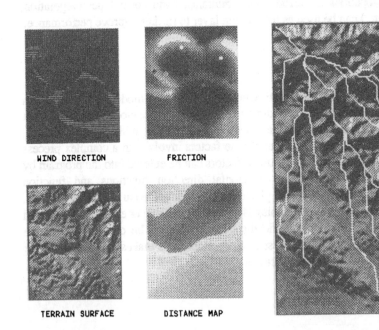

WIND DIRECTION   FRICTION

TERRAIN SURFACE   DISTANCE MAP

Fig. 5 Distance Mapping Using Wind Patterns and Eleveation Data

Fig. 6 The Shortest Paths from the Peak to a Set of Locations

Symbolic Spatial Reasoning on Object Shapes for Qualitative Matching

Erland Jungert

FOA (Swedish Defence Research Establishment)
Box 1165, S-581 11 Linköping, Sweden
telephone + 46 13 11 83 37
fax + 46 13 13 16 65
e-mail jungert@lin.foa.se

Abstract: Using qualitative methods for spatial reasoning about object shapes is of interest in, for instance, matching. Normally matching is based on methods that require massive computational resources since most such methods work on pixel level. The method described here is qualitative and can be used to investigate objects for convexities and concavities where the position of these concavities and convexities relative each other are also of interest. Furthermore, information about local extreme points will also become available as well as various types of sub-forms. The approach that will be discussed here is based on symbolic projections. In symbolic projections strings of objects and their relations are generated from projections of the objects down to the coordinate axes. This technique is developed further in this paper so that relational strings made up by sequences of angles are used in the matching process together with projections of slopes from parts of the object contours.

1 Introduction

Spatial reasoning is mainly used for determining various types of object relations in applications where otherwise methods requiring massive computational resources are the only alternative. For this reason spatial reasoning has become subject to intensive study in order to determine those object relations that use less heavy computational methods. Various approaches to spatial reasoning have been proposed of which some are qualitative. One such qualitative method employs symbolic projections, first suggested by Chang et al. [1]. Further work on this approach, for various types of applications, has been carried out, see e.g. Holmes and Jungert [2], who showed how symbolic projections can be used for path planning in digitized maps. Lee and Hsu [3] used symbolic projections for similarity retrieval. A theoretical base for the method has been presented by Jungert and Chang [4]. Lately, in an extension to symbolic projection, which is an important cornerstone of the work discussed here, Jungert [5] showed that symbolic projections also can be used for determining directions and distances. Other approaches to symbolic projections have their origin in work done by Allen [6] whose method was originally developed for reasoning about time. In this category of methods for spatial reasoning work by Freksa [7] can also be mentioned. Methods based on set theory are suggested by Egenhofer [8] and by Smith and Park [9]. Frank [10] discusses qualitative methods for determining distances and directions.

This list is by no means complete. The work is mentioned simply to illustrated some important approaches of special interest.

The work discussed below deals mainly with qualitative spatial reasoning on shapes of objects that will eventually be used for matching geographical objects that may originate from many different sources, such as digitized maps, satellite images or aerial photos. The method that is proposed here is based on symbolic projections [1] which is a qualitative technique. One main aspect of symbolic projections is that the objects and the interrelations between them are formalized as strings. This concept is developed further so that sequences or strings of angles and angle relations of type 'larger than', 'equal to' and 'smaller than' can be used. The latter is a complement to another type of projection strings that have their origin in slope projections, suggested by Jungert [5]. Slope projections play a fundamental role in reasoning about shape in that they support the determination of various characteristics of the objects. This includes determining whether the objects are concave or convex at certain points. Other features of the objects that can easily be determined are the local extreme points of the objects which here are denoted as north, south, west and east. It is also possible to estimate the relative angle sizes of the points, i.e., whether they are obtuse, acute or right-angled. Furthermore, curvatures of primitive sub-forms of the objects can be determined as well.

Earlier matching methods have mainly been quantitative and based on methods that require massive computational power. For this reason methods based on qualitative spatial reasoning may prove useful as well. Among the conventional methods on matching the work by Borgefors [11] can be mentioned. In her work distance transforms play an important role in the matching process and the method is very robust. Another method that is partly related to this work is reported in Terano et al. [12]. Their method is based on Fuzzy sets and allows matching of simple objects such as fruits and tools belonging to a certain type such as a knife. Whether the qualitative matching method that will be discussed here can become robust or not is not yet clear. However, so far no real qualitative methods for reasoning about shape is known to exist. For determining, for instance, concavities only numerical methods have been used. Generally, there exist methods for determining object features, but they are mainly quantitative.

2 Basic assumptions and symbolic slope projections

The main objective in this work has been to develop qualitative methods for determining various features of geographical objects that can be used for the analysis of object shapes and to form a basis for qualitative object matching.

The basic assumptions are concerned with how the objects and their various features are interpreted:

1. The start of the reasoning process is always in the uppermost (left) corner of the object.

2. The objects are always traversed counter clockwise, i.e. the inside of the objects is always on the left side of the path.
3. The angle of a convex point is always on the inside of the object, see figures 1 a and b.
4. The angle of a concave point is always on the outside of the object, see figures 1 c and d.
5. The angles are either right-angled, acute (figures 1 a and c) or obtuse (figures 1 b and d).

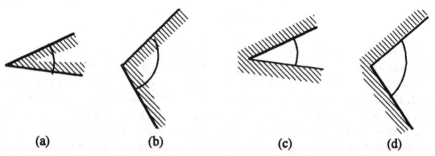

(a)   (b)   (c)   (d)

Fig. 1. The positions of the angles in convex points, a and b, and concave points, c and d. the angles in a and c are acute while they are obtuse in b and d.

In the original approach to symbolic projections, the projections were perpendicular to the coordinate axes. However, it turned out that this projection technique could not be used for cases where points and lines were used and where it is necessary to determine whether a point is to the right or left of a line that is not perpendicular to the coordinate axes. For this reason Jungert [5] introduced the slope projections to overcome this problem. It turns out that this extension of symbolic projections is quite useful in reasoning about object shapes. The basics of the symbolic slope projection method is demonstrated in figure 2 which shows that the point $p(x_3,y_3)$ is to the right of the line segment 'l' that starts in $p(x_1,y_1)$ and ends in $p(x_2,y_2)$.

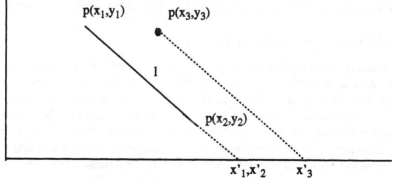

Fig. 2. The basic idea of symbolic slope projections.

The corresponding projection string that can be generated from figure 2 thus becomes:

$$U_s(l): x'_1 = x'_2 < x'_3$$

where

$$x'_1 = x_1 - y_1/k$$

and k is the slope coefficient for the line segment. Similarly x'_2 and x'_3 become:

$$x'_2 = x_2 - y_2/k$$
$$x'_3 = x_3 - y_3/k$$

The label U in front of the string indicates that the objects are projected down to the x-axis. U has the subscript 's' indicating that the projection is of symbolic slope type. The direction of the slope projection here is the line 'l' which is found in parentheses after U. Henceforth the latter information is left out because the slope directions are evident. It is now a simple matter to see that since x'_1 and x'_2 are less than x'_3, the latter must be situated to the right of the line segment. For further explanations of the slope projection method see [5]. Normally, the '='-signs are left out from the projection strings.

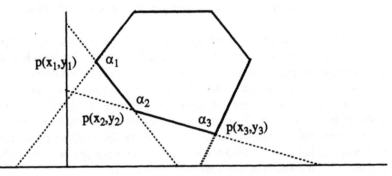

Fig. 3. The principles of slope projections applied to qualitative reasoning on shape.

3 Determining symbolic shapes

Basically reasoning about shape to determine what are here called symbolic shapes can be done by projecting all object sides down both to the x- and y-axes, as illustrated in figure 3. The purpose of this type of projections is to determine a qualitative measure of the angles of each particular vertex. Another feature that must be determined for all vertices is whether they are extreme points or whether they are convex or concave. This is performed by projecting the two lines connected to an object point to both coordinate axes in accordance with the rules that will be discussed below. Finally, when the type of all vertices have been determined it will be possible to tell whether the object is convex or not, i.e. if the object contains any concavities and where they are situated. This also includes the types of the angles and whether they are extreme points or not. However, as will be demonstrated later, these features do not suffice for matching. For this further information is necessary, which will be discussed subsequently.

Reasoning about the vertices to determine their features is based on the incoming and outgoing lines of each vertex. An illustration of the basic principles of this projection technique is found in figure 4 where the angle between the two lines is α_{j+1}. The incoming line is $p_j\text{-}p_{j+1}$ and the outgoing is $p_j\text{-}p_{j+2}$. Projection is made down to the x-axis along or parallel to $p_j\text{-}p_{j+1}$ and to the y-axis perpendicular to that line. The following projections can then be identified from figure 4:

$$U: x'_j\, x'_{j+1} < x'_{j+2}$$
$$V: y'_{j+2} < y'_{j+1} < y'_j$$

where x'_j corresponds to

$$x'_j = x_j - y_j/k$$

and in the perpendicular projection to the y-axis

$$y'_j = y_j + kx_j$$

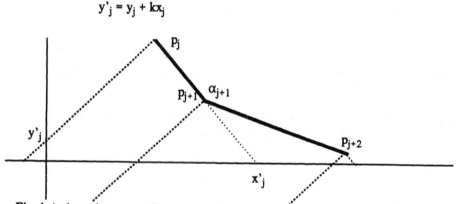

Fig. 4. An incoming contour line $p_j\text{-}p_{j+1}$ and an outgoing $p_{j+1}\text{-}p_{j+2}$ and their projection lines

However, since it is simpler to talk about p_j instead of x'_j and y'_j, the former will subsequently be used in the projection strings. This will not cause any problem since the context is clear. Observe also that the perpendicular projection to the y-axis uses the same slope coefficient. Hence, the U and V strings become:

$$U: p_j\, p_{j+1} < p_{j+2}$$
$$V: p_{j+2} < p_{j+1} < p_j$$

Now the angle α_{j+1} for the point p_{j+1} can be determined from the following reasoning technique. Since, the inside of the object is to the right when the contour is traversed from p_j through p_{j+2}, a left turn at p_{j+1} means that the object is convex. Furthermore, a right turn means a concave point, while a straight line cannot occur. Consequently, since p_j and p_{j+1} are less then p_{j+2} this can be interpreted as p_{j+1} is convex. To determine whether p_{j+1} is acute, obtuse or a right angle we must examine the V string. If

p_{j+1} and p_{j+2} are equal, a right angle can be inferred. If p_{j+2} lies to the right of p_{j+1}, that is $p_{j+1} < p_{j+2}$, then the angle is obtuse; otherwise acute.

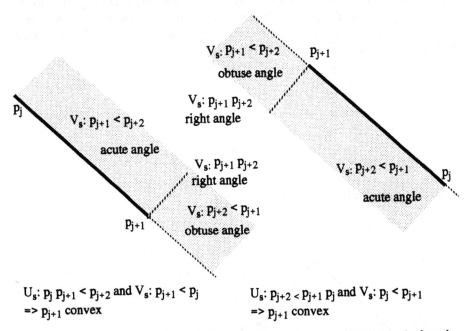

U_s: p_j $p_{j+1} < p_{j+2}$ and V_s: $p_{j+1} < p_j$
=> p_{j+1} convex

U_s: $p_{j+2} < p_{j+1}$ p_j and V_s: $p_j < p_{j+1}$
=> p_{j+1} convex

Fig. 5. Possible out-going directions (shaded areas) for two in-coming lines (p_j, p_{j+1}) where the point p_{j+1} is convex. Slope projection direction is p_j-p_{j+1} towards the x-axis.

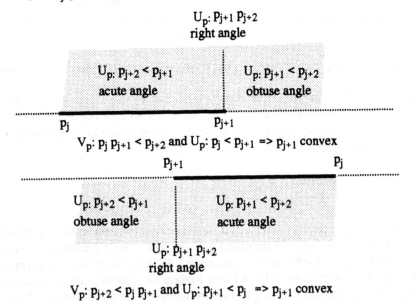

Fig. 6. Possible outgoing directions for incoming horizontal lines (p_j, p_{j+1}) where p_{j+1} is convex.

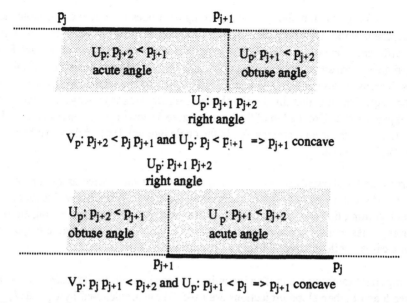

Fig. 7. Possible outgoing directions for horizontal, incoming lines (p_j, p_{j+1}) where p_{j+1} is concave.

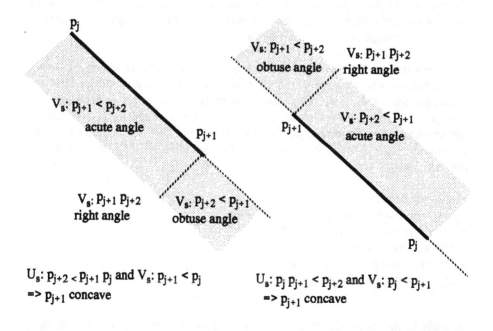

Fig. 8. Possible outgoing directions for incoming lines (p_j, p_{j+1}) where p_{j+1} is concave. Slope projection direction is $p_j\text{-}p_{j+1}$ on the x-axis.

The inferred information about point p_{j+1} in figure 4 does not cover all possible cases. Hence, the method must be generalized. By looking at the all possible incoming lines eight different orientations are possible, which are illustrated in figures 5 through 8. The shaded areas correspond to possible outgoing lines. The shaded areas can be split into two sub-areas, one with an acute angle at the point and the other corresponding to an obtuse angle. The line that divides the shaded area corresponds always to a right angle. The object is situated to the left of the lines in figure 5 showing two convex cases, where the one to the left has an incoming line in the angle interval $[0,\pi]$ while the right one lies in the $[\pi,2\pi]$ interval.

Figures 6 and 7 show horizontal incoming lines where the shaded areas can be interpreted in the same way as in figure 5. The only difference to be observed is that a horizontal incoming line is a special case due to the fact that no slope projection, along the incoming lines, can be made down to the x-axis. The two cases in figure 6 show two convex cases, while the two in figure 7 are concave.

It is important to observe that if an incoming line corresponds to the cases illustrated in figures 5 and 8, then slope projections are used. These are denoted by U_s and V_s. The horizontal cases on the other hand shows alternatives where perpendicular projections are used, which are denoted by U_p and V_p. The rules given in the figures do not in any of the demonstrated cases require any exact calculations of the angles; they are just estimated in qualitative terms. Those cases where the angles are right angles are exceptions, but the correct angle in those cases is just a side-effect.

Extreme points can be determined in two ways, that is either by identifying those projection directions in the slope projection strings for which the projections correspond to crossing lines, a flip over, or by using perpendicular projections, as done here. The four types of extreme points in each group are called east, north, west and south. The east and west points are determined by examining the projections on the x-axis, while the two others are obtained from the y-axis. The definitions of the four cases are presented below:

north/south extreme points:
$V_p{:}p_{j+1} < p_j$ and $p_{j+1} < p_{j+2}$ => south
$Vp{:}$ $p_j < p_{j+1}$ and $p_{j+2} < p_{j+1}$ => north
east/west extreme points:
$U_p{:}$ $p_j < p_{j+1}$ and $p_{j+2} < p_{j+1}$ => east
$U_p{:}$ $p_{j+1} < p_j$ and $p_{j+1} < p_{j+1}$ => west

By applying the above rules to all points in any geographical object in a map or image it is now possible to create a list of three different features which provides a qualitative description of the objects as well. On an overall level it is possible to tell whether the objects are convex or not. It is also possible to tell the position and orientation of concavities including their extreme points. In some cases it will also be possible to identify the actual shape of an object.

A further observation of interest is that for sequences of convexities the U-projections are increasing, that is,

$$U_s: \dots p_i < p_{i+1} < p_{i+2} \dots$$

This goes on until a north or south extreme point is reached, for whereupon a flip over occurs. If the convexity continues after the extreme point, the projections start to increase again after the flip over. The same occurs for sequences of concave points but instead of increasing projections they decrease. Hence, it is possible to determine the type of the points by just projecting the points. The drawback is, however, that no further information about the angles will be available unless the full analysis of the projections is applied to each point of the object.

4 Classification of characteristic points

A number of characteristic points can be identified with the technique that was demonstrated above. Some of these are independent of their orientation, while others can just be considered characteristic in certain orientations. To the former class belong all vertices with a right angle that may be either convex or concave. Two sub-classes exist which depend on the orientation of the objects. One that has incoming and outgoing lines that are parallel to the coordinate axes and a second one that has random orientation. These cases are clearly trivial and do not occur very often but have to be considered nevertheless.

A second class of characteristic points are all vertices that have an angle that is acute, because they can always be considered as extreme points no matter which orientation they have or whether they are convex or concave.

A third class of characteristic points which are rotation dependent are those that have either a line that enters the point either horizontally or vertically. The instances of this class can be found in the table below.

Rotation dependent classes	type	angle	enter	exit
Horizontal entering lines:	convex	obtuse	west,	north to east
		acute	west	north to west
		obtuse	east	south to west
		acute	east	south to east
	concave	obtuse	west	south to east
		acute	west	south to west
		obtuse	east	north to west
		acute	east	north to east
Vertical entering lines:	convex	obtuse	north	south to east
		acute	north	north to east
		obtuse	south	north to west

	acute	south	south to west
concave	obtuse	north	south to west
	acute	north	north to west
	obtuse	south	north to east
	acute	south	south to east

The information in the above table can be extracted directly from the projections used in section 3.

5 Sub-forms

The existence of both concavities and convexities can be determined according to the reasoning technique that was described above. When the shape of an object is determined, both concavities and convexities are important since they can be looked upon as sub-objects that may have characteristics of their own. This section is concerned with the identification of such sub-forms.

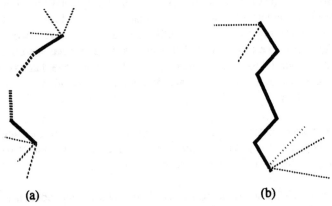

(a)   (b)

Fig. 9. Elementary sub-forms of convex (concave) (a), and zig-zag type (b).

There exist just two basic sub-contour types of which the first one is a curve that can be either convex or concave and the second is a zig-zag contour. These two types, which are illustrated in figure 9, can, depending on their start and end-points, be described as:

- Convex sub-forms that may have start and end-points of concave type that have either right, acute or obtuse angles.

- Concave sub-forms that may have start and end-points of convex type that have either right, acute or obtuse angles.

- Zig-zag sub-forms that have a start and an end with two points of equal type that are either convex or concave and with arbitrary angles.

The simple sub-convexities and concavities can be combined, thus creating a class of sub-forms with twelve different instances. The simple contour parts can have a point that merges two parts such that the angle of that point is either obtuse or acute and in some cases also a right angle as well. The point is called a passage that is either of A, O or R type depending on whether the angle is acute, obtuse or a right angle. Hence, if the R-passage is excluded, the most common eight instances are:

- convex, convex with O-passage
- convex, convex with A-passage
- concave, concave with O-passage
- concave, concave with A-passage
- concave, convex with O-passage
- concave, convex with A-passage
- convex, concave with O-passage
- convex, concave with A-passage

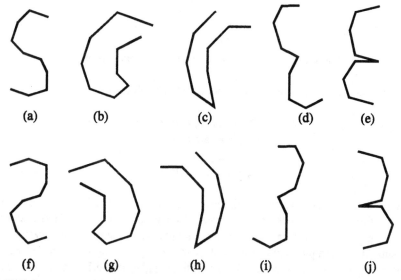

(a)   (b)   (c)   (d)   (e)

(f)   (g)   (h)   (i)   (j)

Fig. 10. Sub-contours of type convex to concave with O-passage (a) and (b), convex to concave with A-passage (c), convex to convex with O-passage (d), convex to convex with A-passage (e), concave to convex with O-passage (f) and (g), concave to convex with A-passage (h), concave to concave with O-passage (i) and concave to concave with A-passage (j).

Some of these elementary sub-form combinations will sometimes result in shapes that are well-known, for instance, convex to convex with an O-passage will in many cases have the shape of an S. Furthermore, the convex to convex with an A-passage is similar to a hand-written E. Of course, the orientation of these shapes may vary and sometimes the shape of these combinations degenerates. For instance, the S shape can be folded at its passage point and thus it may look more like a U but in those cases it will also have an extreme point. The instances of the above class of combined elementary sub-forms are illustrated in figure 10.

These sub-form combinations can, of course, be combined further in any arbitrary way. For instance, it is possible to talk about convex to convex with A-passage followed by a single concavity with an O-passage, followed by a zig-zag contour with an A-passage etc. From this it will be possible to identify sequences of patterns describing the full contour of the object in a qualitative way.

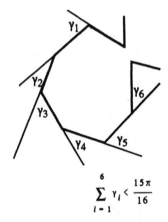

$$\sum_{i-1}^{6} \gamma_i < \frac{15\pi}{16}$$

Fig. 11. An illustration to the curve of a convex sub-form.

Zig-zag forms can be classified with respect to their angles, e.g.:

- all angles are obtuse
- all angles are acute
- mixed angles
- right angles are included

A final problem that is concerned with the sub-forms is the curve-length of concavities or convexities, i.e. how much do they curve? The answer to this question is that we do not want an answer that is exact but rather a qualitative measure of the curvature, that is the sum of the complement angles at all points of the sub-form. In other words, the sum of $\pi - \alpha_i = \gamma_i$ over all i in the sub-form. These angles are fairly simple to calculate as will demonstrated in section 7. However, since it is a qualitative measure of the curvature of the sub-form that is of main interest, it is angular intervals that provide the answer. Hence, angles of size $\pi/16$ can be used as equidistant steps in these intervals, that is, if the curvature is less than $3\pi/16$, this implies that is larger than $2\pi/16$. This is illustrated in figure 11.

6 Illustrations

To illustrate the method two examples are given, see figure 12. The first object is entirely convex while the second one contains three concavities.

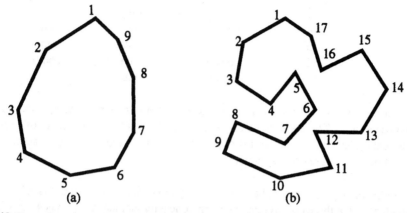

Fig. 12. Examples of two objects that are subject to qualitative shape analysis where (a) is convex and (b) has three concavities.

The traversal of an object produces a table of the type that can be seen below for each of the given objects

Object a.

description:	k	type	angle	extreme point type
	1	convex	obtuse	north
	2	convex	obtuse	
	3	convex	obtuse	west
	4	convex	obtuse	
	5	convex	obtuse	south
	6	convex	obtuse	
	7	convex	obtuse	east
	8	convex	obtuse	
	9	convex	obtuse	

Object b.

description:	k	type	angle	extreme point type
	1	convex	obtuse	north
	2	convex	obtuse	
	3	convex	obtuse	west
	4	convex	obtuse	south
	5	concave	acute	north
	6	concave	obtuse	east
	7	concave	obtuse	south
	8	convex	right	north
	9	convex	right	west
	10	convex	obtuse	south
	11	convex	acute	east
	12	concave	acute	west

13	convex	obtuse	
14	convex	obtuse	east
15	convex	obtuse	north
16	concave	acute	south
17	convex	obtuse	

Object a is clearly convex since all its vertices are convex. Furthermore, all its angles are obtuse as well. There are four extreme points, which is the least number of extreme points that can occur. In a sense this object can be considered as being round or close to round but it is not entirely symmetric.

The second example, object b, has both convex and concave point and it is fairly simple to identify the three concavities. The first concavity includes the points 4 through 8. The start and end points of a concavity are convex but must be included as well since the lines between the last convex point and the first concave point and the last concave point and the first convex point are part of the concavity. The two other concavities are identified in the same way. Of course, the analysis of the object can be carried out the other way around, i.e. the convexities can be of concern instead, that is points 16 through 5 and so on. A convex to concave sub-form with an A-passage can be identified in the interval 16 through 8. The elementary sub-form 16 through 5, which is convex, has a curvature less than $25\pi/16$.

7 Qualitative matching

The reasoning technique can be extended further so that it can be used for matching as well. To do this we need calculate angles or rather relative angles, thus creating strings very much like projection strings. In these strings three operators will be used, i.e. $\{<, =, >\}$. The angle-strings, hence, differ from the original projection strings in that they also include the "larger than"-operator. This is obviously necessary since an arbitrary angle can be followed by a larger angle. To generate these angle strings we need to identify methods for determining angles or at least relative angles.

An angle between an entering and an exiting line is easily determined. Two cases exist that can be applied both to convex and concave points in the polygon. This is illustrated in figure 13.

The angle α_j is associated with a point p_j of the polygon while β_j is the slope projection angle of the incoming line to point j. North/south extreme points are exceptions, because the projection directions of the outgoing slope projection lines change at these points, in other words, there is a flip over. The point angles will be expressed in terms of slope angles. The method is based on simple geometry and does not require any extra explanation. Below the angles for the existing cases are given. Since it is the relative angles that are of interest we need just consider the difference $\alpha_j - \alpha_{j+1}$. However, the expression can be simplified and for that reason ξ_j has been introduced and it is easy to see that for non-extreme points $\alpha_j - \alpha_{j+1} = \xi_j - \xi_{j+1}$.

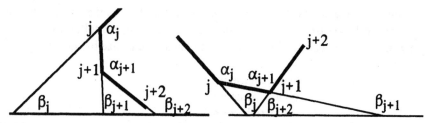

Fig. 13. The projection and contour angles of a convex object corresponding to non-extreme points (a), and a north/south extreme point (b).

It is then a simple task to show the relationships between α_j and α_{j+1} for various types of vertices, as in the expressions (6) through (8) below.

α_j, α_{i+1} convex, non-extreme points
(1) $\alpha_j = \beta_i - \beta_{i+1} + \pi$
(2) $\alpha_{j+1} = \beta_{i+1} - \beta_{i+2} + \pi$
(3) $\xi_j = \beta_j - \beta_{j+1}$
(4) $\xi_{j+1} = \beta_{j+1} - \beta_{j+2}$
(5) $\alpha_j - \alpha_{j+1} = \beta_i - \beta_{i+1} - (\beta_{i+1} - \beta_{i+2}) = \xi_j - \xi_{j+1}$
(6) $\xi_j - \xi_{j+1} = 0 \Rightarrow \alpha_j = \alpha_{j+1}$
(7) $\xi_j - \xi_{j+1} > 0 \Rightarrow \alpha_j > \alpha_{j+1}$
(8) $\xi_j - \xi_{j+1} < 0 \Rightarrow \alpha_j < \alpha_{j+1}$

α_j convex, non-extreme point, α_{j+1} convex, north/south extreme point
(1) $\alpha_j = \beta_i - \beta_{i+1} + \pi$
(2) $\alpha_{j+1} = \beta_{i+1} - \beta_{i+2}$
(3) $\xi_j = \beta_j - \beta_{j+1}$
(4) $\xi_{j+1} = \beta_{j+1} - \beta_{j+2}$
(5) $\alpha_j - \alpha_{j+1} = \beta_i - \beta_{i+1} + \pi - (\beta_{i+1} - \beta_{i+2}) = \xi_j - \xi_{j+1} + \pi$
(6) $\xi_j - \xi_{j+1} + \pi = 0 \Rightarrow \alpha_j = \alpha_{j+1}$
(7) $\xi_j - \xi_{j+1} + \pi > 0 \Rightarrow \alpha_j > \alpha_{j+1}$
(8) $\xi_j - \xi_{j+1} + \pi < 0 \Rightarrow \alpha_j < \alpha_{j+1}$

We do not intend to cover all combinations here since it is quite simple to determine the values of all combinations of $\xi_j - \xi_{j+1}$ from the observation that for any non-extreme point

$$\alpha_{j+1} = \beta_{i+1} - \beta_{i+2} + \pi$$

and for any north/south extreme point

$$\alpha_{j+1} = \beta_{i+1} - \beta_{i+2}$$

independently of whether they are convex or concave.

There are also some shortcuts that will occur here and there in the object. The short-cuts occur for those cases where, for instance, an obtuse angle is followed by an acute angle or the reverse. All such short-cuts can be found in the table below:

$$
\begin{aligned}
&\text{acute, obtuse} => < \\
&\text{obtuse, acute} => > \\
&\text{acute, right} => < \\
&\text{obtuse, right} => > \\
&\text{right, acute} => > \\
&\text{right, obtuse} => < \\
&\text{right, right} => =
\end{aligned}
$$

These rules work for all combinations of convex and concave angles. The following two remaining cases cannot, clearly, be defined by using the above short-cuts; instead the angles must be calculated as described in the beginning of this section:

$$
\begin{aligned}
&\text{acute, acute} \\
&\text{obtuse, obtuse}
\end{aligned}
$$

By applying the methods for determining qualitative relations between angles, as discussed in section 6 to figure 12a, the following angular string can be created:

$$\alpha_1 < \alpha_2 > \alpha_3 > \alpha_4 < \alpha_5 > \alpha_6 < \alpha_7 < \alpha_8 \; \alpha_9 >$$

Since the angles themselves are of less interest, they can be ignored. Hence, just the relations are left such that equal relations in sequence may be put together and replaced with a coefficient corresponding to the number of equal relations, i.e.:

$$1 <, 2 >, 1 <, 1 >, 2 <, 1 =, 1 >$$

This simplification can be taken even further; in other words, the operators can be ignored leaving just the coefficients:

$$-1, 2, -1, 1, -2, 0, 1$$

This gives a sequence of figures where a negative integer corresponds to the number of <-relations in the sequence while a positive integer corresponds to the number of >-relations. =-relations are 0 but the problem is that an =-sign cannot be dropped easily. The solution to this is to replace $2 =$ with 0,0. This will not cause too much trouble since the probability that there are three sequential angles that are equal is very small and it is less likely that $3 =$ exist, at least for geographical objects. We believe that this way of handling the =-relations will not cause any major difficulties.

The advantage of this sequence of figures is obvious and is especially useful in matching. Matching can simply be accomplished by comparing two sequences to see if the number sequences are equal. This is trivial if we have two objects with the same orientation given in the same resolution. However, if the objects that should be matched with the map object are somewhat rotated, this matching is not so simple. This problem is illustrated in figure 14 where 14a is considered to be the original or model object to which other objects should be matched and where these objects could be rotated to some degree. (Observe, that the object in figure 14 is equal to the object in figure 12.) Figures 14b through d demonstrates this. From figure 14b we can now generate the following sequence:

-1,0,1,-1, 2, -1, 1, -1

In this case matching does not fail just because the first elements are not equal since a fairly simple algorithm can be designed to deal with this. In this algorithm the first element in the second sequence will be matched with the first element of the model sequence which are equal. After this the second element is matched with the next element in the model's sequence and so on until there is a failure or a correct match of all elements in the sequence. This means that the model sequence might end first but then the next element is the first in the sequence. In case of failure the process starts with the first element again and matching starts at the point right after the last unsuccessful matching in the model sequence. This goes on until there is a hit or until the first element has reached the last element of the model sequence.

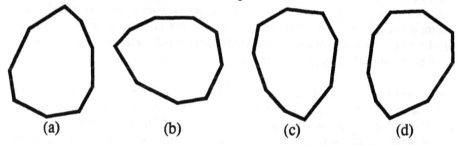

Fig. 14. (a) the model object and (b), (c) and (d) the same object with different orientation.

An interesting observation here is that all the illustrations in figure 14 have different orientations but c and d have nevertheless equal sequences since they both have the same start point.

8 Discussion and conclusions

A crucial problem that has not been thoroughly discussed has to do with the robustness of the matching method. Matching map-objects with objects in arbitrary orientation and scale, coming from various sources, such as sensors, must be permitted. So far, it has been assumed that the information from such sources is stable and without any distortions. Furthermore it has also been assumed that this information has been properly processed and represented in the same data structure as the original map data objects.

It is common that sensor information is not without distortions and for this reason this problem must be studied further. Another problem that must be considered is that the map objects may have been generalized so that their resolution is not sufficient for the matching process. A third problem, which can be harder to deal with, is that reality may have changed since the map was created. These problems, clearly, require further research. However, if the techniques described above are used as an indexing method, then it is simple to prune off such solution candidates that differ too much from the target object. However, a main goal is to improve the method so that it becomes less sensitive to distortions in the objects and hence more robust, Whether this is possible is at present too early to say.

Two further observations, concerning qualitative matching, which demonstrate the complexity of this problem can be made:

- there are infinitely many objects corresponding to a qualitative object description,

- a single object may have a large number of different qualitative descriptions, especially if distortions and different orientations are allowed to occur.

There are also a few other interesting object types to which the method may be applied that, however, may have less bearing on geographical objects but nevertheless may be of interest. Symmetrical geometric objects such as rectangles or octagons where all the angles are equal belong to this category. However, it cannot be determined whether a rectangle has sides which are all equal in length. We can just tell whether the angles are equal with the methods presented here. On the other hand, there are simple mathematical methods for calculating the length of the object sides.

The power of the method at this stage is that it provides a description of the objects and their characteristics that is both independent of scale and invariant against the orientation of the object. A possible usage of the method would be to integrate it in a query system where the goal is to give the user a set of solutions corresponding to, e.g., drawings given by the users. In such a query system the lack of robustness will clearly play a minor role since it will be the user who makes the final selection.

In this work we have not been concerned with objects with holes. However, a hole is just another object about which we can reason in the same way as any other type of object. This means that if a hole is present, the method discussed here can be applied in the same way as it can be applied to the external contour of that object.

A question that has not been dealt with and which is quite important is whether two concavities with an intermediate convexity are all part of the same single concavity or whether they are two separate concavities with a convexity in between. This is really an important problem that needs further study. However, it seems quite clear that more information, including the proportions between the various sub-forms, must be made available during the reasoning process.

What here has been called a method for determination of object shapes or alternatively a technique for generalization of qualitative shape descriptions is clearly not just a single method but instead a set of methods applied to the objects to provide a general characterization of the objects.

References

1. S.-K. Chang, O. Y. Shi and C. W. Yan, Iconic indexing by 2D strings, IEEE transactions on Pattern Analysis and Machine Intelligence (PAMI) 9 (3), 413-428 (1987)

2. P. D. Holmes and E. Jungert, Symbolic and Geometric Connectivity Graph Methods for Route Planning in Digitized Maps, IEEE Transaction on Pattern Analysis and Machine Intelligence (PAMI) 14 (5), 549-565 (1992)

3. S. Y. Lee and F. J. Hsu, Picture Algebra for Spatial Reasoning of Iconic images represented in 2D C-string, Pattern Recognition Letters 12, 425-435 (1991)

4. E. Jungert and S.-K. Chang, An Image Algebra for Pictorial Data Manipulation, To appear in Computer Vision, Graphics and Image Processing (CVGIP): Image Understanding.

5. E. Jungert, The Observer's Point of View: An Extension of Symbolic Projections, In: A. U. Frank, I. Campari and U. Formentini (eds.) Theories and Methods of Spatio-Temporal Reasoning in Geographic space, Springer 1992 pp 179-195

6. J. F. Allen, Maintaining knowledge about Temporal Intervals, Communications of the ACM, Vol. 26 (11), 832-843 (1983)

7. C. Freksa, Using Orientation Information for Qualitative Spatial Reasoning, In: A. U. Frank, I. Campari and U. Formentini (eds.) Theories and Methods of Spatio-Temporal Reasoning in Geographic space, Springer 1992 pp 162-178

8. M. J. Egenhofer and K. Al-Taha, Reasoning about Gradual Changes of Topological Relationships, In: A. U. Frank, I. Campari and U. Formentini (eds.) Theories and Methods of Spatio-Temporal Reasoning in Geographic space, Springer 1992, pp 196-219

9. T. Smith. and K. K. Park, Algebraic Approach to Spatial Reasoning, International Journal of Geographical Information Systems 6 (3), 177-192 (1992)

10. A. U. Frank, Qualitative Spatial Reasoning about Distances and Directions in Geographical Space, Journal of Visual Languages and Computing 3 (4), 343-371 (1992)

11. G. Borgefors, Hierarchical Chamfer Matching: A Parametric Edge Matching Algorithm, IEEE Transaction on Pattern Analysis and Machine Intelligence (PAMI)10 (6), 849-865 (1988)

12. T. Terano, S. Masul, S. Kono and K. Yamamoto, Recognition of crops by Fuzzy Logics, Preprints of 2nd IFSA Congress, Tokyo, 20-25 July, 474-477 (1987)

Reasoning About Spatial Structure in Landscapes with Geographic Information Systems

Claude W. Saunders

Argonne National Laboratory
Argonne, IL 60439

Abstract. Geographic Information Systems offer a new means to store, transform, and present cartographic information. Persons involved in landscape ecology, planning, and management use these systems to transform and examine data pertaining to a given geographical region. Unfortunately, there is often little correspondence between the language of landscape analysis and the operations provided by most Geographic Information Systems. This paper describes a language based on the landscape element abstractions commonly used by landscape architects and ecologists. With this language, a task specific landscape definition may be expressed in terms of matrix, patch, and corridor elements. The landscape definition is then used to guide the interpretation of cartographic data. This paper presents an extension to an existing Geographic Information System which provides the ability to locate instances of specified spatial structures in cartographic data. This extension represents an initial step towards being able to reason directly about the spatial relations amongst elements of interest in a landscape. The definition of a landscape analysis task is given in the landscape language and solutions generated by our system are presented.

1 Introduction

The wealth of information contained in cartographic data poses many challenges for those who develop methodologies and computational aids for its storage, analysis, and presentation. The advent of Geographic Information Systems (GIS) has provided a basic architecture for dealing with cartographic data. Persons involved in landscape ecology, planning, and management use these systems to transform and examine data pertaining to a given geographical region.

Unfortunately, there is often little correspondence between the language of landscape analysis and the operations provided by most GIS. Users must now map their problems onto operations provided by the GIS. For example, to create a map overlay which shows the proximity of agricultural regions to rivers, one might use the following sequence of GIS data transformation operations: "Isolate Landuse-map assign 1 to agricultural for temp-map1. Isolate Waterbodies-map assign 1 to rivers for temp-map2. Add temp-map1 to temp-map2 for Resultant-map." This indirect, procedural method of analysis has prevented the use of GIS in some problem domains.

This paper investigates the role of landscape structure concepts in facilitating GIS interaction and analysis. Central to the paper is the notion that analysis of data should be carried out in the language domain of landscape structure, not in the language domain of GIS architecture. One assumption underlying this approach is that the spatial structure of a landscape is a sufficiently powerful organizing principle on which to base analysis tasks. This paper describes a mapping from low level cartographic data to the higher level language elements of landscape structure. It presents an extension to an existing GIS (macGIS) which provides the ability to locate instances of specified spatial structures in a landscape. The extension is called LRM (Landscape Reasoning Module).

This paper, in proposing and implementing a mapping from low-level GIS data to high-level primitives of a landscape language, hopes to facilitate many analysis tasks. Analysis should no longer require mapping a task onto a sequence of system specific operations. The landscape language allows a more direct statement of the problem through a description of desired landscape configurations. This is the notion of declarative versus procedural interaction. One declares the problem and allows some interpreter to evaluate it and generate a procedural solution. In short, the interaction with a GIS can be in terms of elements of the problem domain, not primitive GIS operations.

2 Landscape Structure

The notion of a landscape has several different definitions depending on the field of study and the scale being considered. This paper considers the landscape as that defined by the fields of Landscape Architecture and Landscape Ecology. Here a landscape is a heterogeneous land area of the scale typically seen in aerial photography.

Landscapes can be examined in terms of structure, function, and dynamics [6]. This paper restricts itself to landscape structure, although function and dynamics do remain a consideration. The structure of a landscape involves the spatial relationships between its basic elements. Spatial characteristics, eg. size and shape, of the individual elements are also considered. Landscape function involves the causal relationships between and within elements while landscape dynamics examines the evolution of structure and function over time.

Researchers in landscape ecology and landscape architecture have developed a language in which one can describe landscape structure. The language is tailored to describing surface ecosystems at the landscape level. The primitives of the language are based primarily on the spatial structure. Three primary elements of a landscape are identified: the matrix, patch, and corridor.

The matrix is the dominant element of a landscape. Needless to say, the definition of dominant is highly context sensitive. Three characteristics are used by Forman to identify dominance. The most obvious characteristic is total area. The matrix of a landscape is generally the type of land occupying the largest percentage of area in a region. The next most important characteristic is connectivity. The matrix usually exhibits high connectivity. While this is not an absolute requirement, one can see that a matrix with very low connectivity is perhaps better described as a collection of patches. The third characteristic is degree of control over landscape dynamics. The matrix is the land type which dominates the flow of materials and energy through a landscape. It generally provides the ability of a landscape to recover equilibrium after perturbation.

Patches of varying types are scattered within the matrix. These patches can be "natural", such as lakes and tracts of forest, or can be "introduced", such as farm fields and landfills. Corridors are linear elements of a landscape such as rivers, hedgerows, and roads. Each of these element types has defining characteristics. While there are numerous variants of these element types, such as the network, ring, and peninsula, this paper will restrict itself to the three basic types. These types are as defined in *Landscape Ecology* [6]. In addition, the landscape as a whole has characteristics which represent the spatial relationships among the primitive element types [6, 8, 9].

Quantitative spatial measures of a landscape serve to summarize the spatial data. Given a set of spatial characteristics, one can compare landscapes, or correlate with other data sources. The spatial characteristics utilized depend on the particular analysis task. This paper utilizes a small, working subset of characteristics and makes no attempt to represent them as universal.

3 An Analysis Task

Before discussing the implementation of the operational landscape language, we present an analysis task and its expression in the language. The analysis task is based on three raster GIS maps. The first map, LANDUSE, gives detailed information about the use of land for the Mt. Pisgah region. Among the data are such items as housing, heavy industry, mining, picnic grounds, timber harvesting, etc. The second map, VEGETATION, shows the distribution of forest, grassland, agricultural land, and orchards. The third map, WATERBODIES, simply displays all open water in the region.

The task is to locate two orchards suitable for an experiment in pest control. A natural predator will be introduced into the "test" orchard and the other orchard will serve as a control. A number of restrictions apply to the suitability of the orchards. First, they must be at least 500,000 square feet in size. Second, they must be at least 10,000 feet apart to avoid the possibility of the natural predator migrating from the test orchard to the control orchard. Third, the orchard chosen for the test must be at least 2,000 feet from any source of noise, such as mining, timber harvest, and heavy industry. Finally, since the natural predator nests at night on river snags and feeds elsewhere during the day, the test orchard must be within 2,000 feet of a river. This goal is expressed as follows:

```
(def-data-layer orchard
     creation-script=(("isolate VEGETATION assign 1 to 1 for ORCHARD.")
                      (load "ORCHARD"))
)

(def-data-layer high-noise
     creation-script=(("isolate LANDUSE assign 1 to 25 1 to 26
                                             1 to 29 for NOISE")
                      (load "NOISE")))
```

```
(def-data-layer river-layer
    creation-script=(("isolate WATERBODIES assign 1 to 1 for RIVERS.")
                        (load "RIVERS")))

(def-element corridor  river
    data-layer=  river-layer
    constraint-list=((> width (50 "feet")) (< width (700 "feet")))
)

(def-element patch  orchard-patch
    data-layer=  orchard
    constraint-list=((> area (500000 "sq feet")))
)

(def-element patch  noise-source
    data-layer=  high-noise
    constraint-list=((> area (700000 "sq feet")))
)

(def-landscape pest-control-test
    component-list=((corridor river) (patch orchard-patch) (patch noise-source))

    constraint-list=((orchard-patch ?op1); test orchard
                        (orchard-patch ?op2); control orchard
                        (river ?r)
                        (! (> (min-dist ?op1 ?op2) (10000 "feet")))
                        (! (< (min-dist ?op1 ?r) (2000 "feet")))
                        (! (for-all noise-source
                             (> (min-dist ?op1 noise-source) (2000 "feet")))))
)

(define landscape-instances (generate-landscapes pest-control-test 'full))

(send landscape-instances generate-GIS-map)
```

The first three expressions (def-data-layer) specify the three "types" of land (or water) which are pertinent to the analysis task. The text between double quotes in the creation script is actually the procedural analysis language of macGIS [7]. While these expressions utilize the procedural language of macGIS, they do not consider any spatial properties, they simply define a type of land on a cell by cell basis. The "isolate" operator simply selects one or more cell types in the data layer and generates a recoded map. In this case we assign 1 to the desired land types, thereby generating a binary data layer. A library of task independent data layer definitions may be built up and referenced as needed. As such, custom data layers do not need to be created for every analysis task.

The next three expressions (def-element) specify the landscape elements of interest to our analysis task. Each expression defines a specialization of one of the three landscape element types: corridor, patch, and matrix. Each element type has pre-defined semantics; the constraint list serves only to further refine the definition. For this analysis, we do not use the matrix landscape element type. The data layer in which instances of this definition will be found is also specified.

The def-landscape expression specifies the desired landscape configuration which represents the analysis task. The component-list specifies the previously defined landscape element types that are of interest to this analysis. The constraint list specifies the desired number and spatial configuration of these landscape elements. The final two expressions generate a list of landscape instances which meet the constraints, and produce a map for each.

An explanation of the language syntax is, unfortunately, beyond the scope of this paper. In general, the language follows the model of functional languages such as lisp and scheme.

4 Implementation

This section describes the core of the Landscape Reasoning Module, a landscape language interpreter. The interpreter has been implemented in MacScheme with SCOOPS, an object-oriented language extension. The interpreter processes expressions of the landscape language one at a time. Data is exchanged with macGIS via raster map data files. The language expressions are divided functionally into three groups: definition of landscape, landscape element, and data-layer classes; instance generation; and instance query. Expressions of the first group serve to define a working set of landscape definitions for the analysis task. Expressions of the second group invoke interpretation of the GIS data subject to the landscape definitions. Expressions of the third group allow the results of the interpretation to be queried.

The first step of spatial analysis with GIS involves developing a set of working landscape definitions and specifying the relevant GIS data-layers. In LRM, these definitions are supplied via the expressions def-data-layer, def-element, and def-landscape. These expressions define sub-classes of a pre-defined object class hierarchy. In other words, they define a number of object "types" for the system.

The power of the object-oriented approach is its ability to facilitate the grouping of data and operations into objects representative of real-world entities. This resulted in its selection as a basis for representing landscape definitions. A class-based inheritance approach is taken, so the methods and data structures of the parent class are inherited, requiring the user only to specify information unique to a sub-class.

The landscape as a whole is an object. Each element of the landscape is also an object itself. Each landscape element is derived from a corresponding data-layer object. Figure 1 shows the class hierarchy upon which objects in the system are based. This class hierarchy is a set of pre-defined object classes, specializations of which are used to define a particular landscape.

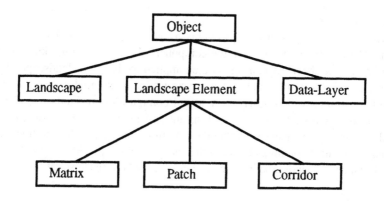

Figure 1. Landscape Class Hierarchy

Message passing is the mechanism by which objects communicate with each other. Each object has a set of methods, or procedures, which defines the messages it will respond to. The internal state of objects may not be directly manipulated, but may be set or read by sending messages. Message passing can also define the flow of execution control within a set of objects. In LRM, for example, an initialize message sent to one object often results in a propagation of initialize messages to other objects.

The second step of spatial analysis involves interpreting the GIS data given the set of working landscape definitions. Interpretation of the GIS data is invoked via the generate-landscape expression. This expression generates instances of defined sub-classes. Each instance is an interpretation of the GIS data subject to the definitions of the sub-class.

Recalling the example analysis tasks of Section 3.0, we assume the user has entered a set of landscape definitions. The interpreter now has an environment of task specific sub-classes. The user is free to request the generation of instances of any of these sub-classes. Analysis normally proceeds by requesting that the system find all possible instances of the landscape sub-class via the generate-landscapes expression. Class definition and instance generation are the core of the Landscape Reasoning Module.

The data-layer class provides an interface between the macGIS data files and the needs of the landscape element classes. The three landscape element classes provide the functionality necessary for finding instances of themselves in the associated data-layers. For example, the patch class contains a connected component algorithm [2], which, when instantiated, locates an instance of a patch in the specified data-layer. The corridor class utilizes a shrink-propagate algorithm [12], in conjunction with the connected component algorithm. The matrix class utilizes a modified connected component algorithm.

The landscape class provides a means for defining a desired landscape configuration, and a means for holding the results of finding an instance of that configuration. The user, when creating a sub-class, supplies the desired configuration in the component-list and constraint-list. The generate-landscapes procedure utilizes these lists to generate instances, of the landscape sub-class, which contain landscape elements meeting inter-element constraints.

The generate-landscapes procedure begins by generating all possible instances of landscape element types specified in the component-list. An algorithm known as satisfy, commonly taught in AI courses, is then applied to the set of landscape element instances and the constraint-list [13]. The constraint-list is essentially a pattern of variables and constraints which describes a spatial configuration of landscape elements. This pattern is matched against the set of instances. Satisfy finds all possible bindings of variables over the instances such that the pattern is consistent. The result is a list of lists. Each list contains bindings between the variables of the pattern and the landscape element instances. A landscape instance is then generated for each set of bindings.

5 Solution to Analysis Task

We now present the solution to the analysis task introduced in section 3.0. Figures 2, 3, and 4 show graphically the instances of landscape elements found in the respective data layers. These instances are shown prior to applying any intra or inter element constraints.

Figure 2 shows the candidate orchards numbered for discussion purposes. The intra-element constraint on minimum orchard size results in orchards 1, 4, 6, 7, and 9 being eliminated by the process of landscape element instantiation. Orchards 2, 3, 5, and 8 are the remaining patches to be considered by the inter-element constraints of the landscape definition. Figure 3 shows the various sources of man-made noise in the region. Here also, the smaller patches are eliminated by an intra-element area constraint.

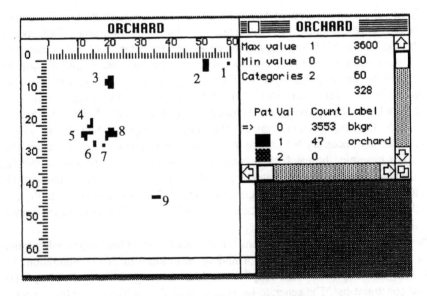

Figure 2. Map of All Orchards

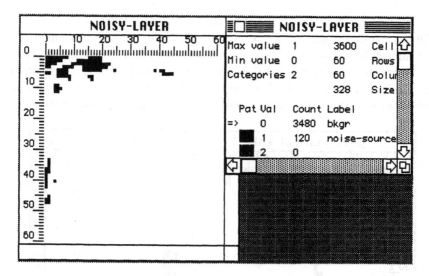

Figure 3. Map of Noise Sources

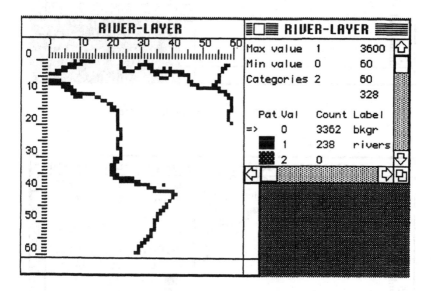

Figure 4. Map of Rivers

Figure 4 shows the rivers of the region. There are four separate rivers, one of them a single cell of the map. The shrink-propagate algorithm used for instantiating corridor elements breaks the rivers into six separate runs and eliminates some short segments. This occurs even though the constraints on river width should result in the four rivers of Figure 4 being accepted "as-is". The utility of the shrink-propagate algorithm, as implemented here, degrades as the width of rivers in a map approaches the grid resolution. Here we have several segments of the river with a cell width of one.

The generate-landscape-instances expression produced four unique solutions. Figures 5 through 8 show the four landscape instances generated. The first solution, Figure 5, shows the test orchard patch (?op1) adjacent to a river with the control orchard patch (?op2) in the upper right. The two orchards are at least 10,000 feet apart and orchard ?op1 is distant from any noise-source, as stipulated by the constraints.

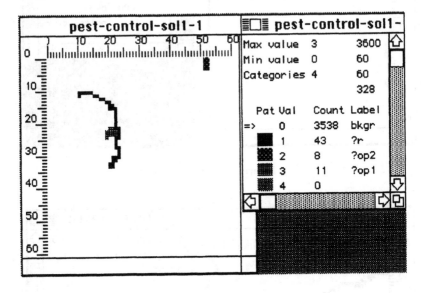

Figure 5. Pest-Control: Solution One

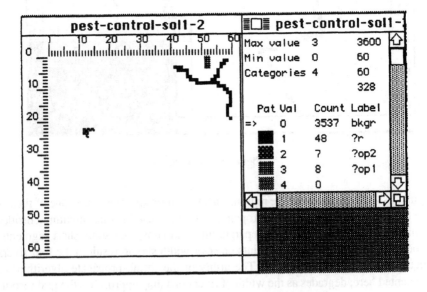

Figure 6. Pest-Control: Solution Two

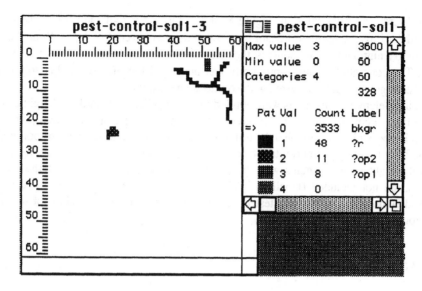

Figure 7. Pest-Control: Solution Three

The second solution, Figure 6, labels the orchard in the upper right as the test orchard ?op1 and selects another orchard as the control ?op2. Again, ?op1 is near a river as stipulated by the constraints.

The third solution, Figure 7, contains the same orchards as the first solution. The labels for ?op1 and ?op2 were switched, another river was selected, and the constraints remained satisfied.

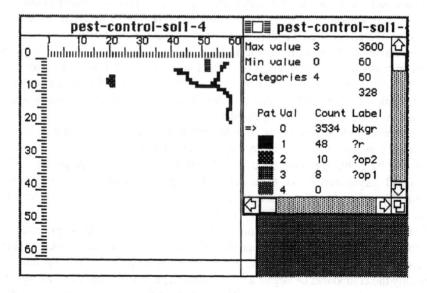

Figure 8. Pest-Control: Solution Four

The fourth and final solution, Figure 8, shows the upper right orchard as the test orchard ?op1 again. Another orchard has been chosen as the control ?op2. While ?op2 is close to a number of noise sources, the constraint on noise-proximity only applies to ?op1. The constraint on noise-proximity has prevented the inverse of this solution from being generated, as occurred with solutions one and three

The inter-element constraints were applied to four candidate orchards and eight candidate river segments (due to the behavior of the shrink-propagate algorithm). There are twelve possible ways to label the four orchards with ?op1 and ?op2, corresponding to P(4,2), or number of 2-permutations of 4 elements. There are eight possible ways to label the eight river segments with ?r. The noise sources, while a part of the constraints, are not included (labelled) in the final map. This means that 96 uniquely labelled landscapes (12*8) were tested against the inter-element constraints, of which only four survived.

6 Evaluation

Our efforts will be examined first in terms of descriptive adequacy, then procedural adequacy. First we consider the syntax of the operational landscape language. The current syntax is quite crude. Our research focussed primarily on the underlying mechanisms used to generate instances of the language elements (patch, matrix, corridor). The text language used to define landscape configurations is a first-pass attempt, and proved adequate to test out the program as a whole.

An advantage of the syntax, which is driven by the underlying object oriented approach, is the modularity of definitions. Since definitions of data-layers, landscape elements, and landscapes are modular, libraries of these expressions may be built up in files and drawn upon as needed. The definitions can be loaded in any order, with all resolution of references taking place at the time of instance generation.

The semantics of the various object classes in the landscape class hierarchy are based primarily on the definitions provided by Forman in *Landscape Ecology* [6]. Here he defines the matrix, patch, corridor, and landscape in terms of structure, function, and dynamics. Portions of the structural definitions were extracted and used as a basis for implementing the operational semantics.

The semantics of the patch were straightforward. A patch can be instantiated by any connected component in the data. In our language, a corridor is made distinct from a patch only by width. A more precise semantics should include other properties such as extent and curvilinearity. The main difficulties here are the development of precise quantitative definitions of the concepts and algorithms for computing these characteristics from the raster representation of the elements.

The semantics of the matrix are somewhat crude. The connectivity requirements of the connected components are temporarily relaxed, if specified, and the largest single component is selected as the matrix. While Forman does point out that the matrix is generally the component of largest area in a landscape, it is unclear as to whether the connectivity relaxation is appropriate. The implementation of a landscape as a collection of landscape elements parallels the definitions well.

The class hierarchy of our landscape language currently represents a minimal, high level taxonomy of landscape objects. The hierarchy could be extended with additional pre-defined specializations of the matrix, patch, corridor, and landscape. The corridor class, for example, could be refined to include sub-classes for closed and open curves, eg. "ring" and "open-ended" corridor types. Another specialization, the peninsula, could potentially become a sub-class of patch. Specializations of the landscape class might include sub-classes such as "network" and "checkerboard". The class hierarchy provides a natural structure for incorporating these additional specializations. For each addition, algorithms for instantiation must be developed.

This leads us to a consideration of the complexity of computing landscape instances from their definitions. The computing time required for locating landscape instances is occupied primarily by the connected-component algorithm [2], the shrink-propagate algorithm [12], algorithms that evaluate spatial characteristics [8, 9], and the satisfy algorithm[13]. The connected-component algorithm is linear in the number of cells of the GIS data layer. The shrink-propagate algorithm is $O(kn)$, where k is the number of cells by which the component is to be uniformly contracted and expanded, and n is the number of cells in the component's bounding box.

The intra-element spatial characteristics are computed for every connected component in every data layer. The characteristics used in this paper are all derived from area and perimeter which are both linear in the number of cells of the bounding box. The inter-element spatial characteristics are computed on demand when a constraint expression is encountered by the satisfy algorithm. Computing whether two components are adjacent is linear. Computing minimum distance between two components is $O(nm)$, where n and m are the number of cells in each bounding box respectively.

Satisfy is an exponential algorithm in general. As used in this paper, we can more precisely characterize the complexity. Consider a generalized version of the inter-element constraint expression: ((type1 ?a) (type1 ?b) (type2 ?c) (! (constraint ?a ?b)) (! (constraint ?b ?c))). We are assuming that the list of desired types and variables, eg. (type1 ?a), will precede all constraints. The constraint expression is matched against a list of landscape element instances of the various types, eg. ((type1 inst1) (type1 inst2) (type1 inst3) (type2 inst1) (type2 inst2)). All possible bindings of variables to instances which result in a consistent constraint expression are computed.

The process of finding all bindings can be broken into two steps. First, find all bindings of variables to instances prior to applying constraints. Second, apply the constraints to each binding set. The number of binding sets produced by the first step corresponds to the product:

$$\prod_{i=1}^{|types|} P(n_i, r_i)$$

There will be one term in the product for each type in the constraint expression. Each term is as follows: given r expressions of a single type, eg. (type1 ?a), and n instances of that type, eg. (type1 inst1), there will be $P(n,r)$ possible bindings. The second step tests each binding set against all the constraints (worst case). The result is:

$$\left(\prod_{i=1}^{|\text{types}|} P(n_i, r_i) \right) |\text{constraints}|$$

The first factor can grow to be quite large. The run time of LRM on a given analysis task was dominated by the satisfy algorithm.

7 Conclusion

The focus of our research has been almost exclusively on the spatial structure of landscapes. In terms of the operational landscape language, the definition of a data-layer with a creation script is the only point at which non-spatial information is considered. The remainder of the process, namely landscape and landscape element instantiation, is based purely on spatial characteristics. A means by which to incorporate non-spatial data would allow analysis to consider the function and dynamics of a landscape, not just the structure.

We have taken a first pass at defining an operational language for reasoning about landscape structure. The primitives of this language derive their semantics from the fields of landscape ecology and landscape architecture. The implementation of the language primitives is made possible by data structures and algorithms from GIS, artificial intelligence, image processing, and object-oriented design. The guiding principle is that problem solving should take place in the task domain, not the domain of the underlying data representation. By providing an operational landscape language, landscape definitions used during analysis may be explicitly represented. The automatic mapping from GIS data to instances of the definitions allows subsequent analysis (intra and inter element constraints) to take place in the domain of the landscape.

The notion of reasoning with a "scene" of objects instead of an image (data-layer) is given an precise treatment in *A Logical Framework for Depiction and Image Interpretation* by Reiter and Mackworth [11]. Reiter and Mackworth address the problem of mapping from image to scene specifically in the context of map interpretation. Here they define a taxonomy of map image objects, the two major components being chains and regions. This corresponds closely to the vector representation of GIS. Also defined is a taxonomy of scene objects, the two major components being linear-scene-objects and areas. Subtypes of these components are entities such as road, river, land, and water. This corresponds closely to the language of landscape structure presented in this paper. The map image is represented as a set of formulas in first order logic. Knowledge about spatial relations in the scene domain is also cast into formulas of first order logic. Finally, a set of depiction axioms defines a mapping between map image objects and scene objects. An interpretation of the map is then formally defined as a model of the map image, scene, and depiction axioms.

In terms of GIS, each data layer can be considered an image. The data layer is a static representation which uniformly records the source data. No emphasis is placed on particular regions or components of the source data. An exception to this would be the relational records associated with points, lines, and polygons in the vector representation. It is assumed, however, that this data is entered without prior knowledge of specific analysis tasks.

A scene is defined to be an interpretation of the images, or data layers, in the context of a specific analysis task. Here one focuses on the content of the images. In the case of landscape ecology, one is interested in the composition and spatial configuration of landscape elements such as patches, corridors, and the matrix. Similarly with land-use planning. Essentially, one is isolating components of the image and reasoning about the inter-relationships.

An important observation of many GIS data representation and transformation operations is that they treat the map as an image. Granted, the attribute values of the data layer represent higher level properties of the region, but they are associated directly with components of an image, not components of a scene. As such, deriving new information requires operations on the image components. In this case, the cells of the raster GIS data representation. Analysis in this situation is a process of mapping a problem onto operations on images. A landscape is a structure of interacting elements, a scene if you will. Representing it only as a collection of cells, or vectors, results in problem solving taking place in the domain of the image, not the domain of the scene.

In closing, we have taken a first pass at developing a high level tool for landscape analysis with Geographic Information Systems. A number of challenges were encountered while attempting to raise the level of user interaction from image operations to scene operations. Research into integrating advanced computing techniques with Geographic Information Systems and landscape analysis has proved to be an area rich with avenues for further exploration.

References

1. Albert, T. M. 1988. Knowledge-Based Geographic Information Systems (KB-GIS): New Analytic and Data Management Tools. Mathematical Geology. Vol. 20, No. 8, pp. 1021-1035.

2. Ballard, D. H. and Brown, C. M. 1982. Computer Vision. Prentice-Hall, Inc., New Jersey.

3. Burrough, P. A. 1986. Principles of Geographical Information Systems for Land Resources Assesment. Clarendon Press, Oxford.

4. Cowen, D.J. 1988. GIS Versus CAD versus DBMS: What are the Differences? In Introductory Readings in Geographic Information Systems. Taylor and Francis, New York, pp. 52-61.

5. Dangermond, J. 1983. A Classification of Software Components Commonly Used in Geographic Information Systems. In Introductory Readings in Geographic Information Systems. Taylor and Francis, New York, pp. 30-51.

6. Forman, R. T. T., and Godron, M. 1986. Landscape Ecology. John Wiley and Sons, New York.

7. Larsen, K. and Hulse, D. 1989. macGIS - A Geographic Information System for the Macintosh: The User's Guide. University of Oregon, Eugene, Oregon.

8. Li, H. 1989. Spatio-temporal Pattern Analysis of Managed Forest Landscapes: A Simulation Approach. Doctoral Dissertation, Oregon State University.

9. Lounsbury, J. F., Sommers, L. M., and Fernald, E. A. 1981. Land Use: A Spatial Approach. Kendall/Hunt Publishing Company, Dubuque, Iowa.

10. Peuquet, D. J. 1984. A Conceptual Framework and Comparison of Spatial Data Models. In Introductory Readings in Geographic Information Systems. Taylor and Francis, New York, pp. 250-285.

11. Reiter, R. and Mackworth, A. K. 1989. A Logical Framework for Depiction and Image Interpretation. Artificial Intelligence 41, pp. 125-155.

12. Rosenfeld, A. and Kak, A. C. 1976. Digital Picture Processing. Academic Press, New York.

13. Tanimoto, S. L. 1987. The Elements of Artificial Intelligence. Computer Science Press, Inc., Maryland.

14. Webster, C. 1990. Rule-based Spatial Search. International Journal of Geographical Information Systems. Vol. 4, No. 3, pp. 241-259.

XII. Posters

An Extended Model for Handling Topological Relationships
T.Y. Jen, P. Boursier (LRI, Université Paris Sud, Paris, France)

Metaphors as a Means to SpatializeInterfaces
A. Tepper (GMD, Sankt Augustin, Germany)

Spatial Relation-Based Representation Systems
D. Papadias (University of Athens, Athens, Greece)

Revealing Spatial Structure of Point Patterns by Quadtree Statistics
L. Pasztor (Research Institute for Soil Science and Agricultural Chemistry, Budapest, Hungary)

Hiearchical Road Networks: A Framework for Efficient Wayfinding
A. Car (Technische Universität Wien, Vienna, Austria)

Classification of Data Quality: Three Case Studies
H. Stanek (Technische Universität Wien, Vienna, Austria)

Modeling Water Use Efficiency in Agricultural Irrigation Practices Through a GIS-Based
Decision Support System
A. Gisotti, R. Loguercio Polosa, N. Sciacovelli (Technopolis CSATA, Bari, Italy)

Object-Oriented Modeling for Map Generalization
LI Weischeng (Space Center for Satellite Navigation, Queensland, Australia)

Analysis of Cognitive Processes Used in Route Descriptions
A. Gryl, G. Ligozat (LIMSI-CNRS, Orsay, France)

Springer-Verlag
and the Environment

We at Springer-Verlag firmly believe that an international science publisher has a special obligation to the environment, and our corporate policies consistently reflect this conviction.

We also expect our business partners – paper mills, printers, packaging manufacturers, etc. – to commit themselves to using environmentally friendly materials and production processes.

The paper in this book is made from low- or no-chlorine pulp and is acid free, in conformance with international standards for paper permanency.

Lecture Notes in Computer Science

For information about Vols. 1–639
please contact your bookseller or Springer-Verlag